陈佳洱院士

陈佳洱院士与夫人周维金教授

北京大学院士文库

陈佳洱文集

陈 佳 洱 著

北 京 大 学 出 版 社
北 京

图书在版编目(CIP)数据

陈佳洱文集/陈佳洱著. —北京:北京大学出版社,1999.5
(北京大学院士文库)
ISBN 7-301-04065-2

Ⅰ.陈… Ⅱ.陈… Ⅲ.① 加速器-研究-文集 Ⅳ.TL5-53

中国版本图书馆 CIP 数据核字(1999)第 10716 号

书　　　名:	陈佳洱文集
著作责任者:	陈佳洱
责 任 编 辑:	周月梅
标 准 书 号:	ISBN 7-301-04065-2/O · 430
出　版　者:	北京大学出版社
地　　　址:	北京市海淀区中关村北京大学校内　100871
网　　　址:	http://cbs.pku.edu.cn/cbs.htm
电　　　话:	出版部 62752015　发行部 62754140　编辑室 62752021
电 子 信 箱:	zpup@pup.pku.edu.cn
排　版　者:	高新特公司激光照排中心
印　刷　者:	北京大学印刷厂
发　行　者:	北京大学出版社
经　销　者:	新华书店
	787 毫米×1092 毫米　16 开本　32.25 印张　532 千字
	1999 年 5 月第一版　1999 年 5 月第一次印刷
定　　　价:	60.00 元

北京大学资源集团出版基金资助出版

谨以此书献给北京大学校庆 100 周年

《北京大学院士文库》编委会名单

主　任：陈佳洱
副主任：王义遒
委　员：(按姓氏笔画为序)
　　　　王　选　甘子钊　巩运明
　　　　侯仁之　赵亨利　姜伯驹
　　　　徐光宪　翟中和

序

最近，北京大学出版社告诉我，北京大学资源集团设立出版基金，资助出版一套《北京大学院士文库》，为北京大学的中科院院士和工程院院士每人出一本学术专著或学术论文集，以记载他们为祖国的科学技术事业所作出的贡献。北大出版社邀我为这套书写个序。

考虑到我较长时间在中国科学院工作，为科学家树碑立传，把他们的伟业记载下来并留传给后人，自然是我应该大力支持的事情。同时，我也曾在北大学习过，这些院士中有的就是我过去的老师，他们对我精心培育的情景，使我终生难忘；有的曾是我的同学或同事，我们之间有着非常深厚的友谊，他们为科学事业无私奉献的精神，给我留下了极为深刻的印象，至今历历在目。无论从工作上考虑还是从师生、同事情义出发，我都愿意为这本书写个序。

我认为，北京大学出版社出版《北京大学院士文库》这套书，是一件非常有意义的事。

首先，《北京大学院士文库》将为我国科学技术文献宝库增添新的内容。北京大学是我国一所著名的高等学府，也是世界上一所有影响的大学。它不仅为国家培养了大批栋梁之材，而且为国家提供了大批重要的科技成果，成为我国一个重要的科学中心。在这所大学里聚集了一批我国最著名的专家和学者，其中仅就自然科学而言，就有中科院院士和工程院院士30人。他们中既有学识渊博、造诣精深、蜚声中外的老专家、学者，也有一批成绩卓著，近年来为祖国科学技术事业作出过重大贡献的中年学者。他们在我国科学技术发展史上占有重要的地位，是我国科技大军中的中坚力量。现在，北大出版社把他们的科学技术著作收集起来，集中出版，无论是他们当年成名之作，还是新发表的学术专著和学术论文，都将为我国科学技术文献宝库增添重要的内容。

其次,《北京大学院士文库》还将为我国科学技术事业的发展提供宝贵的经验。这套学术文库不仅完整地记载了这些学术大师的发明和创造,而且还生动地描绘了他们在不同历史时期为科学事业奋斗的历程。他们以亲身的经历,丰富的史料,独特的见解,深奥的思想,总结了科学技术发展的规律。例如,科学家最需要什么样的支持,在什么样的条件下最容易出成果等。这里既有成功的经验,也有失败的教训;既有成功的喜悦,也有受挫的苦恼。有的院士还从他们的切身感受出发,对我国科技人才的培养,科技体制的改革提出了很好的建议。这些都为我们科技管理部门和科技管理工作者,特别是为我国制定有关的科技政策,提供了很好的经验和借鉴。

第三,《北京大学院士文库》不仅是一套科学技术著作,而且是一套富有教育意义的人生教科书。这套文库详细地记载了这30位科学家的学术成就,也如实地记载了他们的人生经历。他们不仅学问好,而且人品好。他们的一生是在爱国主义旗帜下,为科学事业奋斗的一生。他们通过自己的勤奋努力,走了一条成功之路。他们的成功经验无论对年轻人,还是对一切有志于献身科学事业的人,都有极好的教育意义。

最后,我向这30位院士为祖国科技事业作出的贡献表示衷心的感谢!对《北京大学院士文库》的出版表示热烈的祝贺!也希望能有更多的科学家的学术著作和传记问世,因为科学是推动我们社会发展的强大动力。

中国科学院院长

周光召

1996年10月

序

北京大学出版社决定编辑出版《北京大学院士文库》,这件事情很有意义,我非常赞成。

从世界高等教育的发展看,教师是大学的核心,他们构成学校的基调。世界一流大学都具有很强的教师阵容,拥有一批世界公认的学术权威和知名学者。正是他们能够培养出世界公认的优秀人才。其中一部分毕业生能够成为当代世界政治、经济、文化、科学领域里的杰出代表。同时,他们能够取得重大的科研成果,特别是在基础研究方面,能取得具有划时代意义的科研成果。

在中国科技、教育界,院士是最高学术水平的象征。他们对国家科学技术的发展起着相当重要的作用。北大是拥有院士最多的大学,北大人一直为此而自豪。北大的几十位院士可分为两部分,一部分是老院士,他们在中国科学院成立之初就因为各自取得的成就而成为最早的一批院士(当时称学部委员)。这些老院士德高望重、学风严谨、蜚声国内外,为北大乃至中国的科学技术和文化事业的发展作出了奠基性贡献。他们当中有理科的王竹溪、叶企孙、江泽涵、许宝騄、周培源、胡宁、段学复、饶毓泰、黄昆、张青莲、黄子卿、傅鹰、汤佩松、李继侗、张景钺、陈桢、乐森璕等教授。北大的盛名,在很大程度上是与这些堪称大师的第一代院士的名字联系在一起的。这一长串院士名单,奠定了北大在中国学术界、科学界的地位。谈起他们,像我这样的后辈无不怀有敬仰之情。他们像一块块强力磁铁,吸引着一代代中华学子到燕园求学,在他们的教诲、指导、影响下,新中国急需的大批优秀人才源源不断地从北大培养出来,成为社会主义建设的栋梁之材。当院士文库推出的时候,这些老院士当中已有不少人离开了我们,但他们为北大、为国家建立的功勋,他们的英名将永远为人们铭记!

北大的学术生命是长青的,继第一批院士之后,80年代、90年代,北大又一批理科教师,其中许多是建国以后培养出来的,成为中国科学院院士和中国工程院院士,他们可以说是北大那些与新中国风雨同舟、不畏清贫、不怕艰险、为教育和科学事业执着奉献的中年教师的代表,是今日北大的骨干依靠力量、学术中坚。

人类就要进入 21 世纪,北大也即将迎来建校 100 周年,当此世纪交替之际,北大雄心勃勃地提出:到 21 世纪初叶建成世界一流的社会主义大学。这是一个需要为之付出极其艰苦努力的、振奋人心的目标。以院士为代表的一流教师队伍是我们实现这一目标在学术上的最重要依托。有这样一支老年、中年教师队伍,再加上我们正在迅速成长起来的生气蓬勃、富有想象力和创造力、奋发向上、成为北大未来希望所在的青年教师,我们的目标是一定能够达到的。

院士们的工作成就,有很多都是在相当困难的条件下取得的,他们的奋斗精神和他们的成果一样,都是我们建设世界一流大学的宝贵财富和源泉。为院士出版文集,将他们的代表性学术成果或成名之作结集出版,是对院士们成就的肯定,也将使人们从他们的奋斗足迹中,得到某种启迪和鼓舞。院士文库将为我校的学术宝库增添重要的内容,成为哺育青年学生成长的极好教材。

北大出版社的决定得到了北大资源集团的热情支持,他们出资建立北大资源集团出版基金,资助院士文库的出版。我作为北大校长和一个院士、一个教师,要向北大出版社和北大资源集团为学术专著的出版和学校建设所作的努力表示敬意!

北京大学校长
中科院院士
陈佳洱

1997 年 1 月

目　录

低能粒子加速器与陈佳洱 ……………………………………（1）

第一部分　回旋加速器研究

EXPERIMENTAL STUDIES OF THE CENTRAL REGION
　　OF THE CYCLOTRON ………………………………（3）
ON THE CENTERING OF CYCLOTRON ORBITS BY
　　MEANS OF HARMONIC COILS ……………………（58）

第二部分　静电加速器及基于串列静电加速器的加速器质谱计研究

DESIGN OF 4.5 MV ELECTROSTATIC
　　ACCELERATOR ……………………………………（73）
PROGRESS AND APPLICATION OF THE 4.5 MV
　　ELECTROSSTATIC ACCELERATOR ………………（86）
北京大学 6 MV 串列静电加速器的运行及其应用 ………（95）
加速器质谱计的原理、技术及其进展……………………（104）
BEAM CHARACTERISTICS AND PRELIMINARY
　　APPLICATIONS OF THE TANDEM BASED AMS
　　AT PEKING UNIVERSITY …………………………（111）
ACCELERATOR MASS SPECTROMETRY AT PEKING
　　UNIVERSITY: EXPERIMENT AND PROGRESS ………（118）
北京大学加速器质谱计研究与应用进展…………………（127）

第三部分　束流物理与脉冲化技术研究

静电加速器脉冲化装置及有关特性………………………（137）
THE BEAM DYNAMICS OF A SINE WAVE SWEEPER ……（151）

BUNCHING CHARACTERISTICS OF A BUNCHER USING
　　HELICAL RESONATORS ……………………………（164）
BUNCHING SYSTEM FOR THE STONY BROOK TANDEM
　　LINAC HEAVY-ION ACCELERATOR ………………（176）
600 kV 强流纳秒脉冲加速器中的前切割与后聚束系统 ………（186）
HIGH-EFFICIENCY TWO-HARMONICS BEAM
　　CHOPPER ………………………………………………（194）
A DOUBLE-DRIFT HAMONIC BUNCHER FOR 4.5 MV VAN
　　DE GRAAFF ACCELERATOR …………………………（202）
离子束 ns 脉冲测量用可变栅距同轴靶 …………………………（213）
束流相图仪的标定 …………………………………………………（222）
HIGH-POWER MULTI-ELECTON-BEAM CHERENKOV
　　FREE-ELECTRON LASERS AT mm
　　WAVELENGTHS ………………………………………（229）
强流束传输中束晕与混沌的 Poincare 图象与 Lyapunov
　　指数分析 …………………………………………………（241）

第四部分　射频离子直线加速器研究

螺旋线波导直线加速器简介 ………………………………………（253）
螺旋波导加速腔的不载束高功率试验 ……………………………（271）
螺旋波导加速腔载束全功率试验 …………………………………（277）
AN INTEGRAL SPLITRING RESONATOR LOADED WITH
　　DRIFT TUBES & RF QUADRUPOLES ………………（287）
A RFQ INJECTOR FOR EN TANDEM LINAC HEAVY
　　ION ACCELERATOR …………………………………（298）
重离子整体分离环高频四极场(RFQ)加速结构的研究 …………（310）
RFQ 加速器同时加速同 q/M 正负离子的设想 …………………（320）
EXPERIMENTAL STUDIES ON THE ACCELERATION OF
　　POSITIVE & NEGATIVE IONS WITH A HEAVY
　　ION ISR RFQ ……………………………………………（325）

第五部分 射频超导直线加速器研究

FIRST OPERATION OF THE STONY BROOK
  SUPERCONDUCTING LINAC ……………… (335)
1.5 GHz 铌腔 RF 超导的实验研究 ……………… (347)
THE PROGRESS OF SUPERCONDUCTING CAVITY
  STUDIES AND R&D FOR LASER DRIVEN RF
  GUN AT PEKING UNIVERSITY ……………… (371)
DESIGN AND CONSTRUCTION OF A DC HIGH-
  BRIGHTNESS LASER DRIVEN ELECTRON GUN …… (379)
THE HIGH-BRIGHTNESS ULTRA-SHORT PULSED
  ELECTRON BEAM SOURCE AT PEKING
  UNIVERSITY …………………………………… (385)
DESIGN STUDY ON A SUPERCONDUCTING
  MULTICELL RF ACCELERATING CAVITY
  FOR USE IN A LINEAR COLLIDER ……………… (393)

第六部分 粒子加速器综述

THE PROGRESS OF LOW ENERGY PARTICLE
  ACCELERATORS IN CHINA ……………………… (409)
RECENT ION SOURCE DEVELOPMENT IN CHINA …… (425)
超导重离子直线加速器 ……………………………… (439)
PROGRESS OF RFQ AND SUPERCONDUCTING
  ACCELERATORS IN CHINA ……………………… (452)

发表的论文和学术报告总目录 ……………………… (470)
自　述 ……………………………………………… (486)
难忘的游戏——代后记 ……………………………… (489)

低能粒子加速器与陈佳洱

陈佳洱长期致力于粒子加速器的研究与教学工作。他是一位理论素养和实验技能兼备,熟悉多种粒子加速器,发展比较全面的加速器物理学家,同时又是北京大学技术物理系的骨干教师和学术带头人。他还先后担任过教研室主任、系副主任、开放实验室主任、研究所所长和北京大学副校长、校长等职。他在实验室建设、课程设置和编写教材、培养青年教师和指导研究生等方面辛勤工作了四十多年,先后发表学术论文及研究报告等计一百余篇。为在我国创建核物理专业,发展加速器学科,培养和输送原子能、加速器及国防科技人才方面做出了重要贡献。

一　等时性回旋加速器中央区的物理研究

陈佳洱早期从事回旋加速器中央区的研究。由离子源引出的束流经过中央区加速后通常大幅度地衰减为初始值的10%～20%。然而衰减的机制是什么?如何减少离子丢失?这是一个重要而难解的课题。中央区里离子成分复杂,速度又低,各种非线性的效应明显,不能作一般性的处理。陈佳洱在英国卢瑟福研究所访问期间,从实际出发,利用微分探针逐一确定各种离子成分的轨道并将实验诊断与理论分析及计算机模拟相结合,逐圈分析各轨道上离子的高频相位分布、轨道曲率中心分布、自由振动的包络及频率等,终于发现束流大幅度衰减的主要原因是:在对射频相位敏感的电聚焦的作用下,粒子在头上3～5圈中,其自由振动的宏包络与中心区的接收相空间失配。基于这一认识,通过逐圈控制离子的高频相位与轨道中心等参量将束流的传输效率提高了三倍以上[1]。在分析离子轨道中心随加速器半径变化的过程中,陈佳洱注意到有一种不随高频电压变化的定向移动,经分析确认它就是具有三个扇形聚焦磁极的等时性回旋加速器所特有的越隙共振。陈佳洱还证实可用振幅不随高频电压变化作为诊断越隙共振的实验判据并可运用一次谐波有效地抑制振幅的增长[2]。这些成果在1965年荷兰召开的回旋加速器会议上发表,引起了同行们的兴趣和重视。那时候发展的用正弦波电位器调整谐波场、控制离子轨道中心的方法,后来也被国内研制的回旋加速器所采用[3]。

二 4.5 MV 静电加速器的设计、建造和 2×6 MV 串列静电加速器的改建及应用

(1) 4.5 MV 静电加速器是我国第一台自行设计、建造的单级静电加速器。陈佳洱作为项目负责人领导研制组首次摆脱以往的仿制模式,吸取国内外先进经验,从优化物理参数入手进行设计。他们系统地优化了高压电极形状、高压柱结构以及各项设备的光学参量等,还根据日后物理研究的需要,增加了束流脉冲化等新的功能[4]。经优化的结构机械性能稳定,电场分布均匀。运行时高压柱端部的最大振幅小于 0.08 mm,空载时的端电压最高可达 6.2 MV,是我国单级静电加速器中最高的。加速器的运行表明它的性能比过去仿制的好得多,不但束流十分稳定,而且在离子种类、能量和束流脉冲化性能方面优于国内同类设备。这台加速器的建成填补了我国单色中子在 3.5~7 MeV,16~20 MeV 等能区的空白,成为北京大学中子与裂变物理实验室的基础设施[5]。目前这台加速器不仅为国内高校包括北京大学、清华大学、四川大学以及原子能研究院等有关研究单位提供实验用束流,还接待了来自台北中研院物理所、俄国杜布纳研究所和镭学研究所以及美国劳伦斯伯克利研究所等访问科学家进行实验(图 1)。

(2) 2×6 MV 串列静电加速器是由英国牛津大学转来的 EN 加速器改建而成的国内高校中最大的一台加速设备。1986年以来在陈佳洱的主持下完成了这台老设备的重装、改造及部分设备国产化的工程,配置了绝缘气体处理系统、电控系统、高真空系统及四条束流管线。经改建后,加速器的合轴精度、真空性能和束流强度等明显高于原有设备的水平。这台加速器的稳定运行,为我国开展基于重离子束的核分析技术创造了新的条件,并已初步做出一批好的工作[6]。例如,首次利用碳轰击氚的核反应建立高灵敏的氚分析方法;首次使用氯、溴离子弹性反冲检测

（ERDA）建立从氢到氧的多元素分析方法以及利用重离子背散射（HIRBS）研究高温超导材料与衬底之间的扩散行为和 Co/Si 多层膜退火的反应动力学等。这项重装和改建工程，前后用了不到两年的时间和近 100 万元人民币，使这台价值 200 万美元、建造于 60 年代的老设备，焕发出新的生命。国家教委就此项工程组织的验收评议中指出，这是"一个投入少、效益高的范例"(图 2)。

(3) 加速器超灵敏质谱计（AMS）是近年来串列加速器应用的一个热点。1988 年以来，陈佳洱主持了国家自然科学基金重大项目"为地球科学应用的加速器质谱计的研制与建立"。他与李坤、郭之虞教授合作，与有关人员一起反复论证，瞄准国际先进水平，确定了高起点的设计方案，成功地在改建后的 2×6 MV 串列静电加速器上建立了北京大学加速器质谱计[7]。这台质谱计采用了国际先进的快交替注入测量系统、全静电化的束流聚焦元件和宽孔径的平顶传输系统，有效地避免了分馏效应（使同位素比如 $^{14}C/^{12}C$ 发生变化的一种效应），保证了 ^{14}C 的测量精度优于 1‰；它通过多个宽孔径的高分辨率分析器，在保证测量本底低达 10^{-15} 的水平下使碳束的束流输运效率高达 25% 以上；此外，还配置了自动化的转靶系统和功能齐全的计算机数据获取系统，可方便地测量 ^{12}C、^{13}C、^{14}C 等同位素以及它们的丰度比，碳样的测量能力可达到每天十个以上。1993 年由国家自然科学基金委员会主持的鉴定认为"它的建造和完成为我国 AMS ^{14}C 年代学研究填补了空白，其性能指标在国内处领先地位，已赶上国际水平"。这台质谱计自鉴定以来已先后为二十多个单位进行了 ^{14}C、^{10}Be 和 ^{26}Al 的测量，并已获得了一批在考古地质学和环境科学、生物科学上有重要意义的结果[8]。1996 年以来该质谱计承担了国家重大项目"夏、商、周断代工程"的任务，并进一步将测年精度提高至 ±40 年以下（即优于 0.5‰），达到当前国际 AMS ^{14}C 测年的先进水平，为"夏、商、周断代工程"的进展做出了重要的贡献（图 3,4）。

三　束流脉冲化物理的研究

直流束的脉冲化技术在利用飞行时间法的物理研究中有着重要的应用，也是提高直线加速器和回旋加速器的束流俘获效率和束流品质的一项关键。但通常沿用的早期理论不能正确预言脉冲波形变化等重要性能。1973～1984年间陈佳洱对束流脉冲化技术进行了深入、系统的研究,将束流光学的理论和方法拓展到束流的群聚、切割和输运过程,取得若干重要成果,包括用二维相空间的理论和方法研究束流的群聚和切割过程,补充了过去传统理论的不足,有效地预言了脉冲化束的波形、束流利用效率以及束流品质与束流初始能散及初始发射度的关系。这些都已为后来的实验测量所证实[9];陈佳洱还与所指导的研究生联合提出了新颖的双谐波切割器原理,试验证明其效率比常规正弦波切割器高约47%[10]等。这些成果已在国内外的一些重要应用中取得良好效果。例如,为北京师范大学低能核子所研制的一套高效聚束装置,能以不到10 W的功率将350 keV连续氘束压缩到1 ns[11](图5)。1986年国家教委鉴定认为应用该装置的快中子物理研究"已做出了当代国际水平的工作"、"时间分辨率达到国际先进水平"。又如为美国石溪大学的超导直线加速器所研制的脉冲化系统,成功地将64 MeV的硫离子压缩到100 ps[12],束流的总效率高达60%以上,达到当时国际先进水平;陈佳洱还将束流脉冲化的研究成果制成软件,由计算机根据实验需要算出参数并直接控制加速器各有关设备,运行起来既精确又方便。石溪的同行们把这种运行方式称之为"陈氏模式",一直沿用至今。石溪物理系主任、美国前政府核科学顾问委员会主席P. Paul教授1990年还来信赞扬陈佳洱所做的贡献。

四　直线加速器前沿技术研究

1984年以来,陈佳洱开展了直线加速器前沿技术包括射频四极场(RFQ)加速器和超导加速腔的研究。RFQ是近年来强流离子加速器发展的前沿之一,在医用离子加速器、新型微电子器件与材料、核废料处理、加速器驱动洁净核能源、爆炸物检测和热核聚变研究等方面都有广泛而重要的应用。然而常规的四扇型RFQ结构复杂,体积大,不宜于加速重离子。为了克服这些缺陷,陈佳洱于1984年间与方家驯教授联合提出并发展了一种由分离环共振线激发的新结构,即整体分离环结构(ISR) RFQ。与常规的RFQ相比,它具有尺寸小、电稳定性好、工作频率低、调频范围宽、适宜于加速重离子等优点[13]，受到国际同行的重视。1992年离子注入用的样机试验成功[14]，经国家自然科学基金委员会主持的成果鉴定认为：该结构很有特色，主要性能指标达到国际同类结构的先进水平，填补了国内在RFQ加速技术上，特别是实验和工艺研究上的空白，为我国发展这类加速器技术提供了有益的经验和技术基础[15]。此后北京大学的研制小组即在国家自然科学基金重点项目的支持下，设计、建造了能量为1 MeV的重离子ISR RFQ加速器（图6）。1994年陈佳洱与于金祥教授等联合建议在一个RFQ中同时加速正负离子，并在实验上取得了同时加速正负氧离子的成功，结果大大提高了RFQ腔的束流效率，并为向一种半导体同时注入任意比例的两种离子创造了条件[16,17]。这在国际上还是首次。为了更好处理现有RFQ结构中聚焦与加速作用之间的矛盾，1998年陈佳洱又与方家驯教授联合提出一种分离作用RFQ的新建议。这种RFQ在加速相对论速度较高的离子时，比常规RFQ有更高的加速梯度[18]。

另一项是1.5 GHz射频超导铌加速腔的实验研究。低温射频超导加速腔不仅具有很高的电效率而且能提供低发射度、低能散度的高亮度粒子束，对于发展高平均功率的自由电子激光器和下一代超高能加速器都具有重要意义。国际加速器界对此极为重视。国内过去因种种原因在这方面长期处于空白状态。1987年以来，陈佳洱运用他在石溪从事超导重离子直线加速器的经验积极推进国内射频超导技术的发展，领导课题小组从无到有地建成了具有较高水平的射频超导实验室，包括建立温度低至2K的低温冷却系统、超高真空系统、超净表面处理设备、高稳定度的微波功率与测量系统以及数字化计算机控制与数据采集系统等。此外，还发展了一套有特色的表面处理工艺，使铌腔的加速梯度在Q约为10^9的状态下稳定地提高到12.6 MV/m[19]（图7）。国家教委和国家高技术计划（863）专家组的鉴定认为该项成果达到了国际先进水平，并且"进度快、成果突出、投资少、效益高"，迅速改变了我国在射频超导加速器领域的空白状态。该项研究于1992年获国家教委科技进步一等奖（甲类）。此后陈佳洱又与赵夔教授合作，进行国家高技术计划（863）课题"国产超导加速腔的研制"以及国家自然科学基金重点项目"高亮度短脉冲光阴极电子枪"[20]的研制等。在以上诸项研究中，陈佳洱获得了国家科技进步二等奖,国家高技术计划先进个人奖和光华科技奖。此外,在广泛研究的基础上,陈佳洱还主编了高校专业教科书"加速器物理基础"[21],为国内高校加速器课程所选用,并获部级优秀教材一等奖。

陈佳洱与国际加速器和物理界建立了广泛的合作交流关系,多次应邀作国际会议特邀报告,受聘担任欧洲粒子加速器会议等国际系列会议顾问委员会委员或大会的执行主席,还主持举办了"第五届国际离子源会议"、"北京自由电子激光国际研讨会"、"北京加速器质谱国际研讨会"等国际会议等。陈佳洱现任国际杂志"粒子加速器"远东地区编委,亚太物理学会联合会理事长等职。

参 考 文 献

[1] C.E. Chen, P.S. Rogers. Experimental Studies of the Central Region of the Cyclotron. RHEL/R 116 (1965), U.K.

[2] C.E. Chen, P.S. Rogers. On the Centering of Cyclotron Orbits by Means of Harmonic Coils. NIRL/M/76, (1965), U.K.

[3] Chang Hong-jun, Yi Guan-qi et al. The INR 120-cm Cyclotron Conversion Project. Proceedings Japan-China Joint Symp. on Accelerator for Nuclear Science and their Applications, University of Tokyo, 1980, Japan, p. 101

[4] Chen Chia-erh & Zhang Ying-xia et al. Design of 4.5 MV Electrostatic Accelerator. Proceedins Japan-China Joint Symposium on Accelerator for Nuclear Science and their Applications, University of Tokyo, 1980, Japan, p. 19 见 北京大学静电加速器小组, 4.5 兆伏静电加器, 全国加速器技术交流论文选 编,(1980) p. 43

[5] Chen Jiaer, Zhang Ying-xia et al. Progress and Applications of the 4.5 MV Electrostatic Accelerator. Proceedings Japan- China Joint Symposium on Accelerator for Nuclear Science and their Applications, University of Osaka, 1993, Japan, p. 174

[6] 陈佳洱,于金祥,李认兴,韦伦存,巩玲华等, 北京大学 6 MV 串列静电加速器的运行及其应用."原子能科学技术" Vol. 27,5(1993)401

[7] Chen Chia-erh, Li Kun, Guo Zhiyu et al. Beam Characteristics and Preliminary Applications of the Tandem Based AMS at Peking Uninversity. Nucl. Instr. & Meth. in Physics Research, B 79(1993)624

[8] 郭之虞,李坤,刘克新,鲁向阳,陈佳洱等. 北京大学加速器质谱计研究与应用 进展. 自然科学进展,Vol. 5, 5(1995)513

[9] 陈佳洱. 静电加速器脉冲化装置及有关特性. 全国加速器技术交流论文选编 p. 75

(1980)(北京大学内部研究报告 BDJ-B-1/1978) 陈佳洱. 离子在高频切割器中的运动. 高能物理与核物理 Vol. 4 3(1980)401

[10] G. S. Xu, Z. B. Qian, C. E. Chen High Efficiency Two Harmonics Beam Chopper. Rev. Sci. Instrum. Vol. 57 5 (1986)795

[11] Chen Chia-erh, Guo Zhiyu, Zhao Kui Bunching Characteristics of a Buncher Using Helical Resonators IEEE. Trans. NS-30, 2(1983)1254

[12] J. M. Brennan, Chen Chia-erh et al. Bunching System for the Stony Brook Tandem LINAC Heavy Ion Accelerator IEEE. Trans. NS-30 4(1983) 2798

[13] Fang Jiaxun, Chen Chia-erh An Integral Splitring Resonator Loaded with Drift Tube and RFQ IEEE. Trans. NS 32 5(1985)2891

[14] Chen Chia-erh, Fang Jiaxun Li Weiguo and et al A 26 MHz Prototype Integrated Split-ring Resonator and the Full Power Test. Proceedings Japan-China Joint Symposium on Accelerator for Nuclear Science and their Applications, University of Osaka, 1993, Japan, p. 52

[15] 陈佳洱，方家驯，李纬国，潘欧嘉，李德山. 重离子整体分离环高频四极场(RFQ)加速结构的研究. 自然科学进展, Vol. 4, 3(1994)271

[16] 于金祥，陈佳洱等. 核物理动态 Vol. 13, No. 2(1996) 34

[17] C. E. Chen, J. X. Fang, J. F. Guo, W. G. Li, D. S. Li, X. T. Ren, J. X. Yu. Experimental Studies on the Acceleration of Positive & Negative Ions with a Heavy Ion ISR RFQ. Proc. 5th European Particle Accelerator Conf., June 1996, Sitges (Barcelona), IOP Publishing, Bristol & Philadelphia p. 2702

[18] 陈佳洱 方家驯. 分离作用 RFQ . 专利申请号：98119331.5

[19] 赵夔，王光伟，王莉芳，陈佳洱等. 1.5 GHz 铌腔 RF 超导的实验研究. 强激光与粒子束 第四卷 第一期, 1992 年 2 月, p. 15~32

[20] Chen Chia-erh, Zhao kui, Zhang Baocheng, Wang Lifang, Geng Rongli, Xu Zengquan. The Progress of Superconducting Cavity Studies and R&D for Laser Driven RF Gun at Peking University. Proc. 5-th Japan-China Joint Symposium on Accelerators for Nuclear Science & their Applications, Oct. 1993, Osaka, Japan, p. 244

K. Zhao, R. L. Geng, L. F. Wang, B. C. Zhang, J. Yu, T. Wang, G. F. Wu, J. H. Song and J. E. Chen. Design and Construction of a DC High Brightness Laser Driven Electron Gun. Nucl. Inst. & Meth. in Phys. Res. A, 375(1996) 147

Kui Zhao, Yin-E Sun, Bao-cheng Zhang, Lifang Wang, Rongli Geng, Jiaer Chen The High Brightness Ultra-short Pulsed Electron Beam Source at Peking

University. Nucl. Inst. & Meth. in Phys. Res. A, 407(1998)322
[21]《加速器物理基础》原子能出版社, 1993 (562页, 合作者: 方家驯, 李国树, 裴元吉, 郭之虞)

图1 4.5 MV 静电加速器

图 2 2×6 MV 串列静电加速器

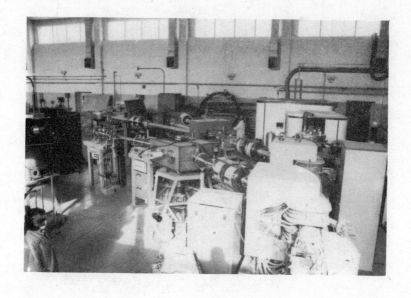

图 3 北京大学加速器质谱计束流线

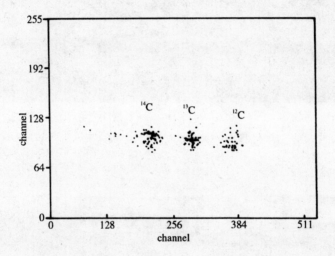

图 4 碳同位素的二维质谱图

图 5 螺旋波导装置

图 6 整体分离环 RFQ 结构

图 7　陈佳洱在实验室

第一部分　回旋加速器研究

EXPERIMENTAL STUDIES OF THE CENTRAL REGION OF THE CYCLOTRON[1]

Abstract

Measurements of the efficiency of ion acceleration have been made on a small fixed-frequency cyclotron. Losses in the first parts of the orbit have been analysed for the fundamental and third-harmonic modes of acceleration. Some ways of improving focusing have been studied. The measurements are part of the design work for the A. E. R. E. Variable Energy Cyclotron.

1. Introduction

An analysis of beam motion and losses in the central region of the cyclotron has been made by means of the measurements described below. Since most of the losses are incurred during the early part of the trajectory, interest has been concentrated on the first few orbits. The experimental system on which the measurements were performed is typical of most isochronous cyclotrons at present, consisting of a source and puller system at the centre of a sector-focus magnet system and a single dee and dummy-dee main accelerating r. f. system. This small working model was set up as a scale model of the central region of the Variable Energy Cyclotron being constructed by the Rutherford Laboratory for the A. E. R. E. at Harwell and the results obtained are primarily applicable to that machine. However, insofar as the measurements reveal some basic limits of this type of accelerating system, they indicate the features to be improved if beam current output, efficiency and quality are to be improved.

---

[1] Coauther: P. S. Rogers, Variable Energy Cyclotron Group, Rutherford High Energy Laboratory, Chilton, Didcot, Berkshire. Reprinted from the RHEL/R 116(1965), U. K.

2. Description of the Model Cyclotron

The cyclotron on which these measurements were made has been described by Clark[1], and earlier (with different pole pieces) by Snowden[2]. Ion trajectories geometrically similar to trajectories in the V.E.C. can be obtained by scaling the dee volts and magnetic field. The scaling factor is such that an ion in the model which has reached a radius R_m has performed the same number of turns as an ion which has reached a radius R_f in the V.E.C.; R_m and R_f are radii such that the magnetic modulation (flutter) and hence the magnetic focusing forces are approximately the same in the model and the V.E.C. The height of the ridges has been arranged so that the ratio of the radii is the same as the ratio of the mean gap widths of the model and the V.E.C., a ratio of 0.633. The requirement of equal turns leads to a scaling relation between the magnetic field, dee voltage and charge-to-mass ratio employed in the model (B_m, V_m, N_m/A_m) and the corresponding quantities in the V.E.C. (subscript f) such that

$$\left(\frac{B^2}{V} \cdot \frac{N}{A}\right)_m = 2.5 \left(\frac{B^2}{V} \cdot \frac{N}{A}\right)_f \tag{1}$$

The frequency of the R.F. dee supply of the model is fixed at 20 Mc/s. which restricts the sorts of ion which can be simulated. Most of the measurements have been made accelerating hydrogen ions.

Since the description of the machine by Clark various changes have been made. The dee geometry has been altered to scale to the V.E.C. dee in the ratio of the gaps. Twelve pairs of orbit trim coils potted in resin and water-cooled have been installed. Upper and lower coils are separately fed by stabilised supplies. The pulse duration of the R.F. power supply has been increased to 0.5 ms. Harmonic coils with a system for injecting a first harmonic magnet disturbance of known amplitude and phase have been fitted[3].

The frequency of the R.F. supply is crystal-controlled and the R.F. power supply is pulsed at about 1% to reduce power and radiation. 1 mA of protons can be accelerated out to 7 inches and 2.55 MeV, or 0.5 mA to 7.5 inches and 3 MeV. The maximum magnetic field is 13 kg, and the maximum dee voltage used is 30 kV. Fig. 1 shows the system of measuring probes in relation to the pole system, the dee and the direction of ion rotation. The central geometry and typical shape of source and puller is shown in Fig. 2. The source is of the hooded-arc type, having a hot tungsten filament with an insulated anti-cathode at the top of the source chimney. Various materials have been used for the source and puller. Carbon is easily broken, small diameter (0.19 inch) copper sources sag, molybdenum is stronger but suspected of sputtering. For the larger diameter source (0.3 inch nominal) on which we ultimately settled, copper is satisfactory.

Various movements of source and puller can be made while the machine is running. The source can be moved across and along the dee gap and also, within limits, up and down in the vertical aperture. The filament can be moved along the axis of the filament tube so as to place it under the hole in the earthed anode. The puller can be moved along and across the dee gap.

The arc supply is fed through a ballast resistor and a current stabilising circuit. The voltage is not stabilised.

3. Measuring Facilities and Techniques

The beam dynamics are deduced from measurements of current and current density at various physical positions, together with the known magnetic field distribution. No energy measurements have been made. Fig. 1 shows the positions in which current measurements can be made. In the N-probe and S-probe positions there are sliding seals permitting target arms to move horizontally and vertically. Calibrated protractors enable the arms to be set radially to the machine centre and there are calibrated scales for radius from centre and vertical position. These two arms have connections for the collection of current. There is also a radius arm mounted on the dee which permits a target to be set at calibrated positions on the dee side of the machine, but current cannot be measured from this probe separately from current reaching the dee and puller. The dee is insulated, so the total current to it can be measured. In addition to these probes there are slides S_1, S_2 on which obstructing fingers or other earthed targets can be placed. The dee voltage is measured by a condenser plate feeding a peak voltmeter while an r. f signal related to the dee voltage can be obtained from a plate on the other side of the dee. The rectified envelope of the pulsed r. f supply can be examined from an oscilloscope monitor point on the dee voltmeter.

Two sorts of probe head are used in the North and South probe arms. Total beam current is measured by a simple insulated plate, tall enough to span the whole of the vertical beam aperture and wide enough radially to collect all of the widest turn. With such a target the position of individual turns is not known unless there is loss from turn to turn, when the turns are indicated as changes of slope in the curve of current versus target radius (Fig. 3(a)). A target which measures beam current density is shown in Fig. 4. This exposes a strip 0.020" wide, to the beam. With this, individual turns can be plotted as peaks. With increasing radius, or at radii where orbit centres are precessing towards the probe regions, are reached where turn spacing is so small that turns superimpose and the current density then rises (see Fig. 3(b)). Both total current and current-density probes are needed, since the beam may be declining with radius for some reason and the only indication on the current-density plot would be a reduced rate of increase of density with radius.

The current can be displayed in various ways. The pulsing duty cycle of about 1% may alter slightly with pulse shape so that an oscilloscope display is needed to convert mean current readings to peak values. Current arriving on the puller and dee is read by a microammeter as this circuit is too heavily by-passed to earth to obtain a pulse picture. Current from the probes is observed by an oscilloscope showing the level during the pulse. The signal from the South probe is fed into an X-Y recorder, the probe being motor-driven with radial position indicated to the recorder as X-position signal. This gives automatic continuous plotting to speed up the radial plot, which needs good discrimination. With the same recording arrangement a vertical scan can be made by hand. Some of the earlier radial measurements were made by hand, e. g. Fig. 14, and these do not show the detail of plots like Fig. 3.

Before making observations on the beam the orbits are centred on the centre of the magnetic system. We adopt the convention that a centred beam is one that would give equal currents in probes at three different azimuths at a radius large enough for turn separation to be negligible. Allowance would have to be made for flutter. In the model the number of turns is small and the turn separation is not negligible. The conventional centre, which would be a close approximation to the centre of the mean circle about which the "scalloped" orbits lie at a large radius such as the extraction radius of the V. E. C. is constructed as follows. Current is received on a target at the radius concerned, generally about the maximum in the model, 6 or 7 inches. The other two probes are used in turn to intercept 50% of the current. The radii from magnet centre of the latter two probes are corrected for increase of radius due to acceleration, approximated by a smooth spiral. The radii are also corrected for 'scalloping' due to the magnetic sector modulation. We then have three points on the circle around the conventional centre from which the centre may be found.

Positive ion beam current measurements on probes may be inaccurate owing to electron emission. During probe measurements with a 1 mA beam at about 2. 6 MeV and 2% duty cycle (52 watts on the target) there was error due to thermal emission of electrons from the hot target. This was corrected by mounting plates horizontally above and below the beam aperture, carried on the target holder. The electron current was intercepted by the plates and fed back to the target. Secondary electron emission gave no trouble. Beam current has been measured with and without the electron capture plates, under conditions which avoid thermal emission. There is no difference within the accuracy of the measurements. Some probe measurements of the first turn show negative signals indicating the presence of electrons at the centre of the machine. These may be due to multipactor effect, secondary emission from parts of the accelerating structure or emission from the source plasma during positive half-cycles of the r. f. The negative signals are azimuthally grouped near the ion source slit and sometimes can be avoided by small changes in probe angle. Unfortunately there is no room for electron capture plates when probing at the centre.

In the N-or S-probe positions there is provision for inserting a phase probe. This consists of a tungsten target shielded within a box with a window formed of a fine grid of horizontal thin wires. The distance between screen wires and target can be controlled so that the beam orbit is able to strike the target without striking the frame on which the grid is carried. The grid gives good screening from r. f. Fig. 5 shows a signal from the phase probe displayed on a sampling oscilloscope. Since the r. f is pulsed, some electronic means is provided to give a stable r. f phase reference for the sampling oscilloscope picture.

Phase measurements can also be made by a modification of the method of Garren and Smith[4](UCRL-10758) Setting the differential type probe at the radius at which it is desired to know the beam, the magnetic field is tuned through the resonant position. Fig. 6 shows such a plot. The point X at which the current density becomes large shows the points at which the phase of the beginning or end of the beam bunch reaches $-90°$ or $+90°$ respectively. The corresponding $-$ve phase point is obscured because of vertical loss at centre. The phase excursion involved,

$$\Delta(\sin\varphi) = \frac{h\pi}{1.8^2} \frac{B}{V} \frac{N}{A} \int_{R_1}^{R_2} \Delta B \mathrm{d}(r^2) \tag{2}$$

where　　$h=$ harmonic number

$B=$ magnetic field (kg)

$V=$ peak dee voltage (kV)

$B=$ magnetic field change (kg)

$r=$ radius (inches)

The formula is subject to the approximation that the synchronous field level does not change between radii R_1 and R_2. Knowing the phase excursion needed to reach $-$or $+180°$, the phase of the beam beginning and end can be found at any radius. Greater accuracy would be obtainable by altering the r. f. frequency, as in the original reference, but this cannot be done in the model.

4. Composition of the Initial Beam

We show in Fig. 7 the form of source chimney and puller with which most results have been taken, together with various relative positions. The phase range of the ions in the beam, and the quality of the beam will be affected by these settings.

During the first measurements to try to centre the beam it was noticed that in some cases the apparent radial width of the turns was very large and it was found that this was due to a mixture of ions in the orbits. The composition of the source output in ion species was investigated in two ways. In one,[5] D. C. conditions were used, avoiding a spread of radii due to ion phase with respect to the r. f. and in the other, a current-density measuring target was used to separate out the individual turns under working conditions.

For the D. C. test the cyclotron source was installed in apparatus used for ion source development, which had a dee system with a steady negative potential. There was an axial magnetic field and a collecting cup and means for measuring the radius of curvature of ions. The peaks of current collected at various radii could be identified by their radius in the constant magnetic field after being accelerated through the constant potential difference between source and dee. There was a puller electrode similar to that used in the model cyclotron, with the added advantage of being able to measure current arriving at the puller separately. With the source power available the maximum proportions of H_1^+, H_2^+ and H_3^+ obtained are as shown in the table.

Table 1 (From Ref. 5) Composition of Source Output

Arc Volts /V	Arc Current /A	Gas Flow st. c. c/min.	Total Ion Current/mA	% H_1^+	% H_2^+	% H_3^+
100	0.45	9.5	18	13	54	33
110	0.40	3.8	18	20	70	10
100	0.40	6.0	23	35	50	15

Other useful results may be found in the reference, but we show only those results which refer to similar conditions to those found to give maximum beams of the various ions in the model. Results were not very dependent on dee voltage or magnetic field. The amount of beam hitting the puller was measured and found to be only about 1% under these D. C. conditions. The table shows that when conditions are optimised for any particular ion there are still large proportions of the other two ions present. The full results show that a beam consisting of 95% of protons can be obtained if increased arc power is used.

Information on the ion species in the model was obtained from plots of current density as a function of radius using the screened strip target mentioned in Section 3. An example is shown in Fig. 8 which refers to a beam of H_2^+ ions accelerated in 3rd harmonic mode. We know from the Table above and reference 5 that by changing the source gas pressure we can increase the contents of either protons or H_2^+ ions. When this is done we obtain changes in the level of the peaks representing the first few turns as shown in the figure. The numbers are settings of a valve in the gas flow line, later correlated with a gas flow meter. In the case of mixed peaks a rough computation of orbit plots done by a method due to Willax[6] with simplified starting conditions shows the order in which one would expect to record turns composed of various ions present in the source output as the probe radius is increased. Figs. 9 and 10 together with the orbit plots in Fig. 11 show this. This knowledge enables a mixed peak to be analysed by shadowing one constituent before it comes to the measuring probe. H_3^+

ions can be distinguished from H_1^+ by the decline in their number when the arc voltage is raised. If the H_1^+ beam is blocked at the centre a plot of H_3^+ can be obtained, at a level of 0.5 mA at maximum radius, Fig. 12. Thus by varying ion source conditions, and an approximate knowledge of the orbits, it is possible to identify beams of unwanted ions.

Some measurements were made to compare the amount of beam coming from the ion source with that lost on the puller under normal operating conditions, with r.f. on the puller. This was done by first placing the puller so as to obstruct the beam emerging from the source. Since it is wide compared with the source it should collect most of the beam and error arising from the redistribution of the electric field when the puller is moved from its normal position should be small. Peculiarities of the model make it necessary to do the measurement using an H_2^+ beam operating on 3rd harmonic mode, since the large H_2^+ content would be lost to the dee and puller and recorded as puller current if a proton beam at fundamental mode was used. The probes cannot penetrate far enough to intercept this H_2^+ orbit. The results of this measurement are as follows:

Total Source Current   9.4 mA

Current on puller   3.2 mA

The operating conditions for these measurements were:

Source Slit 0.060×0.3 inches high

Puller Slit 0.16×0.4 inches high

Source-Puller Spacing   0.13 inches

Dee Voltage   25 kV

Thus the proportion lost on the puller is much more than in the D.C. case. As pointed out in reference 5 this may be due to the influence of electrons extracted from the source during the positive half-cycle, which would increase the proton current at the beginning of the negative half-cycle when the r.f. voltage is low.

These measurements show the difficulties caused by unwanted ion species. Experiments must be designed to remove them or subtract their effect. It should be mentioned that troublesome spontaneous changes of source output sometimes occurred. Some examples are shown in Fig. 13. It is not known whether these are due to the design of our source, or to the effect of the pulsed r. f. dee voltage. The main effect was spontaneous variation of H_3^+ content, which made some proton current density measurements difficult. The presence of these fluctuations was readily detectable in the display of beam current density. In general results were repeatable.

5. Beam Characteristics in the Horizontal Plane

The main possibility for loss in the horizontal plane is collision with the source or puller. The resonances which might be troublesome when $Q_r = 1$, such as the $Q=3/3$ resonance and the gap crossing resonance, can be avoided by keeping the beam properly centred. Excessive centre movement, noted during our measurements and possibly to be attributed to gap-crossing resonance, was easily cancelled by means of harmonic coils. These were controlled by a system which makes the application of harmonics of any desired amplitude or azimuth easy[3].

The orbits of ions ejected from the source at different phases of the r. f. field between source and puller and the orbits of ions coming through the puller with different angles in the horizontal plane both come to a focus after one revolution. The avoidance of loss by striking the launching mechanism depends on giving the ion an acceleration away from the dee which is not small compared with the initial acceleration from source to puller. This loss will also be a function of the initial orbit radius, where we have

$$R = \frac{1.8}{B}\sqrt{(VA/N)} \text{ (inches)} \qquad (3)$$

B=magnetic field (kGs), A=Mass No., N=charge state, V=dee volts (kV).

In the V. E. C. the smallest initial orbit occurs when accelerating protons to an energy of 50 MeV in the fundamental mode. For those ions where it is necessary to extend the limit imposed by the r. f. frequency by acceleration in the third harmonic mode, the smallest initial orbit mode occurs when accelerating ions of $N/A=0.27$ to an energy of 10 MeV per nucleon. In the model centre we simulate these orbits using protons and H_2^+ ions for fundamental and harmonic mode acceleration respectively. To obtain the geometrical scale factor referred to in Section 2 the dee volts and magnetic field in the model have to obey the scaling law given by equation (1) in Section 2. Putting the magnetic field values of 13.3 kGs and 17 kGs for fundamental and third harmonic modes respectively and 80 kV for the V. E. C. dee voltage, we arrive at values of 13 kGs, 30 kV and protons to simulate the smallest V. E. C. fundamental mode orbit and 8.7 kGs., 15.6 kV. and H_2^+ ions to simulate the smallest V. E. C. third harmonic mode orbit.

Using these values, measurements of beam current as a function of dee voltage for voltages near the threshold value were made. Fig. 14 shows these curves. As the voltage is reduced the puller has to be re-adjusted in a direction along the dee gap. The target radius affects the accuracy of threshold determination insofar as phase-slip may occur at low voltages. The threshold voltage in the case of fundamental mode is about a third of the simulation voltage. Curves are shown for operation with and without a dummy dee. For third harmonic mode acceleration the threshold coincides with the simulation voltage if no dummy-dee is used, showing that the beam would be lost against the source with this arrangement. A dummy-dee improves the situation so that the threshold voltage is now 2/3 of the simulation voltage. Some curves of beam current for a bigger range of dee voltage are shown in Figs. 15 and 16.

To find whether the loss in the first turn is against the source or the puller the threshold measurements have been repeated using a smaller value of source chimney diameter, 0.2 inch as against 0.375 inch. The threshold was unaltered, showing that loss is mainly against the puller. Estimates of horizontal loss were made by measuring the current contained in individual turns. For this purpose a target which measures current density is used and we plot current density as a function of radius. Plotting the initial turns for the fundamental and harmonic case enables us to compare the beam total current in the first turn, just before passing the source chimney, with that in the second turn. Such a measurement shows that in the model when simulating the smallest fundamental orbit of the V. E. C. 80% of the current in the first turn is found in the second turn, see Fig. 17. Similar measurements carried out for acceleration in the third harmonic mode using H_2^+ ions show that the ratio of second to first turn currents is about 50% for the voltage 15.6 kV, which gives an orbit size scaled to the smallest V. E. C. orbit in the third harmonic mode.

From Fig. 17 it may be deduced that radial and vertical losses both occur. The curve for third harmonic mode limits to 80% whereas it should go closer to 100% if radial losses only were present. As regards comparative loss of the fundamental and harmonic modes, the radii of the turns as measured on the current-density curves show that the distance of the third harmonic mode orbit from the ion source for 22 kV dee voltage is the same as that of the fundamental mode orbit for 30 kV. This agrees approximately with the curve in the above figure which shows 65% loss for 22 kV voltage, where the ratio of first to second turn current for fundamental mode acceleration at 30 kV is 70%.

The radial quality is difficult to assess in the model because the range of ion phases among ions crossing the gap gives a variation in radius which may be nearly as big as the turn separation, about 0.1 inch at the maximum radius. This is a large uncertainty compared with a betatron oscillation amplitude of, say, 0.2 inches. Since the width of turns is a function of azimuth it is difficult to estimate quality by measuring turn widths. We have not made precise measurements of quality in terms of divergence and displacement as we are unable to find any simple way of doing this at the centre.

Some computations of orbits have been done by Mayhook and compared with results in the model centre[7]. These neglect variation of magnetic field but include the detail of electric field around the source and puller, found by a conductivity paper plot. Orbits have been computed assuming a range of starting angles and phases referred to the ion as it leaves the puller. These are then compared with plots of current density against radius and the starting conditions of those orbits which agree are assumed to be those of the orbits in the model. Agreement is construed as correspondence of radii for the first five turns on both North and South probe positions. For the fundamental mode this gives a phase range of -20 to $+40$ degrees (0 degrees being maximum voltage) and a range of starting angles of 80 to 60 degrees, where 0 degrees is the line of the dee edge. The centres of curvature after three and ten turns have also been computed and from the range of starting conditions, the spread of centres along and across the dee gap are approximately both equal to ± 0.4 after ten turns. The corresponding figures for third harmonic acceleration show the centres increasing in spread between three turns and ten turns; this does not correspond to what is observed in the model.

Comparing the computed phase width with experiment, we obtained a picture from the phase probe and sampling oscilloscope showing a phase width of 70 to 80 degrees. Measurements made by detuning the magnetic field showed a phase width of 90 degrees at 7 inches radius (40 turns) with even larger ranges at smaller radii. (See Fig. 18). The figure shows the loss due to phase slip and improvement made by the orbit coils. A smaller range, about 50 degrees, has been measured at 4 inches radius (8 turns) using third harmonic mode acceleration of H_2^+ ions. Phase slip is more severe on this mode, as the iron field requires more correction than our orbit coils can supply at the level of field required for such operation. Measurements made by magnetic de-tuning may be subject to inaccuracies due to slight shifts of magnetic level, due to hysteresis.

Summarising, we would expect to get from the central region of the V. E. C. 70% of the initial fundamental mode beam past the source and puller and so available for acceleration to larger radii, together with a phase range of about 100 degrees at radius of 4". As shown in the next section, vertical and horizontal loss can not be distinguished at the first turn. By restricting the beam quantity and increasing the vertical focusing the vertical loss can be avoided, in which case the percentage of the initial beam which can be got past the puller rises.

The above results apply to fundamental mode acceleration. On harmonic mode, the transmission efficiency and the phase range for the smallest orbit are not so favourable. This can be partly attributed to gap factor effect which reduces the energy gained at the second acceleration, bringing the orbits nearer to the source and puller. The phase slips faster by a factor of the harmonic number, so it is more difficult to preserve a given phase range out to larger radii. The figures here are 50% and a range of 50 degrees for an H_2^+ beam at a radius of 4″ or 8 turns, subject to the inaccuracies noted above.

Harmonic acceleration is required for heavy ions at low energy. These are subject to loss from gas scattering and stripping. The effect of pressure on a beam of H_2^+ ions has been studied and Fig. 19 shows the loss of beam as a function of pressure for various radii. The problem has been studied theoretically by Clark[8] who expresses the beam transmission exponent as a product of two terms, one of which depends on the cyclotron characteristics; as radius, pressure and dee voltage, while the other depends on the energy, particle and medium. From the shape of our curves in Fig. 19 we calculate the latter exponent to lie between the figures for H_2^+ in N_2 and H_2^+ in H_2 taken from the above reference.

6. Beam Motion in the Vertical Plane

The vertical forces are of four kinds:

(i) A force derived from the slope of the mean magnetic field with radius. "Mean" implies averaged over azimuth.

(ii) A force derived from the sectoral nature of the magnet.

(iii) An electric force due to the different times spent by the ion in the focusing and defocusing parts of the electrical accelerating field.

(iv) An electric force due to the variation of the electric field with time during the passage of the ion.

(v) A defocusing force due to the space-charge.

The focusing force (i) is small at the centre. A suitable slope may be introduced at a small radius either by shaping the iron field or by orbit coils. If isochronism is to be maintained the slope must change at larger radii so that the field increases with radius to match the relativistic increase of mass of the ion. This positive slope gives a defocusing force. The 'flutter focusing' force (ii) is small at the centre owing to the weak modulation of the magnetic field at small radii. Until the radius at which flutter focusing is effective is reached, the electrical forces are predominant. The 'energy-change' force (iii) is always focusing but the force (iv) is defocusing if the ion arrives at the centre of the electric accelerating region before the instant of maximum field (negative phase angles), is zero if it arrives at the instant of maximum and focusing if it arrives later, (positive phase angles).

In the Model Centre the isochronous curve is practically flat, as the Fig. 20 shows. The iron field has a defocusing shape at the centre due to a hole, intended for axial injection, which is only partially filled. The above figure refers to the field as set for proton acceleration. Limited changes of slope can be made by means of orbit coils, as shown in Fig. 21. Fig. 22 shows the rate at which the 'flutter' rises with radius. Fig. 23 shows the iron field and the difference made by the orbit coils when the mean field is set to accelerate H_2^+ ions in the third harmonic mode. The iron field here has a pronounced dip at about 5 inches radius which the orbit coils are unable to correct fully.

A curve of the combined vertical focusing force as a function of radius when the machine is set for proton acceleration is shown in Fig. 24, illustrating the strong effect of ion starting phase in the central region.

The first observations of ion motion in the vertical plane were measurements of beam density within the vertical aperture at different radii. A target exposing a small rectangle, eventually reduced to 0.020 inches high by 0.050 inches radially, to the beam was traversed vertically within the vertical aperture. When the radial current density plots had established the position of individual turns, the vertical current density probing was performed on individual turns. Fig. 25 shows examples of these plots. They show that the aperture is full in the early turns of the trajectory. The vertical scans of current density were repeated turn by turn while altering various parameters. Of these, the most important are beam current, varied by varying the source current, to show the effect of space charge, and phase, varied by altering the magnetic field, to show the effect of electric focusing. Fig. 26 shows how the beam spreads vertically as the beam current is increased by increasing the source are current. Similar measurements on the effect of phase showed different effects at different radii, due partly to the effect of phase-slip.

To simplify the situation and avoid the confusion due to space charge and due to a wide range of ion phases a method of selecting a small range of phases was used. This consisted of placing a plate near the start of the ion trajectory having a notch in one edge (Fig. 27). The radial position was such as to pass a range of phases of $\pm 15°$. The range was checked by threshold measurements, varying the dee voltage. The notch now acts as a source of ions of known small phase range and the vertical trajectory of this small beam can then be traced under various conditions. Because the beam is smaller and limited in phase the vertical movements should be predictable from the forces shown in Fig. 24.

With this smaller proton beam the background of H_3^+ ions, which are accelerated in 3rd harmonic mode, became very troublesome. A radial current density plot was made of the background only, stopping the protons by moving the notch up above the dee edge. This determined where the H_3^+ turns were and a finger was set up in the position shown to intercept them. The H_3^+ turns are wide and some mix with the proton turns, but the finger gives considerable improvement.

Fig. 28 shows a plot of the purified beam examined with a target exposing a 0.50×0.50 inch rectangle to the beam, placed at various vertical levels. The plot at median plane level shows the amplitude coming to a maximum at various radii; these are focal points. In between are minima where the beam is diffuse on the median plane. At the corresponding radii there are maxima on the plot taken at dee level, showing vertical expansion and probable loss to the dee. The plots show the positions of the turns which can then be examined in succession. Fig. 29 shows such an examination. The target has been scanned vertically at the radial position of each turn in succession. The beam can be seen spreading and concentrating as the ions oscillate vertically under the influence of the focusing forces.

Fig. 30 is derived from the previous figure by measuring the half-density points. The phase-selected beam through the notch can be seen to come to a focus between the 7th and the 8th turn. Measuring the centre of symmetry shows variations of the magnetic median plane with radius. The position of the focus can be found with more accuracy by making small vertical movements of the notch, and studying the corresponding movements of the vertical distribution where the beam is nearly focused. At a turn before the turn during which the beam is first focused an upward movement of the notch causes an upward movement of the maximum of the current density distribution; at a turn after the focus the maximum moves oppositely to the movement of the notch. At the second focus the situation is reversed. In Fig. 31 an example is shown of the maximum moving oppositely to the notch movement. In this way one can 'bracket' the focus turn and distinguish between first, second etc. foci.

The experimental curve, Fig. 30, should be compared with a theoretical plot of the vertical oscillations of particles with an assumed range of vertical displacement and deviation, using values of the focusing power Q_v^2 calculated point-by-point from the magnetic field data and the formulae for the electrical vertical focusing for the known phase range of the beam through the notch; $\pm 15°$. A convenient method of constructing the vertical trajectory is an analogue computer. Such a computer has been developed for calculating the beam handling elements in the external beam line of the V. E. C.[9] Each gap crossing is treated as a focusing element of strength Q_v^2 where Q_v is calculated from point-to-point. Any interaction between turns is neglected and the elements are spaced along the horizontal axis of the computer display by an amount proportional to the distance which the ion travels between successive gap crossings. Fig. 32 shows the display. The vertical dotted lines mark the position of the elements while the horizontal lines at the top and bottom mark the limits of the vertical aperture, between the accelerating gap. During the first two turns, where the electric focusing is changing rapidly, the focusing is approximated by two elements per turn, later by one element per turn. The starting conditions assumed are a displacement of $\pm 0.05''$, corresponding to the notch height of $0.10''$, and deviations of ± 24 m/radian. The analogue traces a trajectory corresponding to a set of starting conditions and strengths of focusing elements. The figure is a superimposition of photographs of the trajectories corresponding to ion starting phase of $0°$ and $\pm 15°$. Different phase implies different focusing strength in accordance with the curve of Q_v^2 vs. r Fig. 23. Comparing Fig. 30 with Fig. 32 we see that the mixture of trajectories comes to a mean focus somewhere between the 6th and 7th turns for the theoretical curves, while the experimental curve shows a mean focus between the 7th and 8th turn. The difference may be due to space charge, to the inaccuracy of the electric focusing formulae at the start of the trajectory or the linear approximations used in the computer. However, the agreement is sufficiently close to give confidence in the experimental measurement as a comparative method.

The most important changes in vertical focusing are made by altering the mean magnetic field level. This alters the beam phase and hence the electric focusing force. Fig. 33 shows the effect. These curves are derived from a series of turn-by-turn plots similar to Fig. 29. The changes in vertical losses caused by these changes in focusing can be measured by using a target big enough radially and vertically to receive all the current in a turn. Knowing where the turns are and plotting beam current against radius it is possible to compare current in the first turn with current in the subsequent turns. Fig. 34 shows such curves. These curves may be taken as representing vertical loss only, since the low-energy parts of the beam, which might otherwise hit the puller, have been stopped. From Fig. 34 it can be seen that 77% of the beam current in the first turn is successfully accelerated to maximum radius. If the magnetic field is reduced so that the beam goes towards the focusing (positive) phase this proportion rises to 90%, and it falls to 55% if the field is increased so that the beam goes towards the defocusing (negative) phase. In the case of the two latter plots the beam is lost by phase-slip before/radius is reached; in a practical case the beam phase would be corrected at some radius to avoid this. The effect of space charge can be gauged from Fig. 35; a low beam of about 50 μA has much smaller loss at the centre than a 150 μA beam.

The less important improvement in the vertical focusing at the centre made by altering the rate of change of the mean magnetic field with radius is shown in Fig. 36, which can be compared with Fig. 30. The magnetic distributions used can be obtained by reference to Fig. 21. Changing from the distribution used in Fig. 30 (curve 3 of Fig. 21) to the most defocusing central slope (curve 1 of Fig. 21) changes the number of turns required to bring the phase-restricted beam to a focus by one turn.

The above improvements in efficiency of transmission from the first turn to subsequent turns are largely lost when the whole beam is present. Fig. 37 shows that about 50% of the central beam is retained at maximum radius with the magnetic field tuned to optimum. Fig. 38 shows that not much change is made to the central loss by detuning the magnet to bring the ion phase into the focusing phase. Comparison with Fig. 34 and 35 shows that the central loss is much increased when the full beam is present.

The above measurements refer to a beam of protons accelerated in fundamental mode. H_2^+ ions present in the output of the source are rejected at the second acceleration and H_3^+ ions are obstructed or subtracted as mentioned. Analysis of the vertical losses for harmonic working cannot be obtained so fully, owing to the lower current and current-density of the harmonic beams we can produce. We can accelerate H_3^+ ions in 3rd harmonic mode using the same field level as for fundamental mode proton acceleration, or we can accelerate H_2^+ ions on 3rd harmonic mode by reducing the field to 2/3 of the above value. In spite of the greater number of H_2^+ ions available, many measurements are better made with H_3^+ ions since the field of the model centre magnet needs less correction by orbit coils at the proton fundamental mode level, 13 kg, see Fig. 20 and 23.

By setting a notch as illustrated in Fig. 27 to a radius large enough to cut off all the H_3^+ beam except the range with radii large enough to pass through the notch a pure and phase-selected H_3^+ beam was obtained. Fig. 39 shows a plot of current density vs. radius for such a beam coming through the notch placed at median plane level. The protons make two turns inside the radius at which the notch plate is fixed before striking it, and hence are recorded. If this beam is examined with a target capable of discriminating variations of current density in a vertical direction, i.e. the target exposing a small rectangle to the beam, curves such as Fig. 40 are obtained. Here the target is at median plane level. There is insufficient sensitivity in the system to probe at levels above and below the median plane, so a curve of half-current-density points cannot be obtained. However, the effect of changing ion phase by changing magnetic field is shown approximately. With the magnet tuned to optimum the beam is brought to a focus about the 6th or 7th turn, and this is changed to the 5th or 6th turn if the magnet is tuned 5% low. The focusing gets weaker if the magnet is tuned 5% high. Small variations of field are sufficient to produce an effect since the phase slips three times as fast as on fundamental mode.

Fig. 41 shows another radial scan of the current density of the beam from the notch on median plane, with a different distribution of central magnet field. The current density is low and so the signal obtained is too small to enable a plot of half-density heights to be made, but the position and the number of turns required to focus can be noted. Fig. 42 is a plot of current received on a target wide enough to collect a whole turn but small enough vertically to show a difference between current per turn at different heights. Some current is received at near dee height and this comes to a minimum at the radius corresponding to the focus in Fig. 41.

Harmonic operation does not differ very much from fundamental mode operation in respect of vertical losses of the full beam. If we take the whole beam the vertical oscillations will be incoherent and we can take the ratio between the currents in turns two and three as a relative measure of vertical loss, since there will be no horizontal loss after turn two and there will not be serious phase-slip loss in three turns. By integrating current-density vs. radius plots for cases where turns consisting only of one ion species can be obtained. We obtain about 80% of the second turn current in the third turn, in both cases. This applies to third harmonic acceleration of H_3^+ but we have no reason to suppose it different for H_2^+. Fig. 12 shows the current-density vs. radius plot for H_3^+ third harmonic acceleration.

A direct way in which to move the ion phase in the focusing direction is to apply angular setback by twisting the source-puller system physically against the sense of ion rotation so that the ions arrive later at the dee-dummy gap[10]. In order to keep the orbit centre of rotation on the machine centre it is necessary to move the source and puller into a position further away from the dee, see Fig. 43. Even so, the spread of orbit radii along the instantaneous radius of curvature of the zero-phase ion emerging from the puller, partly due to the phase-range and partly due to the more diffuse and unsymmetric fringing field when the source is twisted, will contribute to worsen the quality of the beam at starting.

The transmission efficiency obtainable with this system is shown in Fig. 44 (a). This refers to a proton beam accelerated at fundamental mode. A notch is set up on the median plane and the whole of the current through the notch is measured as a function of radius. The unwanted third-harmonic beam of H_3^+ was obstructed. To reduce it to an acceptable level two fingers had to be used. This may be due to the worsening of the beam quality mentioned above, although owing to the changed angle of arrival of the beam at the probes, it is not possible to make direct comparison of turn widths. The H_3^+ background is plotted separately by moving the notch plate out far enough to obstruct the proton beam. This background may be subtracted from the mixed plot to give the true proton transmission efficiency. Over 90% of the proton beam from the slit present in the first turn is transmitted to subsequent turns. This may be compared with Figs. 34 and 35 for the source geometry without angular setback. Fig. 44b shows the corresponding current density plot showing the position of individual turns.

Analysis of the focusing similar to that performed on the proton beam with the normal source aspect follows. Fig. 45a shows a turn-by-turn vertical scan of the beam through the notch on the median plane. From this Fig. 45b is constructed, which is a plot of the vertical position (with respect to beam centre) of the half-maximum density points as a function of number of turns. Comparing this with Fig. 30 shows that the beam through the notch is brought to a focus in five turns with angular setback as compared with seven to eight turns without it. Fig. 46 shows the method of identifying the first focus by moving the notch vertically. Turn 5 is just beyond the first focus, turn 10 just beyond the second focus. The improved vertical quality as compared with Fig. 30 is mainly due to a realignment of the source and puller.

The effect of angular setback on the total beam is shown in Fig. 47(a) and (b). About 50%~60% of the central beam is retained at outer radii. This is not much improvement over the figure without angular setback, Fig. 37. In Fig. 47 the H_3^+ background has been removed by a finger. The H_3^+ background is a variable amount, sometimes accompanied by bad beam quality, and is thought to be due to spontaneous source fluctuations.

In the above series of vertical measurements we note that some parameters, particularly ion phase, have a large effect on a small element of the beam but much less effect on the total beam. For a beam of given current and phase range there is a maximum of transmission efficiency, defined as proportion of first turn current successfully accelerated out to greater radii under conditions of zero horizontal loss (dee voltage above threshold and no phase-slip or gasscattering). For our 1 mA proton beam this is about 50%. If either the phase-range or the current-strength is reduced the transmission efficiency improves, and it can approach 100% if both these quantities are made small and the electric focusing is optimised.

7. Conclusion

Our measurements show that a large proportion of a beam of a given ion species emitted from the source-puller system can be accelerated away from the central region to larger radii. For similar orbits third harmonic mode acceleration is almost as efficient as fundamental mode acceleration.

For our particular geometry a large proportion of the beam transmission loss is due to the beam expanding vertically and striking the limits of the vertical aperture. The electrical focusing forces due to the accelerating electric field are powerful at the centre, but since they are phase-dependent it remains a problem how to obtain a wide phase-range in the beam without heavy loss vertically. We have not made an experimental study of the focusing due to the ion source and puller geometry; this offers possible scope for further improvement.

Acknowledgments

We should like to make acknowledgment of the work of others which has contributed to this study. The conversion of the magnet poles to a three-sector system and the setting-up of the fields was done by Dr. D. J. Clark who also made some of the early beam measurements. We had useful discussions with Drs. F. M. Russell and J. R. J. Bennett and the D. C. ion source tests performed by the latter form part of our text. Similarly, discussions with Mr. A. R. Mayhook on orbit computations were of great assistance in planning experiments. Messrs. W. E. Walford and T. P. Parry gave much experimental and constructional assistance. Finally, our work forms a part of the design for the A. E. R. E. Variable Energy Cyclotron and we acknowledge the continual help and encouragement of Mr. J. D. Lawson who is responsible for this project.

References

[1] D. J. Clark, International Conference on Sector-Focused Cyclotrons &. Meson Fac-

tories, C. E. R. N. 63~19, 1963, p. 209.
- [2] M. Snowden, "Sector-Focused Cyclotrons", 1959 Publication 656, National Academy of Sciences-National Research Council, p. 97.
- [3] Ch'en, C. E. and P. S. Rogers, "On the Centreing of Cyclotron Orbits by Means of Harmonic Coils", Rutherford Laboratory Memo. NIRL/M/76, 1965.
- [4] A. A. Garren and Lloyd Smith, International Conference on Sector-Focused Cyclotrons and Meson Factories, C. E. R. N. 63~19, 1963, p. 18.
- [5] Rutherford Laboratory Internal Memo. J. R. J. Bennett, N. I. R. N. S. Cyclotron Design Note 500/50/90. 1964.
- [6] H. A. Willax, Lawrence Radiation Laboratory Report U. C. R. L. 9416, 1960.
- [7] Rutherford Laboratory Internal Memo. A. R. Mayhook and Miss C. Smith. N. I. R. N. S. Cyclotron Design Note 500/50/95, 1965.
- [8] Rutherford Laboratory Internal Memo. D. J. Clark, N. I. R. N. S. Cyclotron Design Note, 500/30/060, 1963.
- [9] R. N. Hansford, A. E. R. E. Report R 4869, March 1965; also, submitted to Nuclear Instruments &. Methods.
- [10] W. I. B. Smith, "Improved Focusing Near the Ion Source", Nuclear Instruments &. Methods, 9(1960) p. 49~54.

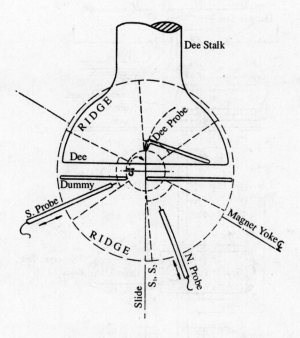

Fig. 1 Plan view of model magnet, dee and probes

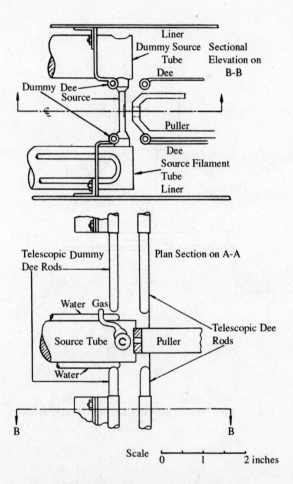

Fig. 2 Central geometry of model cyclotron

EXPERIMENTAL STUDIES OF THE CENTRAL REGION OF THE CYCLOTRON

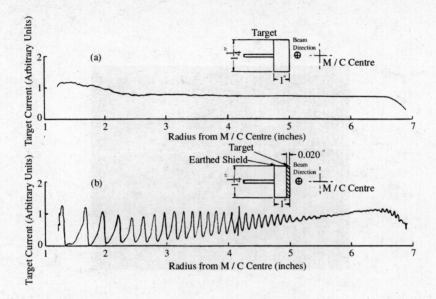

Fig. 3 Typical plots from probes in proton beam
(a) current vs. radius (b) current-density vs. radius

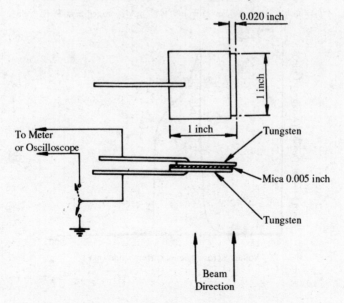

Fig. 4 Target for current density measurements

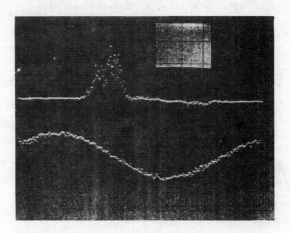

Fig. 5 Beam picture from phase probe
200 mV/cm. vertically, 5 ns/cm. horizontally target at 4″ radius,
$V_{dee}=30$ kV. Proton beam current, 1.25 mA averaged over R.F. cycle

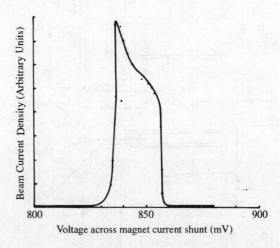

Fig. 6 Beam current density vs. magnetic field level. Proton beam, fundamental mode acceleration. Target radius 7 inches

EXPERIMENTAL STUDIES OF THE CENTRAL REGION OF THE CYCLOTRON

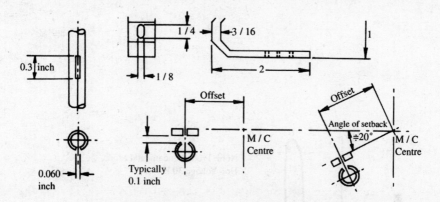

Fig. 7 Source and puller, with relative positions

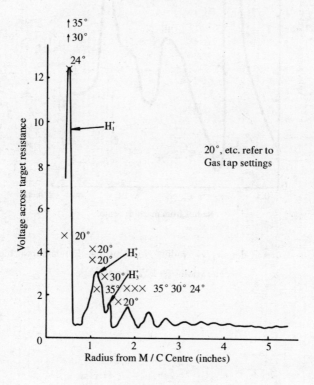

Fig. 8 Beam of mixed ion species

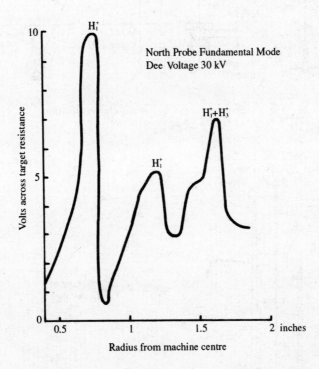

Fig. 9 Current density vs. radius North probe. Fundamental mode acceleration. 30 kV. dee voltage

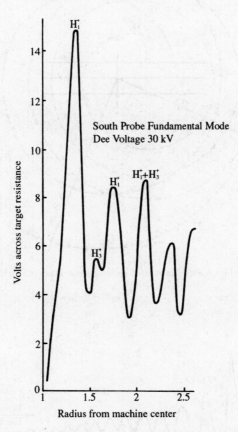

Fig. 10 Current density vs. radius South probe. Fundamental mode acceleration 30 kV. dee voltage

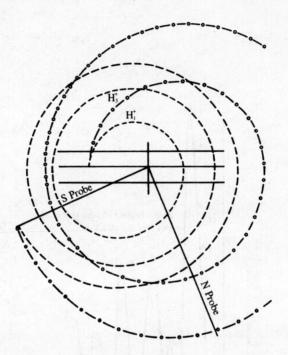

Fig. 11 Ion orbits for fundamental mode acceleration, 30 kV. dee voltage

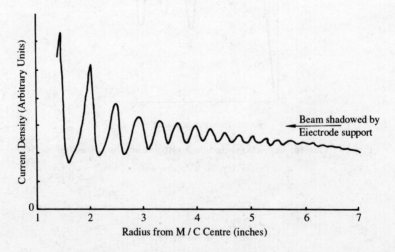

Fig. 12 H_3^+ beam, accelerated in third harmonic mode. Current-density vs. radius (H_1^+ obstructed)

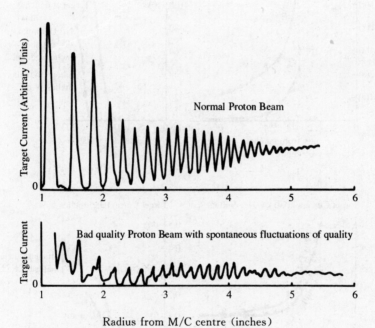

Fig. 13 Examples of spontaneous fluctuation of beam quality

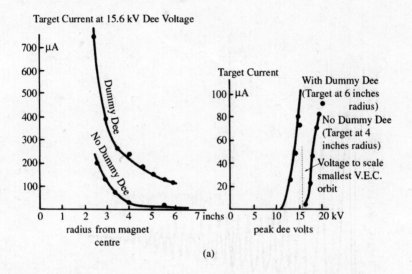

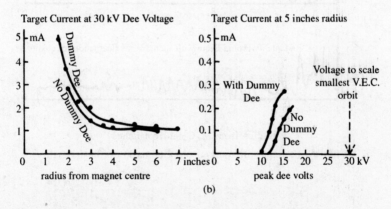

Fig. 14 Beam current vs. radius and dee voltage
(a) H_2^+ beam, 3rd harmonic mode acceleration
(b) Proton beam, fundamental mode acceleration

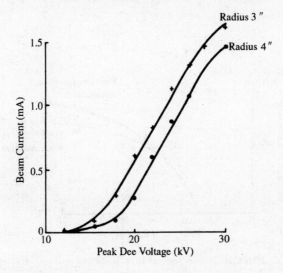

Fig. 15 Beam current vs. dee voltage for proton beam. Fundamental mode acceleration. Arc current 0.4 A.

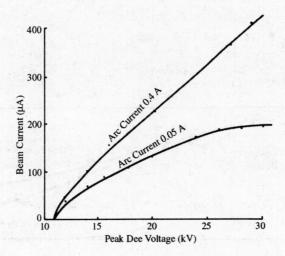

Fig. 16 Beam current vs. dee voltage for H_2^+ beam. Third harmonic mode acceleration. Target at 7 inches radius from M/C centre

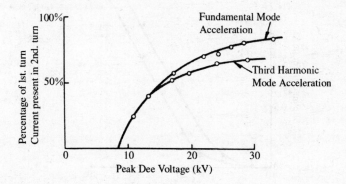

Fig. 17 Ratio of first to second turn beam current vs. dee voltage

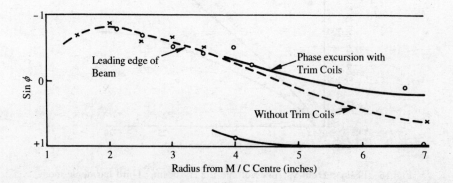

Fig. 18 Phase excursion vs. radius. Proton beam, fundamental mode acceleration

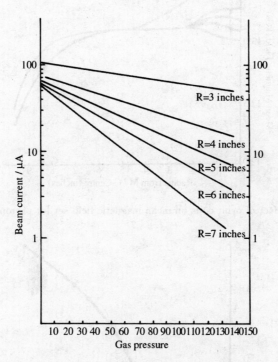

Fig. 19 Beam loss with pressure. Third harmonic mode acceleration. H_2^+ ions. 15.6 kV. Peak dee voltage

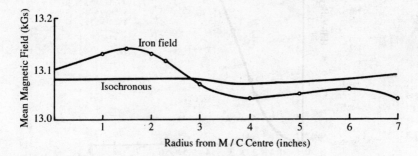

Fig. 20 Calculated isochronous field and measured iron field for proton acceleration

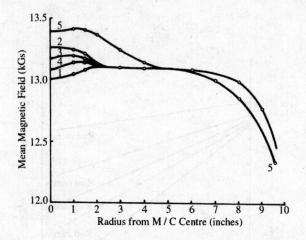

Fig. 21　Effect of orbit coils on mean magnetic field set for proton acceleration

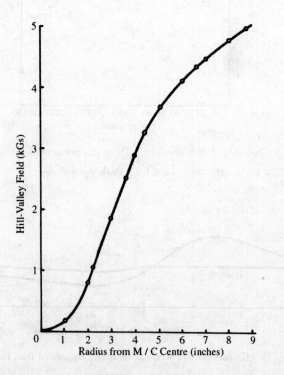

Fig. 22　Magnetic flutter modulation vs. radius, with main field set for proton acceleration

EXPERIMENTAL STUDIES OF THE CENTRAL REGION OF THE CYCLOTRON

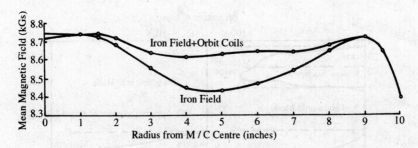

Fig. 23 Effect of orbit coils on magnetic field set for H_2^+ third harmonic mode acceleration

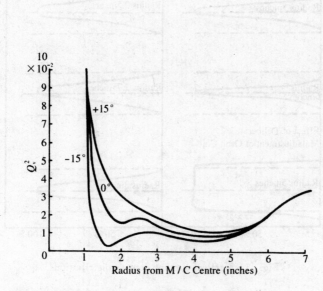

Fig. 24 Total vertical focusing force vs. radius for various ion starting phases

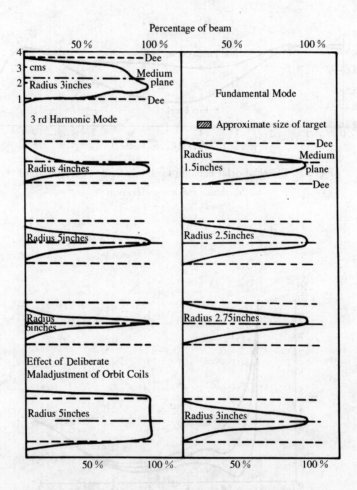

Fig. 25 Vertical current density distributions

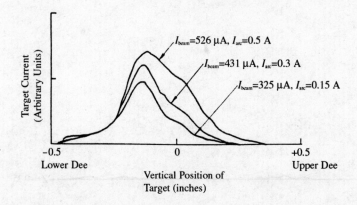

Fig. 26 Effect of beam current on vertical distribution.
Target $0.050 \times 0.020''$ radially

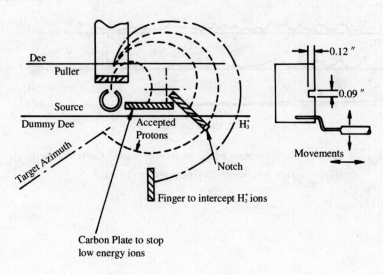

Fig. 27 Arrangement for phase selection of beam

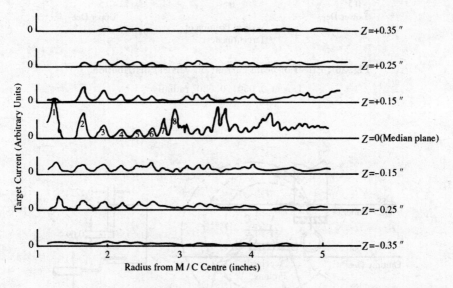

Fig. 28 Proton beam through notch on median plane. Current density vs. radius at various vertical levels of target

EXPERIMENTAL STUDIES OF THE CENTRAL REGION OF THE CYCLOTRON

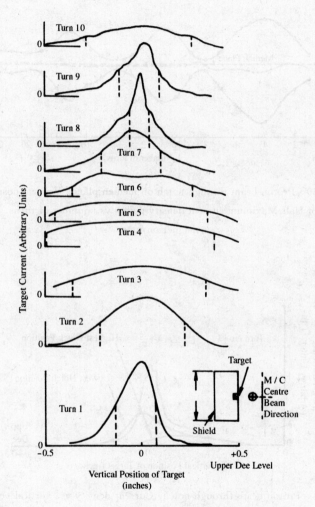

Fig. 29 Proton beam through notch on median plane. Turn-by-Turn vertical scan of current density

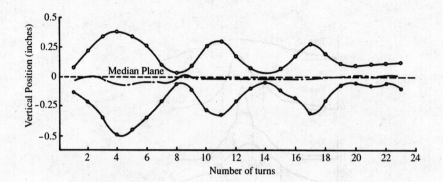

Fig. 30 Proton beam through notch on median plane. Vertical position of Half-Maximum current density points vs. number of turns. magnet on tune

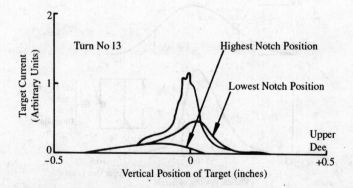

Fig. 31 Proton beam through notch. Current density vs. vertical position of target, with notch height as parameter

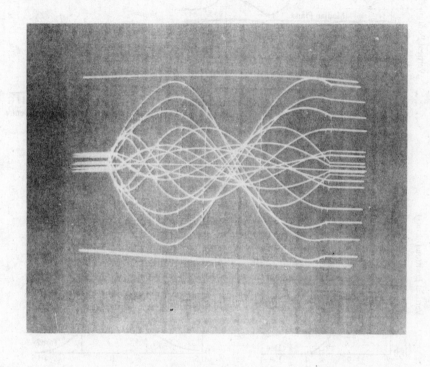

Fig. 32   Trajectories plotted on analogue computer

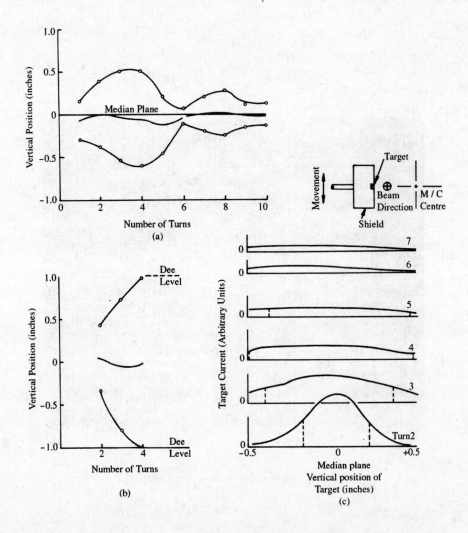

Fig. 33　Proton beam through notch on median plane. Vertical position of Half-maximum current density points vs. number of turns
(a) Magnet tuned 2% low. (b) Magnet tuned 1% High.
(c) Turn-by-turn vertical scan of current density. Magnet tuned 1% high

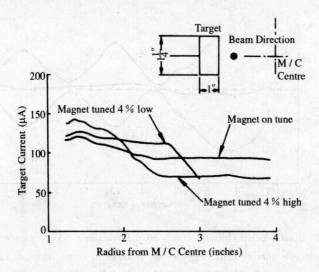

Fig. 34 Proton beam through notch on median plane. Beam current vs. radius with magnet tune as parameter

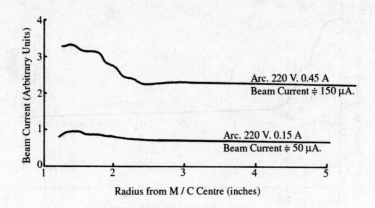

Fig. 35 Proton beam through notch on median plane. Beam current vs. radius with beam current as parameter

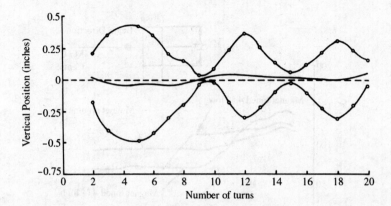

Fig. 36　Proton beam through notch on median plane. Vertical position of Half-maximum current density points vs. number of turns. slope of mean magnetic field with radius De-focusing (Curve 1 of Fig. 21)

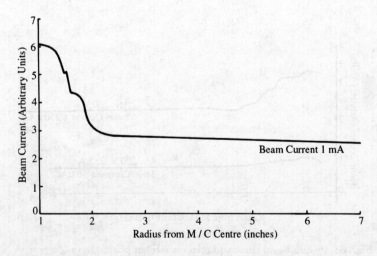

Fig. 37　Total proton beam current vs. radius. Magnet on tune. 30 kV. peak dee voltage

EXPERIMENTAL STUDIES OF THE CENTRAL REGION OF THE CYCLOTRON

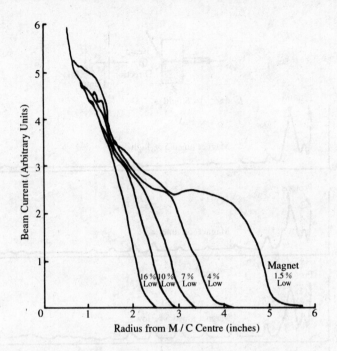

Fig. 38 Total proton beam vs. radius with magnet tune as parameter

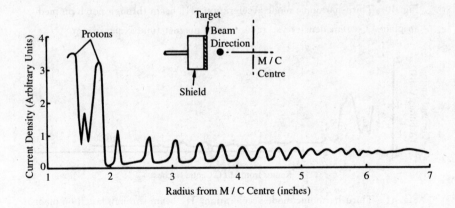

Fig. 39 Third-Harmonic mode acceleration. H_3^+ Beam through notch on median plane. Current-density vs. radius

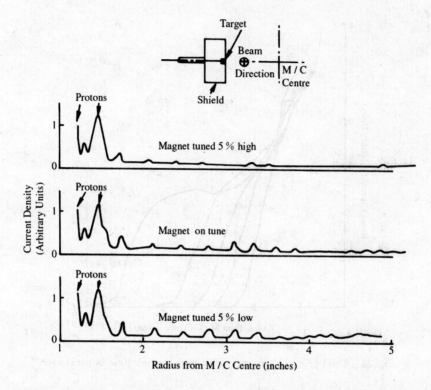

Fig. 40 Third-harmonic mode acceleration. H_3^+ beam through notch on median plane. Current density vs. radius, with magnet tune as parameter

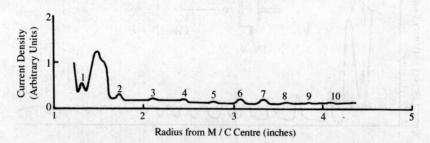

Fig. 41 Third-harmonic mode acceleration. H_3^+ beam through notch on median plane. Magnet on tune. Current density vs. radius, with target at median plane

EXPERIMENTAL STUDIES OF THE CENTRAL REGION OF THE CYCLOTRON

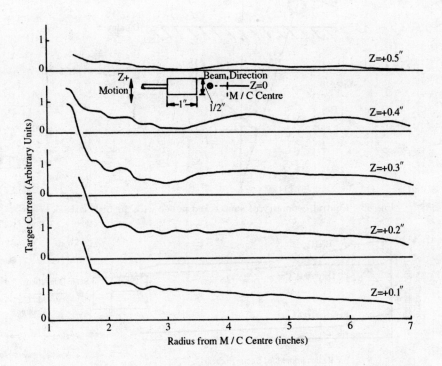

Fig. 42 Third-harmonic mode acceleration. H_3^+ beam through notch on median plane. Current vs. radius with vertical position of target as parameter

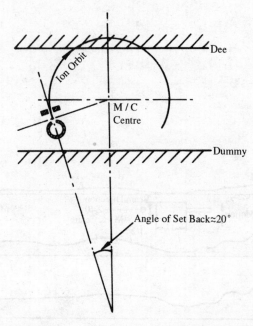

Fig. 43 Central geometry of source and puller with angular setback

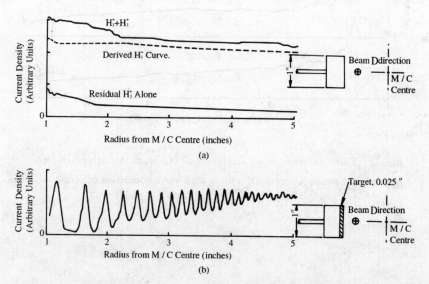

Fig. 44 Source and puller with angular setback. Proton beam through notch on median plane.
(a) Beam current vs. radius. (b) Current density vs. radius

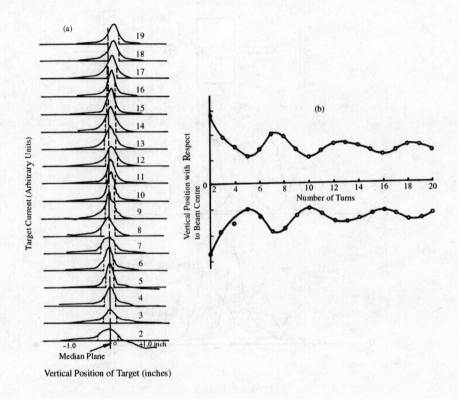

Fig. 45 Source and puller with angular setback. Proton beam through notch on median plane.
(a) Turn-by-turn vertical scan of current density. (b) Distance from beam centre of Half-maximum density points vs. number of turns

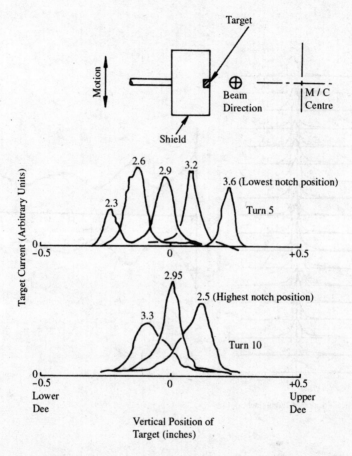

Fig. 46 Source and puller with angular setback. Proton beam through notch. Current density vs. vertical position of target with notch height as parameter (Numbers opposite peaks refer to vertical notch traverse)

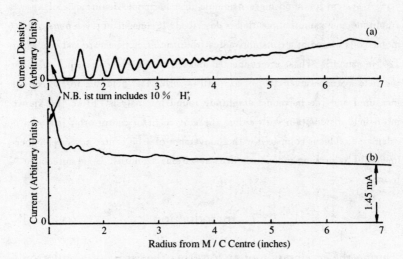

Fig. 47 Source and puller with angular setback. Total proton beam.
(a) Current density vs. radius (b) Current vs. radius

ON THE CENTERING OF CYCLOTRON ORBITS BY MEANS OF HARMONIC COILS[①]

Abstract

A system for imposing a magnetic first harmonic disturbance of known amplitude and azimuthal position is described. It consists of three pairs of harmonic coils placed at a constant radius from the machine centre and spaced by 120° in azimuth. These are connected and fed with current in such a way that the field averaged over 360° of azimuth is zero. The field due to the coils is measured and the harmonic amplitude found by analysis. From the known mean field distribution with radius the movement of centre orbit per turn is calculated. This is compared with a movement of orbit centre measured experimentally by probing individual turns at various azimuths. Good agreement is found.

1. Introduction

During the acceleration of an ion from the centre of a cyclotron to its maximum radius, the instantaneous centre of curvature traces out an oscillatory path near the centre of the machine. This movement is greatest during the first few turns and in the absence of harmonic errors in the field settles down to within about 0.001 inches after some 20 turns. It is desirable that at large radii the centre of curvature should coincide with that of the machine. The factors which control the ultimate position of the centre will now be considered. Firstly, there is a series of movements of orbit centre of curvature at the beginning of the acceleration process, due to the

---

① Coauther: P. S. Rogers, Variable Energy Cyclotron Group, Rutherford High Energy Laboratory, Chilton, Berkshire. Reprinted from the NIRL/M/76 (1965), U.K.

relatively large fractional increase in momentum at each passage through the electric accelerating field. Secondly, any residual first harmonic component in the magnetic field will move the centre. Thirdly, there may be movement due to "gap-crossing resonance". [4] The ion source has to be placed so that the centres of orbits converge on the magnetic centre. However, accurate control by means of ion source movements may be inconvenient. The measurements described here concern a system of harmonic coils which will give easy and accurate control of orbit centres after the source system has been approximately placed.

2. Description of System

Harmonic coils have been added to a small sector-focused cyclotron which is used for central-region studies concerned with the Variable Energy Cyclotron. This model machine has been described in the proceedings of the C.E.R.N. International Conference on Sector-Focused Cyclotrons[1]; immediately after that report orbit coils were added and the central geometry modified. The radiofrequency is crystal-controlled at 20 Mc/s. but by scaling the dee voltage and magnetic field one can obtain orbits which are geometrically similar to orbits in the V.E.C. [2] The scale is such that there will be the same number of ion revolutions out to radius R_m in the model as there are revolutions of the ion being simulated out to radius R_f in the V.E.C.; and $R_m/R_f = 0.63$. In what follows, the orbit radii quoted are radii in the model.

A harmonic coil has been placed in each of the valleys of the three-ridge system. Each coil is of 124 turns on a water-cooled elliptical former; the centre is placed on a radius of six inches. See Fig. 1. The six coils are connected in three pairs of corresponding upper and lower coils wound in the same sense. The three pairs are joined in star connection. Current is fed to the three free ends and the central point of the star is left unconnected; thus the algebraic sum of currents in the three pairs of coils is zero. So, if the coils are all similar, the law of variation of mean magnetic field with radius is not disturbed when the coils are excited.

The free ends of the star-connected coils are fed with current by three linear push-pull amplifiers able to inject or extract about four amperes of current into the coils. The input to the amplifiers comes from voltages derived from a modified sine-cosine potentiometer. The modification consists in replacing the single wiper contact by three wipers arranged so that they lie on radii mutually at 120°. Thus the position of the triple wiper assembly with respect to the direction of the turns on the potentiometer dictates the relative excitation of the three pairs of coils through their amplifiers and hence the azimuthal placing of the imposed harmonic in the cyclotron field. The magnitude of the harmonic can be varied without changing its position by varying the voltage supplying the potentiometer. The system is shown in block diagram in Fig. 2, and the circuit of one of the three amplifiers in Fig. 3.

3. Measurements

The magnetic field due to the coils when excited by known currents was first measured at 15° intervals around the circumference of a six inch radius circle, with angular settings of the sine-cosine potentiometer as parameter. See Fig. 4. The variation of field due to one coil along a radius through the centre of the former was also measured. See Fig. 5. Fig. 6 shows how the induced first harmonic varies with Azimuth Control setting. From these measurements the amplitudes of the various azimuthal harmonics of magnetic field up to the fifth were found by graphical Fourier analysis. See Fig. 7. The azimuthal position of the maximum in relation to the setting of the sine-cosine potentiometer was also found.

The isochronous law of mean magnetic field variation with radius in the model is such that

$k \approx 0$; $Q_r \approx 1$, over the useful area of the machine

The growth of the amplitude of radial oscillations (x) about the equilibrium orbit radius (r_e), and hence the movement of orbit centre, can be calculated by writing

$$\frac{d^2x}{d\theta^2} + x = hr\sin\theta \qquad (1)$$

where $x = r - r_e$ and $h = \Delta B/B_e$, the relative harmonic amplitude. This yields a value of centre movement per turn,

$$x_c = \pi r h \qquad (2)$$

This can be calculated at intervals of radius, knowing the number of turns in the intervals. For our coils, the analysed first harmonic amplitude is 7 gauss at 6 inches, with a variation with radius according to Fig. 5. This gives a calculated centre shift of ± 0.14 inches between radius 3 inches and 6 inches, when using a mean field of 8.7 kg. and 15.6 kV dee voltage, accelerating H_2^+ ions.

4. Residual Centre Shift

Calculation and measurement of the centre shift arising from various causes was made, to see if the harmonic coils would be strong enough. The simple calculation of initial orbit movements assuming that ions experience the peak available accelerating potential difference at each acceleration gives a series of movements along the dee/dummy-dee gap, alternating in direction and diminishing in amplitude. The change in orbit centre from turn to turn, from the series, is

$$x_c = r_1 \{2\sqrt{(n-1)} - [\sqrt{n} + \sqrt{(n-2)}]\} \qquad (3)$$

where n is the total number of accelerations and r_1 is the radius of curvature after the first acceleration. When operating with conditions favourable to accurate centre measurement, i. e. accelerating H_2^+ ion on third harmonic mode, with 15.6 kV dee voltage and 8.7 kGs field, the calculated centre movement is 0.001 inches per turn, at a radius of 6 inches. The number of turns is 18 and the turn separation 0.17 inches. From the 4th to the 18th turn the orbit centre approximately moves 0.05 inches. In the calculation an allowance is made for the change in dee voltage while the ion is passing through the dee-dummy-dee gap. This is only significant on the second acceleration, so here we multiply by a 'gap factor' of less than 1.

When the harmonic coils are unexcited, the movement actually measured from the 4th to the 18th turn, is 0.3 inches in an approximately linear direction, agreeing with $n=0$, $Q_r=1$. If this arose from residual first harmonic, then, using equation (2) this would amount to 10^{-3} of the main field, or 8.7 gauss.

The presence of a third cause of centre shift may be inferred by comparing the centre movement at third harmonic with that when operating at fundamental mode. Adjusted to accelerate protons with 30 kV on the dee and 13 kGs magnetic field, the ions execute 31 turns to reach 6 inches. Assuming the first harmonic fractional error to remain the same, one would expect the measured residual shift to be about twice the figure for the previous harmonic operation over 18 turns. In fact the centre drift is about the same. Movement due to the gap-crossing resonance is independent of volts per turn, but depends on the rate of change of the third harmonic component of the magnetic field disturbance with radius. Here $Q_r \approx 1$ for both third harmonic mode and fundamental mode, so this differential will be the same in both cases. So the gap-crossing resonance might be the cause of our constant residual shift.

5. Performance

Having noted the amount of centre shift which the harmonic coil would be required to correct, the performance of the coils was measured. The amplitude of shift was determined independently of the residual centre drift by plotting orbit centres against position of the harmonic azimuth control (the sine-cosine potentiometer) while keeping the harmonic amplitude control (the current supplied to the sine-cosine potentiometer) constant. The method of measuring orbit centres was as follows. A probe, see Fig. 8, consisting of an earthed plate which shielded all but a thin strip of a second plate was used and the current on this second plate was plotted against radius from the centre of the machine. This enabled individual turns to be identified. For any particular turn the centre was found by intercepting half the current received on the probe by probes at two different azimuths. Allowances were made for the 'scalloping' of the orbits due to the flutter, and for the change in the radius due to any acceleration between the different azimuths. This effectively gave the centre of curvature of the mean orbit just before it struck the current-measuring probe. The accuracy of the centre determination is limited by the accuracy of reading the probe radius and by the magnetic field regulator. The errors are estimated at about ± 0.020 inch. Fig. 9 is constructed by measuring the direction of centre movement due to the coils and plotting against position of the sine-cosine potentiometer. On this plot are also placed points showing the calculated direction of centre movements obtained from the measured harmonic distribution due to the coils, see Figs. 4, 5 and 7. There is good agreement between the calculated and measured centre movements.

The total centre movement available with this apparatus is ± 0.13 inch, which can be applied in any direction desired by means of the azimuth control. There is a small constant field effect, which also has a distribution with radius, and which is therefore presumably due to slight differences between the three coils or the iron around them. It amounts to about 5×10^{-4} of the main field of the model.

Acknowledgments

We are glad to acknowledge the help of Mr. H. Lane of the Rutherford Laboratory, N.I.R.N.S., who designed the amplifier circuits and Mr. T. P. Parry, who installed the harmonic coils and assisted in the measurements.

References

[1] D. J. Clark, C.E.R.N. 63~19, 1963, p. 209.
[2] J. D. Lawson, NIRL/R/85.
[3] W. L. B. Smith, Sector-focussed Cyclotrons, Nat. Research Council Publication No. 656, p. 187.
 Also, D. J. Clark, Rutherford Laboratory, N.I.R.N.S. CDN. 500-05-024, (In-

ternal Design Note)

[4] M. M. Gordon, Nuclear Instrumets &. Methods, 18, 19 (1962) p. 268.

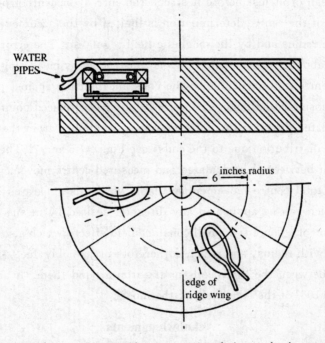

Fig. 1　Coils in position in ridge system, in relation to the dee system

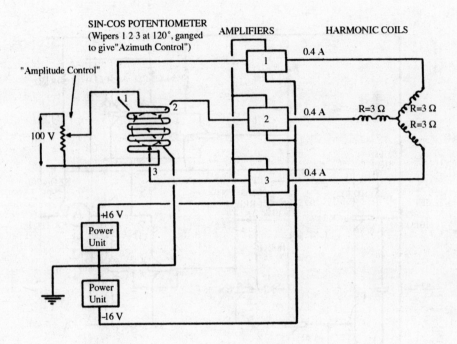

Fig. 2 Block diagram of system for applying harmonics

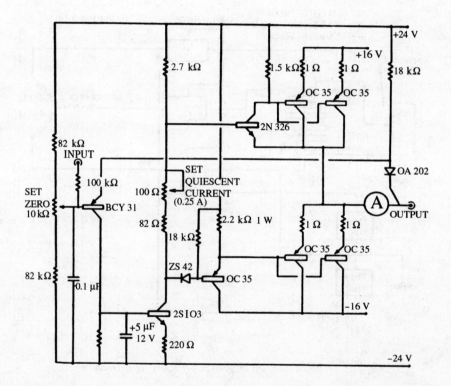

Fig. 3　Amplifier circuit

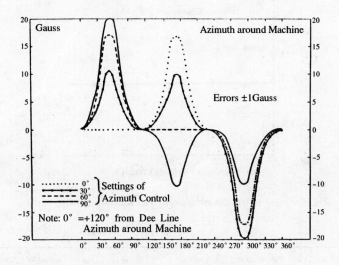

Fig. 4   Azimuthal distribution of harmonic coil field vs. azimuth control settings

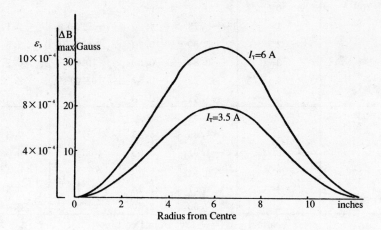

Fig. 5   Radial distribution of harmonic coil field

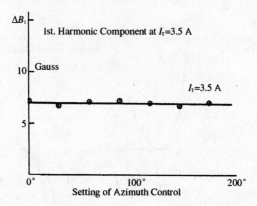

Fig. 6　First harmonic component vs. azimuth control settings

$$B = a_0 + \sum a_n \cos(n\theta) + \sum b_n \sin(n\theta)$$

Setting of Azimuth Control	Harmonic Number n	a_n	b_n	Phase relative to Machine (Note 1.)	Amplitude (gauss)
0°	0	0	—		0
	1	−5.73	4.20	−53°30′	7.1
	2	5.80	−0.10	91°	5.8
	3	0.84	−0.32	110°40′	0.9
	4	2.28	−0.57	103°10′	2.52
	5	−1.41	0.78	108°40′	2.47
90°	0	0.5	—		0.25
	1	4.39	5.83	37°	7.3
	2	−1.91	6.05	162°30′	6.35
	3	−0.27	−0.40	75°50′	0.48
	4	2.67	−2.18	129°40′	3.48
	5	0.53	2.40	12°20′	2.46

Note 1.　0° = +120° from Dee Line

Fig. 7　Harmonic analysis of harmonic coil field

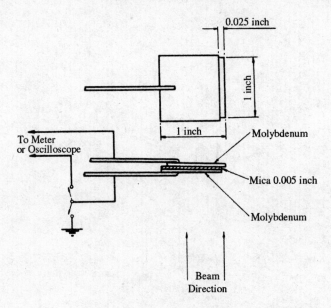

Fig. 8 Beam current probe

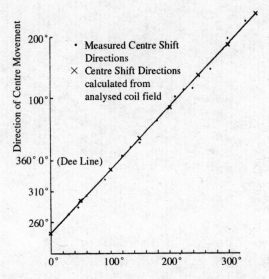

Fig. 9 Centre movement vs. azimuth control settings

第二部分 静电加速器及基于串列静电加速器的加速器质谱计研究

DESIGN OF 4.5 MV ELECTROSTATIC ACCELERATOR[①]

Abstract

A 4.5 MV electrostatic accelerator was designed and is being constructed in the Department of Technical Physics, Beijing University. Protons and multiply charged ions as heavy as Ar are to be accelerated with an intensity of $10^{11} \sim 10^{14}$ p.p.s. RF Choppers and buncher are installed inside the terminal electrode to provide a pulsed beam of 2 ns. with a repetition rate of 3 or 9 MHz. The general layout of the accelerator and main features of the high voltage generator, accelerating and differential pumping tube, terminal equipments, and beam transport system are described.

1. Introduction

In order to meet the extensive need for various appllications of light and heavy ion beams as well as for research in low energy physics, it was decided in 1976 that a heavy ion 4.5 MV single stage Van de Graaff accelerator is to be built as the first facility for the Department of Technical Physics, Beijing University. This accelerator will also act as an injector to a post linear accelerator consisting of helix-loaded resonators for future extension of ion energy. The performance parameters designed for the above purposes are as follows:

[①] The Electrostatic Accelerator Group. Written by Chen Chia-erh, presented by Zhang Ying-xia. Reprinted from the proceedings 1-st. Japan-China Joint Symposium on Accelerators for Nuclear Science &. their Applications, University of Tokyo, Japan, 1980, p. 19~29.

Table 1

Ion Species	Energy /MeV	Intensity /p.p.s.	Pulsing Beam
P. d	4.5	10^{14}	1~2 ns, 3 MHz
α, C^{2+}——O^{2+}	9	10^{13}	
Ne^{3+}——Ar^{3+}	13.5	10^{12}	2~3 ns[a], 9 MHz
Ar^{4+}	18	10^{11}	
Emittence	$7 \cdot \sqrt{Q/A}$	mm · mrad · $MeV^{1/2}$	
Energy dispersion	$\leqslant 10^{-3}$		

a 1 ns after futher compression by post buncher.

In the course of designing the efforts to get the necessary performance have been made mainly on 5 items namely, the high voltage generator, the accelerating and the differential pumping tube, the terminal equipments, the control system and the beam transport system. Main features of these items will be described seperately in the following sections.

2. The High Voltage Generator

The high voltage terminal, the insulating column and the pressure tank are schematically shown in Fig. 1.

Profiles with complicated curves have been adopted for the high voltage terminal electrode in various labs[1,2] to minimize the surface field strength. However, an electrode consisting of simple smoothly joined spherical and cylindrical surfaces was adopted in our design on the technical ground. The field in the latter case will be concentrated at the joint between the surfaces where the change of radius of curvature takes place. The field strength there could be as high as twice that of the mean value. In order to optimize the field distribution, extensive calculations were made to evaluate the effect of geomatrical parameters on the surface field[3], including the ratio of radii R/r_0, r_0/r (see Fig. 1), the height of the neck of the terminal electrode h_1 and the distance between the top of the terminal electrode and of the pressure tank g, etc. The calculated re-

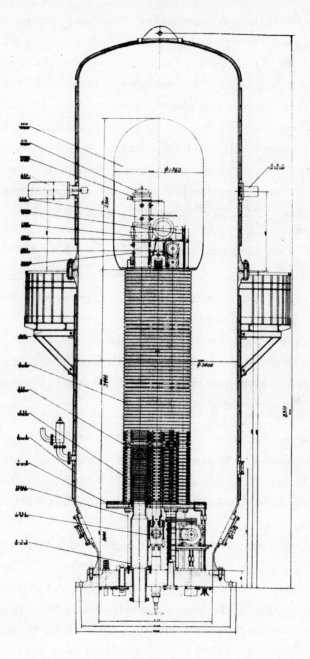

Fig. 1

sults indicate that the maximam field on the terminal electrode can be minimized by taking $R/r_0 \approx 2$, and it decreases with increasing h_0, h_1. On the other hand, the highest field of the generator may occur at the surface of first few equipotential hoops in the vicinty of the terminal electrode. And the most serious overvoltage conditions are produced by tank to column breakdown, especially by breakdowns to the hoops that are near the terminal, which may result in serious failure of the tube sections and resistors, as has been pointed out by Purser.[2] To minimize the field there, one can make use of the shielding effect of the terminal itself. Yet this calls for a $\frac{R}{r_0} < 2$, $\frac{R}{r} \sim e$ and a shorter neck height h_1 against the earlier requirements. As a compromise $R/r_0 = 1.76$, $\frac{R}{r} = 2.4$, $h_1 = 0.54\ r_0$ were chosen. Besides in contrast with Eastham's[4] stasement, our calculations indicate that oval hoops can be helpaul if an optimun ratio of axes, corresponding to a minimum of peak surface field, is chosen. The value of the optimunm ratio depends on the axial voltage gradient of the column.

The geometrical parameters finally decided (Table 2) ensure the highest field E_{max} to be less then 1.4 times the ideally optimized field between coaxial cylinders $\left(\frac{e}{R} \cdot V\right)$, i.e, $E_{max} < 1.4 \cdot \frac{e}{R} \cdot V$. Taking $E_{max} \approx$ 16 MV/m, the maximum achievable voltage wille be $V_{m[MV]} = 4.2R_{[m]}$ which is in accord with most of single stage Van de Graaff generators. Recently the highest surface field in upgraded FN tandems has been raised up to 20 MV/m, then $V_m = 5.2R$ and a voltage as high as 6 MV might be expected in our case.

Stable operation of the generator depends to a large extent upon a sound structure of the insulating column. As we've got to have a big terminal electrode for housing a lot of equpments, strict demands on a stiff column are inevitable. To meet the requirements, the column is divided into 4 short seetions with a height of 870 mm each. Every section is supported by 4 lucite rods ($D = 146$ mm) which are fixed on two 20 mm thick stainless plates to form a rigid frame. Comparing with a 2.5 MV Van de

Graaff built by the same factory, the gain in rigidity for such a structure was estimated to be 5 times more.

Fig. 2 shows the layout of an equipotential frame.

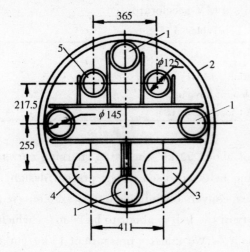

1. insulating column
2. ring hoop
3. accelerating tube
4. differential pumping tube
5. control hole

Fig. 2

Table 2

Pressure Tank
    Diameter $\emptyset=2440$ mm, wall thickness 20 mm
    Height $H=8310$ mm
High Voltage Terminal Electrode
    Radius of cylindrical and spherical surfaces $r_0=680$ mm
    Height of top sphere part $h_0=680$ mm
    Height of bottom neck $h_1=370$ mm
    Total height of the electrode $H_0=2300$ mm
    Gap between top of electrode and tank $g=1000$ mm
    Maximum field at 4.5 MV $E_{max} \leqslant 14$ MV/m.
Equipotential hoops with elliptical section
    Diameter $\emptyset=1000$ mm
    Mojor axis $2a=40$ mm
    Minor axis $2b=22$ mm
    Pitch $p=50$ mm
    Maximum field at 4.5 MV $E_{max} \leqslant 14$ MV/m
Insulating column
    Total height 3440 mm

Static and dynamic stability of the column were evaluated briefly by a simplified model. It can be seen from the results shown in Table 3 that so far as axial and transversal oscillations are concerned, the stabilities are nearly as good as the Oxford's 9 MV accelerator.

Table 3

Axial stable limit	17 t
Eigen frequency for axial oscillation	15 Hz
Transverse displacement under leteral force of 50 kg	1.4 mm
Transverse oscillation under a 3 t load	1.3 Hz

At a charging voltage of about 50 kV the rated charging current is about $400 \sim 500$ μa. Charging belts are used in the present design, but rooms are reserved for future conversion into pelletron system. With a width of 500 mm, the velocity of the belt is about to be 20 m/s, which results in a windage loss of about 4 kW, under a pressure of 16 kg/cm^2 consisting of 25% CO_2 plus 75% N_2 gas.

3. The Accelerating Tube and Differential Pumpimg Tube

This crucial element of the accelerator accounts for most of the limitations in accelerator performance. The voltage gradient of a conventional tube available in our country lies in $1.1 \sim 1.2$ MV/m because of the traditional loading effects. However, a gradient of 1.4 MV/m is required in our design and measures should be taken to get rid of the loading effects as much as possible. As small aperture tubes of NEC have been run successfully at a high gradient level with high beam intensity, a small aperture version is used for this purpose in the design.

To eliminate the back stream of electrons without intercepting the useful beam, the apertures along the tube should be so tailored as to fit the accelerated beam profile as close as possible. Hence diaphragms of $8 \sim 20$ mm in diameter are set along the tube having an accelerating aperture of $\varnothing = 25$ mm.

The accelerating tube, overall length 3.44 m, is devided into four sections. Each section consists of 33 pyrex glass rings, ⌀ 180 mm in inner diameter and 24 mm in height, and stainless steel electrodes, and is jointed altogether by polyvinyl seal. Fig. 3 shows the shape of accelerating electrode with two 120° sector openings on every electrode, the gas conductance of the whole tube is estimated more than 32 L/s. The electrodes are displaced 60° azimuthaly with each other to intercept the electrons. The maximum energy of the electrons can be limitted to less than 150 keV in this case. The use of magnetic traps is also under consideration[5].

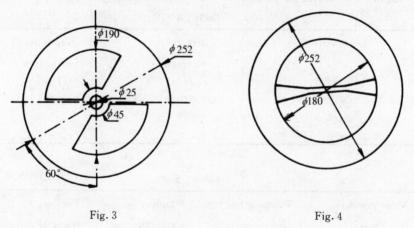

Fig. 3   Fig. 4

Poor gas conductance of a small aperture tube causes a severe limitation to the vacuum inside the tube. The situation would become untolerable in the case for accelerating heavy ion beams where high conductance is important. (see Table 4) To tackle this problem, a differential pumping tube with a conductance of 150 L/s is connected to einzel lens section. By making use of the selecting aperture of the pre-charge selector as a vacuum aperture, the vacuum inside the terminal section and the accelerating tube could be improved by nearly a factor of 7.

The structure of the differential pumping tube is similar to that of the accelerating tube, except the shape of electron intercepting electrode (Fig. 4) is different from that of accelerating electrode. Here electrons are deflected by the field between curved electrodes and limited in energy to less than 400 keV. According to Howe[6] this kind of structure can run safely at a gradient of 1.7 MV/m. with a long life time. Meanwhile, provisions have been made to put turbo molecular and ion getter pumps inside the terminal electrode for further imporvement of the vacuum. The fore vacuum in the latter case can be provided either by the differential pumping tube or by a tank inside the terminal electrode. Table 5 shows the improvements in vacuum by ditterential pumping and terminal pump.

Table 4

Ion Species	Source gas consumption [L/s]	Requirements of conductance [L/s]
N^{2+}	3.2×10^{-3}	>105
Ar^{3+}	2.5×10^{-3}	>250
Xe^{6+}	2.5×10^{-3}	>930

Table 5

Mean pressure [torr]	Pumping by acclerating tube	Pumping by accelerating and differerential tube	Pumping by terminal pumps [400 L/s]
Charge selector	1.2×10^{-4}	3.8×10^{-5}	8.2×10^{-6}
gap lens	9×10^{-5}	1.3×10^{-5}	4.1×10^{-6}
Accelerating tube	4.5×10^{-5}	6.8×10^{-6}	2×10^{-6}

4. The Terminal Equipments

Terminal equipments consist of multiply charged ion source, pre-focusing lenses, charge state selector and beam pulsing device. Fig. 5 shows schematically the general layout.

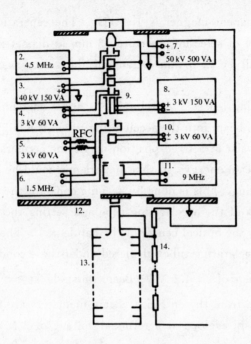

1. ion source 2. R. F choopper 3. einzel lens 4. steering electrode
5. steering electrode 6. R. F chopper 7. extracting electrode
8. pre charge selector 9. pemanent magnet 10. steering electrode
11. buncher 12. gap lens 13. accelerating tube 14. resistance chain

Fig. 5

a) Heavey ion source   A cold cathode PIG source[7,8] with end extraction, which is similar to that of Baumann's[9], was under development. Dimensions of electrodes, discharge parameters as well as electrode meterials are investigated on a test bench. Beams of $100\mu a$ protons, $2.2\mu a$ N^{2+} and $16\mu a$ Ar^{2+} were obtained with a gas consumption <5 cc/h. and a lifetime >50 h. Aimming at a yield of $\sim 10^{15}$ p. p. s. protons and $10^{13} \sim 10^{14}$ pps lighter heavy ions, various efforts have been made to coupe with the requirements of a source operating inside the high voltage terminal.

b) Charge state selector   A cross field filter is used to avoid unneccessary beam loading. The electric field applied by a pair of electrods, 200 mm in length, can be varied up to 4 kV/cm. The magnetic field produced by a permanent magnet is 1.3 kGs. The seperation of Xe^{6+} and Xe^{7+} at the plane of selecting aperture, 3 mm in diameter lying 160 mm downstream, will be 5.5 mm Details of the selector are given elsewhere[10].

c) Prefocusing lenses   An einzel lens and a matching gap lens are used to focus the beam into the accelerating tube. It is highly desirable to keep the beam waist at a fixed position inside the accelerating tube independent of the beam energy, so that diaphragms could be set to intercept the electron loading. This is done by keeping constant the voltage ratio of the entrance lens of the accelerating tube and setting the image waist of the einzel lens at the optical centre of the gap lens[10]. The beam trajectories inside the accelerating tube calculated under these conditions are similar within the range of 2~4.5 MV. Beam waists lie at $\approx \frac{1}{3}$ of the accelerating tube length from the entrance. Beam diameters are no more than 6 mm. Typical beam evelope at a terminal voltage of 3 MV. is shown in Fig. 6.

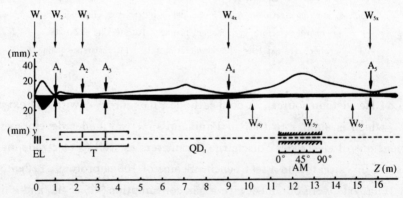

W—Waist QD—Quadrupole lens AM—Analyzing magnet T—accelerating tube
A—Apertures EL—Eingel lens

Fig. 6

d) Prebuncher and RF choppers   Two sets of RF plate chopper and a klystron-like buncher to give either a proton (or deuteron) burst of 2 ns.

with a repetition rate of 1.5~3 MHz or a heavy ion burst of ≈3 ns. with a repetition rate of 9 MHz. Heavy ions will be compressed futher to a width less than 1 ns by a post buncher[11]. Drift distance as long as 800 mm is chosen for the heavy ion chopper to make the chopped beam as nearly parallel to the geomatrical axis as possible[12]. Besides, wire gauzes mounted on the bunching electrodes are used to prevent the emittance in the transverse plane from deterioration by phase dependent focusing of the buncher. The main parameters of the pulsing device are shown in Table 6.

Table 6

RF choppers
    Length of electrode 50 mm, gap 35 mm
    Frequency   1.5 MHz   (4.5 MHz for HI beam)
    Voltage   <1 kV
    Pulse width   37 ns
    Inclination of chopped beam <0.1 mrad (0.5 mrad for HI beam)
Buncher
    Diameter of the electrodes 25 mm
    Length of the intermediate electrode 66~100 mm
    Bunching voltage 3 kV
    Initial energy dispersion 5.6×10^{-3} (1.2×10^{-2} for HI beam)
    Pulse width 2 ns (3.5 ns for HI beam)

5. Beam Transport System

The physical layout of the beam transport system and helical boosters is shown in Fig. 7. The accelerated beam after passing a 90° analyzing magnet ($R = 1200$ mm) and stablizing slits is either transported to the boosters or deflected by a 90° bending magnet and a switching magnet to the experimental area. The post accelerated beam will finally be redeflected back by several bends into the experimental area. A pair of 45° bending magnets ($R = 400$ mm) with a time spread of 0.2 ns are used to deflect the beam directly to the neutron target.

Quadrupole lenes, steering magnets, Faradgy cups as well as beam profile and beam emittance monitors are distributed along the beam line to facilitate the beam transport.

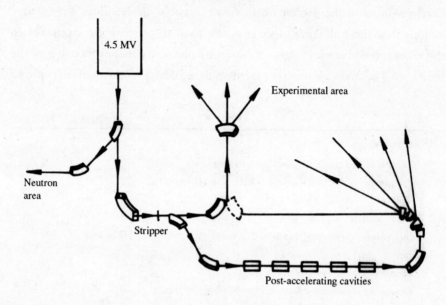

Fig. 7

6. Control System

a) Energy stablizing circuits   Ripples and slow drift of the high voltage are detected either by a pair of capacitive sensors around the terminal or by the image slits after the 90° analyzing magnet. The signals are then transfered to a feedback circuit to control corona triode and the charging voltage source, giving an overall voltage stability of $\approx 10^{-3}$. Generating voltmeter is also used to pick up signals when the beam is off.

b) Control panel   Central console and local penels are used to facilitate the operation of the accelerator. The former controls mainly the high voltage generator, the terminal equipments, the analyzing magnet and the main pumping system etc., while elements on the individual bcam line can be controlled on the spot by local panels. Five racks standing by the side of the central console display all kinds of signals.

7. Concluding Remarks

The accelerator is now being constructed by Xien Feng Electromachinary Works in Shanghai. We expect that the machine will be set up by the end of 1982 to obtain the first beam in 1983.

References

[1] A. Kiss, E. Kaltay et al. H. I. M. 46(1967) 130.
[2] K. I. Purser. N. I. M. 122(1974) 72.
[3] Xu Fang-guan. Internal memo, Beijing University BDJ-B-7/1978.
[4] D. A. Eastham. N. I. M. 108(1973) 593.
[5] Xu Fang-guan. Internal memo, Beijing University.
[6] F. A. Howe. IEEE Trans. NS-14 3(1967) 126.
[7] Yu Jin-xian, Li Ren-xing. Internal memo, Beijing University BDJ-B-5/1978.
[8] ibid BDJ-B-6/1978.
[9] H. Baumann. II conf. on Ion Source (1972), p. 577.
[10] Yan Sheng-qing. Internal memo, Beijing University BDJ-B-3/1978, BDJ-B-2/1978.
[11] Chen Chia-erh. ibid BDJ-B-1/1978.
[12] Chen Chia-erh. Physica Energiae Fortis Et Physica Nuclearis vol. 4, 3(1980) 401.

PROGRESS AND APPLICATION OF THE 4.5 MV ELECTROSTATIC ACCELERATOR[①]

Introduction

The 4.5 MV electrostatic Accelerator at the Institute of Heavy Ion Physics, Peking University has been developed as a monochromatic neutron source after several years operation and improvement. The operational stability and reliability has been satisfactory. The energy dispersion is $<1\times10^{-3}$. The bunching facility installed in the top terminal provides a pulsed beam of 2 ns with an analyzed current of $>1\ \mu A$ in average at a repetition rate of 3 MHz which meets the requirements of neutron physics studies.

The progress and application will be described in the following sections.

1. RF Ion Source

In order to meet the needs of the neutron experiment, a RF ion source with high proton ratio has been installed to replace the previous PIG source.

High quality ion beam is essential for the bunching facility. Especially since the distance between the ion source and the entrance of the accelerator tube is quite long, about 1140 mm (The layout is shown in Fig. 1), in

---

① Coauther: Zhang Yingxia, Wang Jianyong, Gong Linghua, Lu Jianqin, Xue Pingfang, Xie Dalin, Yuan Zhongxi, Quan Shengwen. Reprinted from the Proceedings 5-th Japan-China Joint Symposium on AcceleratorS for Nuclear Science & their Applications, University of Osaka, 1993, Japan, p. 174.

order to get a better beam transmission, the emittance and energy dispersion must be small. An ion source test bench has been set up for this purposes to examine the emittance and energy dispersion of the extracted beam, so as to determine the structural parameters and operational conditions of the RF source.

The structure of the ion source is shown in Fig. 2, the inner diameter of discharging quartz tube is 24 mm, the height is 70 mm. The extractor is made of Al, and is shielded by quartz glass. The RF power generated by

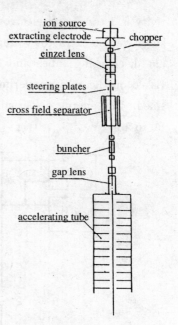

Fig. 1 The bunching facility

a push-pull oscillator is transferred to the discharging tube through inductance coupling. The extracted current of hydrogen ions ranges from 240 μA to 500 μA at the exit of ion source. The ratio of H^+ : H_2^+ is more than 80%.

Fig. 3 shows the schematic layout of energy dispersion measurement assembly where a decelerating lens is used for measuring the beam energy. The sensi-

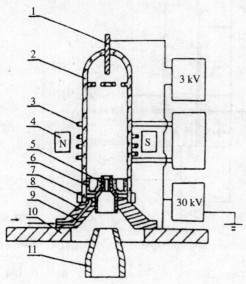

1. probe 2. discharging tube 3. coil
4. magnet 5,6. shielder 7. extractor
8. base of extractor 9. air duct 10. flange
11. post-accelerating electrode

Fig. 2 The structure of ion source

tivity is 1 to several electron volts. The energy dispersion has been measured versus extractor voltages and various magnetic field levels for different discharging tubes and extractors of various dimensions and shape. Based on the above experiments, the structure of the ion source was defined. The energy dispersion for the source being used now lies between 50 to 100 eV.

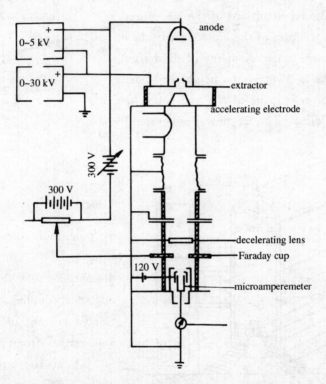

Fig. 3 The schematic layout of energy dispersion measurement assembly

The emittance measurement device is shown on Fig. 4, which has a single-slit and single scanning filament and is controlled by a computer. Phase diagrams have been measured against various extraction voltages, magnetic fields and pre-acceleration voltages. The typical results show that the source emittance lies between 2.3 to 6.6 mm-mrad-meV$^{1/2}$, as can be seen from Fig. 5.

The life time of the source turns out to be more than 260 hours on the accelerator, which is quite satisfactory to meet the requirements of various measurements.

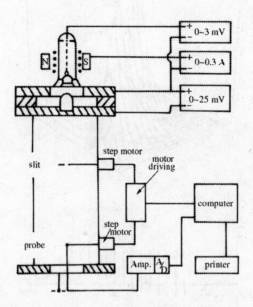

Fig. 4 The emittance measurements device

2. Beam Pulsing System

The 4.5 MV Van de Graaff was designed as a pulsed machine to meet the need of fast neutron TOF spectroscopy.

The beam pulsing system consists of a pair of deflecting plates, to which 1.5 MHz r.f. voltage is applied, and a Klystron type buncher inserted between the deflecting plate and the chopping aperture (Fig. 1), to which bunching voltage of 9 MHz is applied. The pulsed beam with width of 1.8 ns (FWHM) has been obtained at the end of the target at a 3 MHz repetition rate. The mean current of the analyzed beam is $>1\,\mu A$ at a 3 MHz repetition rate. The bunched and chopped proton beam waveform observed on a 50 target assembly is shown on Fig. 6.

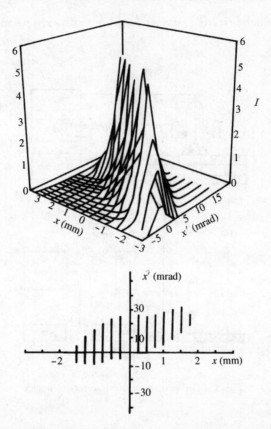

Fig. 5　The source emittance phase diagrams

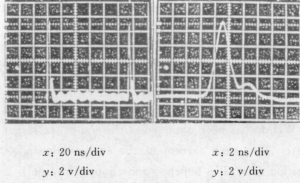

x: 20 ns/div　　　　　　x: 2 ns/div
y: 2 v/div　　　　　　　y: 2 v/div

Fig. 6　The bunched and chopped proton beam waveform

3. The Accelerating Tube and Matching Gap Lens

In 1987, the small aperture accelerating tube with dismountable electrodes made by the Research Institute of Automation was installed in the machine. It appeared that the beam transport efficiency of this tube is very good. A beam currents of 60 μA H^+ could be easily obtained at the exit of the accelerator. However, it could not sustain the terminal voltage as high as it was expected. After long time conditioning, it could reach 3.8 MV in contrast with more than 6 MV if without the tube. After further examining, a great number of sparking traces on the inner wall of the insulating glass ring were found. They were obviously caused by secondary electrons. Actually two 120° sector openings on each electrode were set up to improve the vacuum of the small electrode aperture tube. It was possibly that the opening was too large to shield the electrons completely. This might be the reason why the tube could not sustain high enough voltage.

In 1992. A spiral inclined accelerator tube made by Dowlish Developments Ltd. was installed in the machine to replace the small aperture tube. Since then the terminal voltage has been reached 4.6 MV with this tube. The H^+ beam current at the exit of accelerator tube is about 40 μA.

In the course of beam adjustments, it was discovered that a maximum of the accelerated beam currents occurred at a terminal voltage between 1.5 to 2 MV. The current decreased when the terminal voltage became high. When the voltage reached 4.0 MV, only half of the maximum could be obtained. This phenomena was not in coincidence with the designing calculation. It was found during the later investigation that the insulator of the matching gap lens was damaged. The insulator was punctured when the electrode voltage went up to 5 kV. So the matching effect was lost. In addition, the gap lens voltage came from the first few voltage dividing resistors of accelerating tube near the terminal, and the designed voltage ratio was 130. However this value would change with the terminal voltage especially when the leakage current and corona current as well as beam loading became remarkable. So an adjustable ±30 kV power supply was added to the gap lens, and the effective ratio of N could be changed by adjusting the power supply through a motor. It was found that the beam current obtained at the exit of the accelerator was higher at the N value of 90 rather than 130.

4. The Calibration of the Analyzing Magnets

The digital Hall effect teslameters with a temperature coefficients＜200 ppm/℃, resolution ±0.1 Gauss, accuracy 0.02% and linearity 0.1 were checked with NMR. The field of the analyzing magnet, consisted of two 45° identical sector magnets with a radius of curvature of 400 mm, was measured and monitored by the Hall probes. The energy of the accelerator was calibrated by measuring the excitation of the ^{7}Li(p, n)^{7}Be reaction. The typical result is shown in Fig. 7.

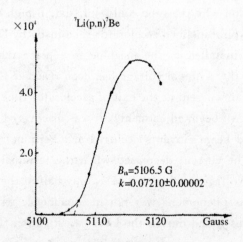

Fig. 7 Energy calibration

5. Neutron Experiments

The mono-energy neutron generated by this machine ranges in

4~7 MeV and 16~20 MeV respectively. The Joint Institute for Nuclear Research in Russia and TsingHua University have reached an agreement with us on using this accelerator to carry out researches on the reactions of (n, p), (n, α). This work is important not only in the light of nuclear structure studies, but also in providing important data on the radiation damage of materials used in fusion reactor. Since 1991, mono-energy neutron has been supplied to carry out the duo-differential cross section measurement for neutron induced reaction. Typical results of the two dimensional spectrum at 4 MeV are shown in Fig. 8, with the angular distribution of particles in Fig. 9. The measurements on the reactions of ^{88}Ni(n, α), ^{65a}Zn(n, α) were carried out too. Moreover, mono-energy neutron of 7.0 MeV was provided to study the performance of shelter material for neutron shielding and the energy calibration of the space radiation detector used in the satellite.

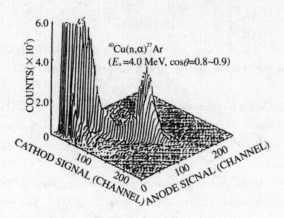

Fig. 8 The two dimensions α spectrum

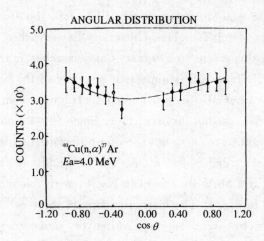

Fig. 9 The α angular distribution (4 MV)

Concluding remarks

This accelerator is now being used as a mono-energy neutron source and providing pulsed beam. Welcome to carry out neutron studies on this machine.

Reference

[1] Proceedings of Japan-China Joint Symposium on Particle Accelerators and their Applications in Nuclear Research and Industry. (1980), Chen Jiaer, Zhang Yingxia et al.

[2] Status of 4.5 MV Electrostatic Accelerators Proceedings of 5th National Conference on Accelerator. (1992), Zhang Yingxia, Wang Jianyong et al.

[3] Energy Dispersion of H^+ Ions From Ion Source, Master Thesis. (1986), Hu Jianpeng.

[4] An Automatic Beam Emittance Measurements Device, Proc. of National Conference on Accelerator Beam Diagnosis and Control. (1986), Yan Shengqing et al.

[5] Emittance Measurements of a RF Ion Source, (Internal Report). Gong Linghua et al.

[6] Measurement of Cross Section at 5 MeV and Angular Distribution at 4 and 5 MeV for Reaction of $^{40}Ca(n, \alpha)^{37}Ar$, Tang Guoyou et al.

北京大学 6 MV 串列静电加速器的运行及其应用[①]

摘　要

北京大学 6 MV 串列静电加速器上建成 4 条束流输运线,加速 ^{1}H, ^{12}C, ^{16}O, ^{19}F, ^{35}Cl, ^{79}Br 等离子,并开展物理实验研究. 首次利用 D(^{12}C,p)^{13}C 核反应建立起高灵敏度的氚分析法;利用共振核反应 ^{1}H(^{19}F, $\alpha\gamma$) ^{16}O 分析材料中氢的深度分布;用重离子背散射分析超导材料 $YBa_2Cu_3O_{7-x}$;用 35 MeV ^{35}Cl 和 45 MeV ^{79}Br 的弹性前冲分析不同材料中的轻元素及测定低能重离子的阻止本领等.

关键词　加速器　加速器质谱　材料科学　离子束应用

　　北京大学 6 MV 串列静电加速器的主体由英国牛津大学提供. 经 22 个月的主体安装及配套设备的设计研制,于 1989 年 8 月调试出束.

　　在该加速器上研制并建立起加速器质谱计[1,2]. 该系统经 ^{14}C 断代研究,对马王堆、大汶口和南庄头等考古遗址样品和澳大利亚蔗糖标准样品的测试,其测量精度已达 1.7%;^{14}C 同位素测量灵敏度达 3×10^{-14},证实系统性能指标达到设计要求,已通过国家自然科学基金委员会组织的专家测试[3]. 加速器质谱计主要应用领域有:(1) 对中国古人类化石及甲骨样品的 ^{14}C 年代测定;(2) 为研究我国干旱、半干旱地区几万年来环境演变的动态过程和发展趋势,从而建立古气候时间序列,需利用 ^{14}C 断代技术;(3) 对深海沉积物与金属锰结核的 ^{10}Be 年代测定;(4) 为研究铝致脑痴呆等生物医学问题而进行 ^{26}Al 的示踪测定.

　　本文还将描述相关系统的研制与改进,以及应用核物理分析方法进行材料科学的研究和低能重离子的阻止本领的测定.

[①] 合作者:于金祥,韦伦存,李认兴,巩玲华,李　坤,卢希庭,江栋兴,刘洪涛. 摘自《原子能科学技术》,Vol. 27, No. 5(1993) 401.

1 北京大学 6 MV 串列静电加速器的改进

1.1 配置控制系统

采用程控技术,实现对真空系统的程序控制及自动保护.程控机根据输入信号及内部程序实现各项功能.当真空进入 2.7×10^{-4}Pa 时,真空测量仪会自动输送"真空好"信号给程控机,收到信号后,程控机自动打开高、低能端的气阀,以便进行调束.若当真空度低于 2.7×10^{-4}Pa 时,程控机将根据测量仪给出的"真空坏"信号自动关闭气阀及电动阀门,并发出警报信号.当需停机时,程控机可按时序切断各种电源.我们还利用光纤传输技术检测高电位下负离子源的工作参量.由于负离子源及预分析器均处于 -60 kV 预加速电压下,近 10 个临测参量需采用遥控及遥测技术,将离子源有关参量送入幅频变换和电流放大后,由光纤传至解调器,在现场和控制室实现负离子源工作参量的监测.还利用计算机计算和监测加速器运行时的部分工作参量,并制成加速器运行动态显示器,从而使这台加速器的运行操作方便可靠.

1.2 建立两器共同的气体处理系统及真空系统的改进

加速器钢筒内用 1∶4 的 CO_2 和 N_2 作为绝缘气体,最高气压约 1.6 MPa.气体处理系统是由我校自己设计的,与 4.5 MV 静电加速器共用.它包括氧压机、过滤器、干燥器、油水分离器和循环干燥系统、储气罐、滑阀式真空泵及管路系统等.该系统可对绝缘气体进行循环干燥、净化处理和回收使用,经循环干燥处理的绝缘气体的湿度降到 10×10^{-6} 以下.该系统几年来的运行情况表明,其性能稳定可靠.

主机和全部束流输运线、物理实验线的真空系统均配以国产分子泵.整个系统真空度高($10^{-4}\sim10^{-5}$Pa 范围).由于实现了真空的程序控制及保护,运行安全可靠,并解决了在牛津大学长期未解决的钢筒充气体时加速管真空下降的问题.

1.3 离子源改进

目前该器装有三台负离子源,分别为 Middleton 型源、上海原子核所研制的 860 型源及双等离子体电荷交换源.前面两台源分别装有 12 和 20 个靶位,可在不破坏真空的情况下,更换靶位,产生各种不同的离子. 860 型源专用于 AMS,用于物理工作的是经我们改进后的 Middleton 型源,一般流强 0.1~20 μA,经 3 年多运行证明该源稳定可靠、寿命长,可满足大部分物理实验的要求.

1.4 四条实验线的建立

我们建立的四条束流输运线分别是:用于测定宇宙成因核素 ^{14}C, ^{10}Be 及 ^{26}Al 等的加速器质谱线(AMS);三束共靶的在线注入及在线分析线;另两条是用于核反应分析(NRA)、背散射分析(RBS)、弹性前冲分析(ERDA)等的通用线.

该器从 1989 年出束以来,经过一年多的试运行,于 1991 年 5 月通过由国家教委组织的验收,现已正常运行使用,端电压 6 MV,端电压稳定度≤±2 kV,达到在牛津大学时的水平. 到目前为止,已经加速了 H,C,O,F,Cl 和 Br 等离子,表 1 是有关的性能参数. 如 ^{12}C, ^{79}Br 等离子的束流强度均高于在牛津大学时的水平,高能端到分析磁铁像缝处的束流传输效率达 80%,分析磁铁后到各实验靶室的传输效率均达到 100%,束斑可达≤3 mm.

表 1　6 MV 串列静电加速器的性能参数

离子	端电压/MV	能量/MeV	靶束流/nA
C^{5+}	5.5	33.07	75
C^{4+}	5.5	27.57	1200
Cl^{6+}	5.5	38.69	70
Br^{6+}	5.0	35.07	120
Br^{9+}	5.7	57.07	23

2　材料科学的研究

6MV 串列静电加速器运行两年来,主要集中于材料科学的研究和低能重离子的阻止本领的测定. 在材料科学的研究中,主要利用了核反应分析、弹性前冲分析、重离子背散射等核物理分析方法.

2.1 核反应分析法

利用 D(^{12}C, p)^{13}C 核反应,我们建立了一个高灵敏度氘的分析方法[4]. 该方法是国内首次采用. 这个核反应有很大的反应截面(~2×

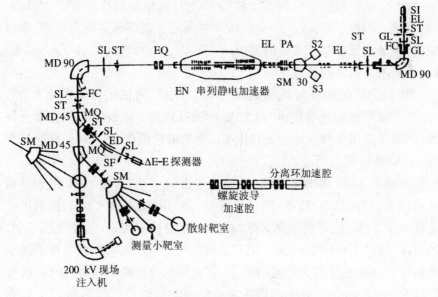

图 1　6 MV 串列静电加速器及束流管线的分布图

S1——溅射离子源；EL——单透镜；ST——导向器；SL——缝隙仪；GL——间隙透镜；
S2——溅射离子源；S3——电荷交换离子源；PA——预加速；EQ——静电四极透镜；
FC——法拉第筒；MQ——磁四极透镜；ED——静电能量分析器；SM——开头磁铁.

$10^{-29} m^2$)并且当 ^{12}C 的能量在 6.8 MeV 到 8.2 MeV 时其反应截面的变化很小(±10%). 分析氘的灵敏度和 $D(^3He, p)^4He$ 的核反应分析的灵敏度相当,达到 $1×10^{14}$ 原子$/cm^2$. 如果对于实验条件加以优化,其分析灵敏度可以超过 $D(^3He, p)^4He$ 分析氘的灵敏度. 该反应对于 Si 中的氘的分析深度可以达到 1.2 μm. 由于使用了非常容易得到的 ^{12}C 离子,而不是使用价格昂贵的 ^3He 气体,因此,成本较低. 利用氘作为氢的示踪同位素,研究了在 SiH_4 和 D_2 气氛中由弧光放电法制备的作为太阳能电池材料的含氢无定型硅的成长机理. 分析结果表明:由此法制备的含氢无定型硅中的氢大部分来自于 $SiH_4(>90\%)$[4].

我们还利用 $^1H(^{19}F, αγ)^{16}O$ 核反应分析了作为太阳能电池材料的 a-Si(H)中的氢的深度分布和含量,以及作为核反应堆材料的 ZrO_2/Zr 中的 ZrO_2 及其界面处的氢的分布和含量,分析结果表明在 ZrO_2/Zr 的界面处氢的含量比 ZrO_2 中氢的含量要高.

2.2 弹性前冲分析法

利用 35 MeV 的 ^{35}Cl 离子束的弹性前冲分析了从氢到氧的多种轻元素. 用金硅面垒探测器, 在其前加有密勒吸收膜(Mylar)[5]. 这是国内首次采用此法, 该方法可以同时分析 H, D, He, C 和 O, 或者同时分析 H, C, N 和 O 元素. 分析的深度分辨率一般好于 20 nm. 对于 C, N 和 O 的分析最小检测限为 $(1\sim2)\times10^{14}$ 原子/cm^2, 而对 H, D 和 He 的分析为 $(4\sim8)\times10^{14}$ 原子/cm^2. 做了在氮气氛中不同的快速热退火条件下对 Co (480 nm)/Si 膜中的 H, C, N 和 O 的含量分析, 结果表明在未退火及较低温度(<500℃)下退火的样品中含有约 $(1\sim2)\times10^{14}$ 原子/cm^2 的 N, 而当退火温度升高时, N 的含量降至方法的最小检测限以下. 还用该方法研究了单晶硅中注入 He 的深度分布, 实验测定的深度分布与 TRIM 计算的结果有明显的差异[5].

另外还建立了用 45 MeV ^{79}Br 和 ΔE-E 探测系统的弹性前冲分析方法[6].

2.3 重离子背散射(HIRBS)法

重离子背散射比起常规的 α 背散射具有其独特的优点. 由于重离子背散射利用较重的离子, 对于比入射粒子重的元素分析的质量分辨率好于 α 背散射, 而且重离子有较大的阻止本领, 因此具有较好的深度分辨率[7].

我们建立了一个以 ^{12}C 离子作为入射粒子、用普通的金硅面垒探测器的重离子背散射方法, 对 YBa$_2$Cu$_3$O$_{7-x}$ 高温超导材料与衬底 SrTiO$_3$ 之间的扩散行为进行了研究.

磁控溅射法制备的高温超导样品 YBa$_2$Cu$_3$O$_{7-x}$/SrTiO$_3$ 的 ^{12}C 的重离子背散射谱示于图 2. 为了从实验谱中提取元素的浓度分布的信息, 我们编写了一个计算机模拟重离子背散射的程序(HIRBS). 同时考虑了入射束流的不稳定性、入射束的有限束斑及探测器的有限立体角引起的能量几何展宽、离子在靶中的能量歧离以及探测器的能量分辨率因素. 通过模拟从 ^{12}C 实验谱而得到了样品中的 Y, Ba, Cu, Ti 和 Sr 的浓度分布示于图 3. 从图 3 可以清楚地看出: Ti 向 YBa$_2$Cu$_3$O$_{7-x}$ 有明显的扩散(扩散深度约为 30 nm)[7].

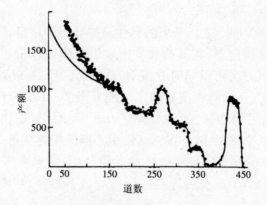

图 2 ^{12}C 重离子背散射的 $YBa_2Cu_3O_{7-x}/SrTiO_3$ 谱

……实验谱；—— 模拟谱；$E(^{12}C)=8.13$ MeV

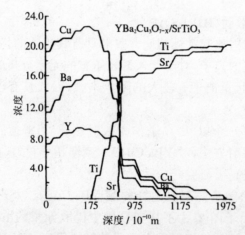

图 3 由 HIRBS 模拟程序计算的 $YBa_2Cu_3O_{7-x}/SrTiO_3$ 样品中的 Y，Ba，Cu，Sr 和 Ti 的浓度分布图

该方法还被用于 Co/Si 多层膜的真空热退火的反应动力学的研究. 结果表明对于 Co/Si 多层膜反应动力学而言，它和 Co/Si 单层膜的钴硅化合物的生成相序是一致的，即 Co_2Si-$CoSi$-$CoSi_2$.

3 低能重离子的阻止本领测定

通过利用弹性前冲法测定样品表面及界面的氢,建立了一个新的阻止本领的测量方法[10]. 该法可用来测量能量在 0.2～1.0 MeV 能区的重离子的阻止本领. 这很有意义,因为在这一能区阻止本领的数据很有限. 测量的 ^{12}C,^{19}F,^{16}O 和 ^{35}Cl 在 Ag,Si 和 Al 中的阻止本领示于图 4. 有些数据与 TRIM-90 的计算值有明显的差别[8].

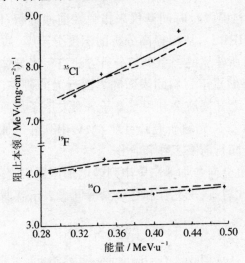

图 4 ^{12}C 离子在 Al,Ag 和 Au 中的阻止本领
与 TRIM-90 计算结果的比较
+——实验点;——实验点拟合;----模拟计算

4 未来的工作

从上面介绍的工作可以看出,6 MV 串列静电加速器无论是对于材料科学的研究,还是对于地球、考古或生物医学等其他领域的研究都是非常有意义的. 我们将进一步提高加速器质谱计的性能,并充分发挥它的效益. 另外,近十年来,在国外的一些实验室里在线注入及辐照对于材料科学及微电子学方面的研究有了长足的进展[9~11]. 一般的在线注入系

统包括一台静电加速器(单级或串列)和一台低能离子注入机.我们已有一台 4.5 MV 的单级静电加速器和一台 6 MV 串列静电加速器,这两台器的束流被引入同一靶室,再将一台 200 kV 的中流强离子注入机的束流引入该靶室,这样就可以做从几十 keV 到几十 MeV 能区的离子注入及辐照的研究和在线分析.另还有广泛的离子种类的选择.现正在加工一个多功能的包括三维转角仪及从液氮温度到 600℃的温度控制系统的新靶室.主要工作将集中于离子注入的材料改性、新功能材料和微电子器件的研究.

我们还将利用飞行时间谱仪来作弹性前冲分析(ERDA)和重离子背散射分析(HIRBS),从而提高分析的深度分辨率、质量分辨率及分析灵敏度.并利用所建立的一系列核分析方法进行半导体和超导体材料的研究,其重点将侧重于材料的表面和界面行为的研究,以及注入改性的研究.另外也计划开展有关利用重离子来激发原子的原子物理研究工作(例如重离子激发 X 光的研究).目前 6MV 串列静电加速器向国内外开放,欢迎国内外同行前来工作和合作.

作者借此机会向参加 6 MV 串列静电加速器实验工作的张征芳、金瑞鑫、张桂筠、沈毅雄、王兆江和任晓棠等同志表示感谢.

参 考 文 献

[1] Chen Chiaerh, Guo Zhiyu, Yan Shengqing, et al. Status of the Tandem Accelerator Mass Spectrometry Facility at Peking University. Nucl Instrum Methods, 1990, B52: 306.

[2] Chen Chiaerh, Guo Zhiyu, Yan Shengqing, et al. The Test Run of AMS System at Peking University. Conf Rec of the 1991 Particle Accelerator Conf, May 6~9, 1991, San Francisco, Vol. 4: 2616.

[3] 国家基金委测试组.北京大学加速器质谱计及〈C-14 断代〉测试报告,北京大学资料,1992.

[4] Wei Luncun, Yang Xihong, Liang Bin, et al. A New High Sensitivity Analysis Method for Deuterium by D (^{12}C,p)^{13}C and Its Application, Nucl Instrum Methods, 1991, B53: 332.

[5] Yang Xihong, Wei Luncun, Li Renxing, et al. Elastic Recoil Detection Analysis of Light Elements in Thin Films Using 35 MeV ^{35}Cl^{6+} Beam. Nucl Science and

Techniques, 1992, 3(3): 175.
[6] 刘洪涛等. 利用 ΔE-E 探测系统的弹性前冲分析方法. 核技术, 1993. (在排印中.)
[7] 韦伦存, 杨熙宏, 梁斌, 等. ^{12}C 重离子背散射方法及其对 $YBa_2Cu_3O_7$ 超导膜界面扩散行为的研究. 核技术, 1992, 15(10):583.
[8] Lu Xiting, Xia Zonghuang, Zhou Kungang, et al. A New Method for Measuring Stopping Power by ERD, Nucl Instrum Methods. 1991, B58: 280.
[9] Allen CW, Funk LL, Ryan EA, et al. In Situ Ion Irradiation/Implantation Studies in the HVEM-TANDEM Facility. Nucl Instrum Methods, 1989, B40/41: 553.
[10] Was GS, Rotberg VH. Michigan Ion Beam Laboratory for Surface Modification and Analysis. Nucl Instrum Methods. 1989, B40/41: 722.
[11] Cottereau E, Camplan J, Chaumont J, et al. ARAMIS: An Accelerator for Research on Astrophysics, Microanalysis and Implantation Solids. Materials Science and Engineering, 1989, B2: 217.

OPERATION AND APPLICATION OF 6 MV TANDEM ACCELERATOR AT PEKING UNIVERSITY

Abstract

Four beam lines are constructed for the 6 MV Tandem Accelerator of Peking University. Various ions such as ^{1}H, ^{12}C, ^{16}O, ^{19}F, ^{35}Cl, ^{79}Br, etc. are accelerated. Some experiments are made, such as deuterium determination for the first time by the D(^{12}C, p)^{13}C nuclear reaction, hydrogen profiling using the resonance reaction ^{1}H (^{19}F, $\alpha\gamma$)^{16}O, backscattering analysis for high T_c superconductor $YBa_2Cu_3O_{7-x}$, ERDA induced by 35 MeV of ^{35}Cl and 45 MeV of ^{79}Br ions for analyzing of light elements in various materials and stopping power measurements of low energy heavy ions and other experiments.

Key words Accelerator   AMS   Material science   Ion beam application

加速器质谱计的原理、技术及其进展[①②]

一 加速器质谱计的兴起

在地球科学和考古学的研究中,长期以来放射性同位素一直是重要的信息来源.对于半衰期在 $10^3 \sim 10^8$ 年的长寿命放射性核素,如果它们不是天然放射性系中的中间核素,那么元素生成时曾存在的原始同位素至今早已衰变殆尽.现存的这类核素的原子,基本上都是后来直接或间接通过宇宙射线引发的核反应而生成的,故这类核素被称为宇宙成因核素.通过对这些核素的研究,可以了解过去漫长时期中地球和太阳系的许多令人感兴趣的过程.几种最常使用的宇宙成因核素列于表 1. 70 年代末期发展起来的加速器质谱计(简称 AMS)技术[1]克服了常规衰变计数法与传统质谱方法的弱点,在宇宙成因核素的研究中发挥了令人瞩目的作用.

衰变计数法的主要缺点是所需样品量大、效率低、本底高. 以 ^{14}C 年代测量为例,衰变法被测样品需含 5~7 g 碳,而 AMS 通常只需要 1~5 mg 的碳就够了.样品量的减少带来了多方面的好处.首先是扩大了样品的选择范围,例如可对一些考古珍品进行年代测定.这方面最突出的例子是最近对耶稣裹尸布的鉴别.其次是使得有可能对地质剖面进行更仔细的分层.此外,样品量少易于实验室化学处理,有利于控制污染,从而使结果的可靠性更高.至于表 1 中的其它核素,半衰期都在 10^5 年以上,衰变率很低,用衰变计数法测量其丰度受到很大限制.例如对 ^{36}Cl,衰变法只能测到 5×10^{-13},但用 AMS 可测到 2×10^{-15}. 故 AMS 大大扩展了可测样品的范围. AMS 的另一个优点是效率高.例如 ^{14}C 测年统计精

---

① 合作者:郭之虞、李坤、陈铁梅.摘自《原子能科学技术》,Vol. 23, No. 6 (1989). p. 76~80.
② 国家自然科学基金资助重大项目,项目编号 9488008.

度欲达1%,衰变法一般要测几十个小时,AMS只用几十分钟.

表1 典型的宇宙成因核素

宇宙成因核素	^{10}Be	^{14}C	^{26}Al	^{36}Cl	^{41}Ca	^{129}I
半衰期(年)	1.5×10^6	5730	7.2×10^5	3.0×10^5	1.0×10^5	1.6×10^7
同位素	^9Be	$^{12}C, ^{13}C$	^{27}Al	$^{35}Cl, ^{37}Cl$	^{40}Ca	^{127}I
同量异位素	^{10}B	$^{14}N^*$	$^{26}Mg^*$	$^{36}Ar^*, ^{36}S$	^{41}K	$^{129}Xe^*$
自然丰度上限	10^{-11}	10^{-12}	10^{-14}	10^{-10}	10^{-14}	10^{-12}
AMS检测下限	10^{-15}	10^{-15}	10^{-15}	10^{-15}	10^{-15}	10^{-14}

* 不能产生稳定的负离子,其干扰可通过采用负离子源消除.

鉴于衰变计数方法的局限性,1970年有人提出采用直接计数方法的建议.随后,在传统的低能质谱计上进行了^{14}C测量的尝试,但终因干扰本底太高而失败.干扰本底系指由被测核素以外的他种离子所引起的本底,包括分子干扰、同量异位素干扰、强峰拖尾及散射干扰.在AMS中,离子被加速到较高的能量,从而使干扰本底有可能得到有效的抑制.现代AMS技术始于1977年.十余年来,AMS技术得到了很大发展,其成果集中反映在一系列国际会议上[2~6].目前世界上的AMS装置已超过30台.AMS的应用领域已遍及地球科学的各个分支以及考古学、物理学、天体物理学和环境化学,近年来又扩展到材料科学与生物科学.关于AMS的应用可参阅文献[7].

二 原理与设备

图1所示为瑞士苏黎世高等工业学院的串列AMS系统布置情况.一般说来,AMS由五个部分组成:(1)注入系统.其离子源大多采用铯溅射负离子源,预分析磁铁应有足够高的分辨率以便有效地抑制高丰度同位素的强峰拖尾干扰.(2)加速器.一般为串列加速器,但也有少数实验室使用回旋加速器或直线加速器.使用串列加速器的好处是,其剥离过程可有效地消除分子干扰.(3)高能分析系统.磁分析器、静电分析器与速度分析器(交叉场速度选择器或飞行时间探测器)均被广泛采用.(4)探测器.使用ΔE-E探测器可将干扰本底与被测核素有效地分离.(5)计算机控制与数据处理系统.AMS的具体设计往往因实验室而异.

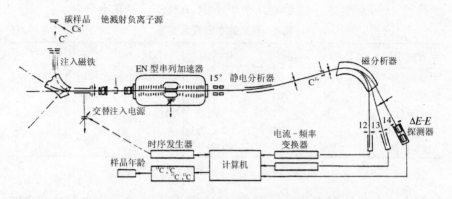

图 1　典型的加速器质谱计配置(瑞士,苏黎世)

AMS 与传统质谱计的主要区别在于,后者只将离子加速到数千电子伏,而 AMS 则将离子加速到数兆电子伏或更高的能量.离子能量的提高带来了多方面的好处.(1) 有可能采用剥离技术来消除分子干扰.(2) 为使用多种手段抑制同量异位素干扰提供了可能,如射程吸收、能损率测量、全剥离以及充气磁铁技术等.(3) 减小了散射干扰.这是由于散射截面随能量升高而减小,且束流发射度在高能下收缩.(4) 使得 ΔE-E 探测器的应用成为可能,而低能区探测器只能对进入的粒子计数,不能分离干扰本底.

AMS 中的分析器主要被用来选择所需的电荷态以及抑制连续动量谱本底.该本底中主要是被测核素的高丰度同位素的离子,它们来自离子源能谱中的强峰拖尾、分子离子经剥离后产生的碎片以及散射与电荷交换过程.消除连续谱本底要使用两个或更多个不同类型分析器的级联组合.由于 AMS 综合使用了多种分析手段,故干扰本底得到了有效的抑制.

AMS 中的另一类本底是污染造成的固有本底.此本底是由外来的而非样品原有的被测核素的同种离子所引起的,一旦引入即无法用分析手段消除,故其防范是十分重要的.污染来源于制样过程中引入的现代样污染、野外污染以及离子源中引入的污染(主要是记忆效应).选择合理的制样工艺、仔细设计离子源、严格取样与制样操作有助于减小污染.

在 AMS 测量中还必须考虑分馏效应.分馏效应是指质谱分析过程

中的质量选择作用,其结果使测量得到的同位素比值偏离于实际值.制样中的化学分馏、离子源中的溅射分馏、剥离分馏和传输分馏都必须给以足够的重视.由于杂散磁场(地磁场等)可使不同同位素的轨迹歧离,若束流通道上某处孔径不足(如分析缝、气体剥离管道等处),就会造成传输分馏.分馏的大小与稳定性关系到测量的精度.为达到尽可能高的精度,最好让所有同位素(如 ^{12}C、^{13}C 和 ^{14}C)都经过同样的加速过程和尽可能共同的传输路径.通常的做法是采用交替注入,令各同位素以固定的周期依次通过加速器,在高能端分析磁铁之后再分开,各自进入不同的法拉第筒与探测器.在不同同位素的共同通道上应避免使用磁四极透镜与磁导向器.为减小分馏,在束流光学设计中要力求实现"平顶传输",即束流接收度较之束流发射度有较大的裕量.这意味着各元件孔径要足够大、各缝要开足、气体剥离管道的通径也要加大.剥离管道的加粗可能导致加速管真空度下降,此时加装循环抽气泵是必要的.各缝开足则可能会引起提取能量稳定信号的困难,苏黎世和英国牛津大学为此发展了用位置灵敏法拉第筒监测高丰度同位素束流的端电压稳定技术.此外,在离子源中,随着溅射过程的延续,样品表面会出现凹坑.这会引起时变的分馏效应.苏黎世与加拿大多伦多大学采用了靶位机械扫描装置,使铯束可以均匀地轰击样品表面,从而避免了凹坑的产生.

尽管采取了以上措施,对测量结果进行本底校正与分馏校正仍是必不可少的.校正用空白样与标准样进行.为减小时变效应的影响,常常把对一个样品的测量划分为许多小区间进行,并在测量未知样的过程中周期地对空白样与标准样进行测量.这也是 AMS 的离子源往往采用多靶位结构的原因之一.

三 性能及其限制

1. 检测下限与样品大小

对于不同的被测核素而言,检测下限取决于不同的限制因素.对 ^{14}C,设计合理的系统可以有效地消除干扰本底,AMS 测年范围目前主要受污染所限.大多数实验室的机器本底在 $3\times10^{-16}\sim10^{-15}$ 之间,约合六七万年.但制样过程中引入的污染可高达 0.1%～1% 现代样,约四万到五万五千年. ^{10}Be 的本底约在 3×10^{-15}. 限制来自 ^{10}B 与吸收体的核反应 H(^{10}B,^{7}Be)^{4}He 所产生的 ^{7}Be 干扰, 以及 ^{9}Be 的散射干扰. 在 ^{36}Cl 的测量中, 本底主要来自两个方面: 同量异位素 ^{36}S 的干扰与高丰度同位素 ^{35}Cl、^{37}Cl 的强峰拖尾. 美国罗彻斯特大学用能损率方法分离 ^{36}S, 用高能端第二磁铁抑制 ^{35}Cl 与 ^{37}Cl 拖尾, 检测下限达到 2×10^{-15}. 能损率方法的分离抑制比与离子束能量有关, 使用端电压高的加速器较为有益. 而 ^{26}Al 的检测下限则在很大程度上受限制于系统的效率, 确切地说, 是受限制于离子源的低流强. 通常铯溅射源所产生的 Al^{-} 流强仅为 $0.1\sim1\ \mu A$, 故在样品丰度过低时, 难于达到可接受的精度.

AMS 所使用的样品大小通常为 $1\sim5$ mg(指元素含量). 在某些情况下, 数十微克的小样品也曾被使用. 但一般说来, 样品量小于 0.5 mg 将导致污染本底上升、精度下降.

2. 精度

AMS 对精度的要求视其应用场合而定. 一般若作为示踪, 总精度能达到 5% 就足够了, 但若用于测年, 自然希望精度尽量高一些. 对 ^{14}C 测年, 1% 的精度相当于 80 年的测年误差. 我们一般所说的精度(精确度)其实包含着两个概念即精密度与准确度. 前者指测量数据相对于均值的离散程度, 后者则指均值相对于真值的偏离程度. 准确度取决于对本底与分馏的控制程度以及校正方法的有效性. 精密度误差依来源可分为内部误差与外部误差. 内部误差主要是被测核素计数的统计误差. 外部误差则主要来自本底与分馏的不稳定性, 反映了 AMS 系统的完善程度, 通常用测量结果的可重复性来衡量. 统计误差的大小主要取决于系统的效率, 目前测量 ^{14}C 现代样可达 0.3%, ^{10}Be 一般可达 2%~3%, ^{36}Cl 做到 1% 没太大困难, 但 ^{26}Al 由于离子源流强难以提高, 一般只能做到 5%~10%. 近年来各实验室为提高 ^{14}C 测年精度做了大量工作. 对于建造完好的系统, 外部误差可达 0.2% 左右, 准确度也基本上令人满意. 苏黎世、牛津、多伦多和美国亚利桑那等大学实验室测量 ^{14}C 的总精度均已达 0.5%. 但对于小样品(<0.5 mg)及老样品(>2 万年), 精度难以提高. 对 4 万年的样品一般只能达到 4%~5%.

3. 效率

效率有两个方面的含义, 一是样品的利用效率, 二是时间的利用效率. 对于小样品与老样品,我们较为关心样品的利用效率. 这是因为其样品中所含的被测核素的原子数较少,统计误差取决于 AMS 的总效率. 该效率等于离子源效率、剥离效率与传输效率三者的乘积. AMS 的总效率一般在 10^{-3} 量级. 对于 ^{14}C, 在理想情况下可达 1%, 而对于 ^{26}Al, 一般仅为 0.02% 左右. 对于普通样品,其中所含的被测核素的原子数足够多,实际上只要达到所需的统计误差,测量也就结束了,尽管此时样品往往尚未用尽. 在这种情况下,我们关心的是缩短测量时间,提高工效. 这时,离子源的输出流强比离子源效率更为重要. 目前许多实验室已实现了 AMS 测量的自动化. 在苏黎世,24 h 之内可以测量 40~50 个未知样品.

四 回顾与展望

在过去的十余年中,AMS 技术取得了巨大的进展. 1978 年的第一次国际 AMS 会议,标志着 AMS 初创阶段的成功. 当时的"第一代"设备都是借用核物理实验用的加速器与束流线,其间确认了 AMS 的可行性,提出了主要的技术问题及其解决方法. 随后到 1984 年的第三次国际 AMS 会议,可以认为是 AMS 的技术建立阶段. 其间 AMS 装置的数量迅速发展,大部分实验室完成了"第二代"系统的建造. 其特征是改用大曲率半径 90°注入磁铁、采用 Ar 气循环剥离、完善高能端分析系统、实现同位素交替注入. 被测核素的种类也大大扩展,并出现了 ^{14}C 测量的商品化专用系统. 离子源技术与制样技术也有较大发展. 在应用领域,特别是地球科学与考古学方面做了大量工作. 1984 年以来则可以认为是 AMS 的技术完善阶段. AMS 的内在机制得到了较为充分的研究. ^{14}C 测量的精度已接近于常规衰变计数法,拓展测年范围的努力仍在继续,在这方面牛津大学正在试验的 CO_2 气体溅射源被寄予厚望. 新的 ^{10}Be 国际标准样品已被建立起来,这为用 ^{10}Be 进行测年创造了良好的条件. 用低端电压小加速器检测 ^{10}Be 与 ^{26}Al 取得了接近于大加速器的好结果. ^{41}Ca 的测量取得了突破性的进展,美国宾夕法尼亚大学不使用浓缩样品已可将检测下限推至 10^{-15} 以下. 一种新的同量异位素分离方法——充气磁铁技术在美国阿贡实验室发展起来,为较低能量下重元素的同量异位素分离提供了有力的手段. 美国 NEC 公司与 US-AMS 公司先后推出了新的商品化 AMS 系统. 其中 US-AMS 公司的"第三代"^{14}C 专用 AMS 系统,采用了同位素重新组合同时注入技术和完善的高能端分析系统,据称精度好于 0.5%,产率可超过每年 3000 个样品[8]. AMS 的应用领域不但在地球科学中进一步扩大,而且出现了向材料科学扩展的趋势. 目前世界上仍有一批 AMS 装置正在建造之中. 中国自己的数台 AMS 装置,在国家自然科学基金会与中国科学院的支持下,也已分别开始建造. 我们期待着他们的成功. 可以期望,AMS 技术将在最近几年中取得更大的进展.

参 考 文 献

[1] Elmore, D. and Phillips, F. M., *Science*, **236**, 543(1987).
[2] Gove, H. E., ed., Proc. First Conf. on Radiocarbon Dating with Accelerators, Univ. of Rechester, USA(1978).
[3] Henning, W. et al., eds., Proc. Symp. on Accelerator Mass Spectrometry, Argohne National Laboratory, USA(1981).
[4] Wölfli W. et al., Zürich, Switzerland, *Nucl. Instrum Methods.*, **B5**, 91(1984).
[5] Hedges, R. E. M. et al., eds., Proc. Workshop on Techniques in Accelerator Mass Spectrometry, Oxford, UK(1986).
[6] Gove, H. E. et al., Niagara-on-the-Lake, Canada, *Nucl. Instrum. Methods.*, **B29**, 1(1987).
[7] 陈铁梅等,海洋地质与第四纪地质,9(1), 103(1989).
[8] Purser, K. H. et al., *Nucl. Instr. and Meth.*, **B35**, 284(1988).

(编辑部收到日期:1988 年 5 月 21 日)

BEAM CHARACTERISTICS AND PRELIMINARY APPLICATIONS OF THE TANDEM BASED AMS AT PEKING UNIVERSITY[①②]

Abstract

A dedicated tandem based AMS facility was constructed and put into operation at Peking University. Preliminary applications to the carbon dating of archaeological samples are performed at a background level of 10^{-14} with an accuracy better than 1.7%. The designed features of the system as well as the measured beam emittance and transmission efficiency are presented. The suppression of background and fractionation effect and the precision of the measurements are discussed.

1. Introduction

To meet the growing requirement of extremely sensitive isotope measurements in China, the construction of the Peking University accelerator mass spectroscopy (PKUAMS) facility was completed after 3 years of joint efforts. The facility was put into operation early this year and had the first batch of samples measured at a sensitivity of 10^{-14}, with an accuracy better than 1.7%.

The PKUAMS consists of four major subsystems: a dedicated ion source, a fast switching injection line, an EN-tandem accelerator and a postacceleration analysis and detection system. The layout is shown

---

① Coauthers: GuoZhiyu, YanShengqing, LiRenxing and XiaoMin, LiKun, LiuKexin, LiuHongtao, ZhangRuju, LuXiangyang and LiBin. Reprinted from the Nuclear Instruments and Methods in Physics Research B79(1993) 624~626.

② Work supported by NSF China and joined by Department of Archaeology PKU and Shanghai Institute of Nuclear Research.

schematically in Fig. 1. The design concepts were reported in previous papers [1~3]. Emphases have been laid on high quality ion beam production, mass independent transmission, high resolution magnetic and electrostatic analyzers and computer control of sequential isotope injection and data acquisition. The practical features and beam characteristics of subsystems are described in the following sections.

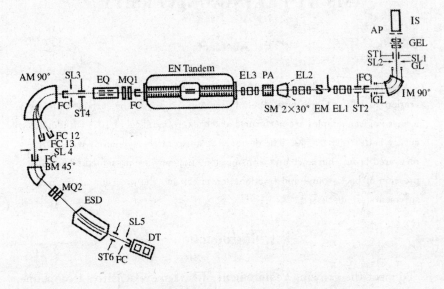

Fig. 1 Schematic layout of the PKUAMS facility

2. Features of the System

2.1 Ion sources

Both an upgraded Hiconex 834 reflection type source and a newly developed sputter source are available for operation. The 834 source, with a sample tray of 12 positions, was used during the commissioning and the measurement of the first batch of archaeological samples. For natural graphite samples a current of 4~10 μA can be extracted. The current ratio of mass number 13 to mass number 12 is 5 times higher than that of natural composition, and it increases with degrading vacuum. This indicates high molecular yields of ^{12}CH. A similar situation is observed for mass 14 due to ^{12}CH$_2$+^{13}CH. The new sputter source developed at SINR [4] is novel in the ionizer design where a spherical geometry is used to get better focusing of the Cs beam. As a result, 150 μA of C$^-$ and 5 μA of BeO$^-$ as well as 0.4 μA of Al$^-$ were extracted. The source has a tray with 20 samples positions. Driven by two stepping motors, it enables rapid change between samples within one minute under remote control.

The emittance of the source is crucial not only for the reason of beam transmission but more importantly in the light of system resolution and fractionation, and it was measured carefully after the injection magnet for related isotopes. The measured emittance ε_n is shown in table 1 for the 834 source which is 4.1 π mm $\cdot$ mrad $\cdot$ MeV$^{1/2}$ for ^{12}C and 3.7 π~6 π for other ions. The corresponding transmission efficiency η through the injection line is also shown. The current dependence of the emittance is not serious for ^{12}C within the range of 4~10 μA; however, the change is remarkable for mass 13.

Table 1 Emittance of related isotopes

	^{12}C$^-$	^{13}C+^{12}CH	^{12}C$_2^-$
I/μA	6.30	0.37	2.46
ε_n/π mm $\cdot$ rad $\cdot \sqrt{\text{MeV}}$	4.07	3.70	1.62
$\eta/(\%)$	60.4	61.1	72.2

As for memory effect, the current of ^{12}C attenuated by a factor of 2$\times$10^{-3} in 10 min after changing the graphite to an Al sample.

2.2 Tandem accelerator

The EN-tandem accelerator from Oxford was reconstructed and put into operation early 1990. A variety of ions from Be to Br have been accelerated successfully for various purposes since then. So far 6 MV is available as the maximum terminal voltage while 3 MV is used with a stability better than $\pm$1.5 kV as a routine for carbon dating where ions with 3 + charge state are selected for high energy mass analysis and detection. The vacuum system ensures a pressure of 10^{-6}~10^{-7} in the accelerating tube.

The transmission efficiency through the tandem is about 60% depending on the ion species, and somehow limited by the stripper.

2.3 Beam line and analyzing elements

The layout of the beam line was worked out using the OPTRYK code. Isotopes are injected into the tandem sequentially along the same orbit by an injection magnet biased with $0 \sim 5$ kV pulsed voltage on the chamber. All lenses and steerer used are electrostatic so as to provide the same focusing condition for all isotopes. In order to approach flat-topping transport, the apertures of the beam elements are fairly large, e.g. 100 mm diameter for the einzel lenses, 50 mm gap 90° injector magnet with a 10 mm slit, etc. Grided lens and curved boundary magnet are also adopted to minimize the emittance growth due to aberrations. Moreover, an aperture of 10 mm in diameter is set right after the source to cut down the beam halo while ensuring a high transmission efficiency for the wanted beam. The beam transmission efficiency measured by Faraday cups along the injection line was $60\% \sim 70\%$.

At the high energy end, both electrostatic and magnetic quadrupole doublets are available. The pole tip of the main analyzing magnet is extended and is incorporated with three position adjustable Faraday cups to enable simultaneous measurement of all isotopes. The transmission efficiency after the main analyzing magnet is $90\% \sim 100\%$, and the total efficiency along the whole AMS line lies in the range of $25\% \sim 40\%$. The feature of flat-topping transport was successfully verified by varying the field strength of related beam optical elements versus target current, and it turns out the flatness for the einzel lens at the tandem entrance and the injection magnet is less than elsewhere so high stability power sources are needed there.

High mass or energy resolution is highly demanding for suppressing the intensive peak tail of abundant isotopes. The resolution R measured for the injection magnet IM, main analyzing magnet AM, bending magnet BM and the electrostatic analyzer are listed in table 2 with related bending angle Θ, radius of curvature r and slit width S. The attenuating factor α

measured as the ratio of the tail of mass 12 at mass 14 to its peak is also listed.

Table 2 Resolution of magnetic and electric analyzer

	IM	AM	BM	ESD
Θ/deg	90	90	45	20
r_{mm}	400	914	1000	3200
s_{mm}	10	5	5	5
R	103	324	119	261
$\alpha \times 10^{-5}$	1.7			

The isobaric background is discriminated by a ΔE-E detector. In principle, the source will not produce negative ions like ^{14}N and ^{26}Mg. However, they can be produced elsewhere. A number of ^{14}N ions was detected once during a ^{14}C test due to a leak at a bellow before the tandem.

2.4 Timer and computer control

The fast switching of isotope injection is controlled by a "timer". The duration for stable isotopes is set to $100 \sim 700$ μs out of the 99.4 ms cycling period, so that stable isotopes can be measured in the high energy end without causing a serious loading effect, which in turn facilitates the essential measurement of isotopic ratios like $^{14}C^{3+}/^{12}C^{3+}$, $^{13}C^{3+}/^{12}C^{3+}$ as well as $^{12}C^{3+}/^{12}C^{-}$. The data acquisition is completed by an IBM PC which is synchronized with the injection and is to be cycled by 500 times in 50 s. The fast switching scheme can compensate to some extent the slow drift and short term variations.

2.5 Fractionation of the system

The fractionation effect was observed during the C^{-} beam tests. The cratering on the sample due to the long period of bombardment reduces the transmission of mass 13 through the injection line by nearly 13% with respect to ^{12}C. The effect of magnetic elements was studied by comparing the transmission through the postacceleration line using a magnetic quadrupole doublet to that of one magnetic plus one electrostatic doublet. The fraction δ^{13}C turns out to be -10.2‰ in the former case and -7.8‰

in the latter case. The result implies that serious fractionation occurred in the process of sputtering and stripping; meanwhile the use of electrostatic elements does ease the system fractionation effect. Actually, the effect also related to the system acceptance. As the width of the slit after the ESD increases from 4×4 mm, 6×6 mm to 10×10 mm, $\delta^{13}C$ decreases from 17.7‰, 6.2‰ to -9.4‰ respectively.

3. ^{14}C measurements

To get reliable results a great number of measurements are taken for each sample. It takes 500 s in total for one measurement which consists of 10 records, each containing data from 500 fast switching cycles. The counts of ^{14}C read from the window of a two-parameter spectrum, and charges of ^{12}C and ^{13}C from the integrated current meter, were recorded with the corresponding ratios between related isotopes. The data are averaged over 10 records. In order to eliminate the error-leading effects like sparking, those records deviating far from the mean value were canceled according to the Chauvenet criterion. In addition, standard and blank samples are measured alter-natively with the dating sample for a number of times, so that the system fractionation effect and the back-ground due to sample or system contamination can be properly corrected.

The results finally obtained for three archaeological samples from Ma Wang-dui, Da Wenkou and Nan Zhuang-tou are listed in table 3.

The precision of the measurement is verified by measuring the specific radioactivity ratio of two standard samples i.e. China radiocarbon Standard and Australia Standard, and the sensitivity for measuring ^{14}C is given by the results of the blank sample measurement.

Table 3 Results of archaeological samples

Sample	β-counting	PKUAMS	Error
Ma WD	2120±90 a	2130±140 a	1.7%
Da WK	5350±90 a	5130±220 a	2.7%
Nan ZT	10510±270 a	10420±310 a	3.7%

4. Conclusion

The beam characteristics of PKUAMS with 834 source meet the basic requirements of AMS measurements while better performance is expected by using the new sputter source with higher brightness, and a new power source for the einzel lens with higher stability. Further studies aiming at higher precision, sensitivity and efficiency are planned. Applications of ^{14}C dating for archaeological and geophysical samples are ready for routine analyses. Measurements with ^{10}Be and ^{26}Al are also in progress.

References

[1] C. E. Chen, K. Li, Z. Guo, S. Yan et al., Nucl. Instr. and Meth. B52(1990) 306.
[2] C. E. Chen, K. Li, Z. Guo, S. Yan et al., Conf. Rec. of the 1991 IEEE Particle Accel. Conf. 1991, San Francisco, p. 2616.
[3] C. E. Chen and K. Li, Proc. 4th China-Japan Joint Symp. on Accelerators for Nuclear Science and Their Applications, 1990, Beijing, China.
[4] H. Z. Si, presented at this Conference (12th Int. Conf. on the Application of Accelerators in Research and Industry, Denton, TX, 1992).

ACCELERATOR MASS SPECTROMETRY AT PEKING UNIVERSITY: EXPERIMENTS AND PROGRESS[1][2]

Abstract

The Peking University Accelerator Mass Spectrometry (PKUAMS) has been put into routine operation. The ^{14}C measurements of archaeological samples with fast cycling injection has shown good results. The new multi-target high-intensity sputtering ion source has been tested and the ^{10}Be measurements were carried out with a new detector in which both the stopping of the intense flux of ^{10}B ions and the identifying of ^{10}Be ions are performed. The ^{26}Al samples were also measured. While various applications show a good prospects of PKUAMS, a further upgrade is desirable.

1. Introduction

The Peking University Accelerator Mass Spectrometry (PKUAMS) project, as a major project of NSFC, was started in 1988.[1] The PKUAMS facility is based on an EN tandem with dedicated ion source, beamline and detector. After optimization the present layout of PKUAMS beamline is shown in Fig. 1. The facility was put into operation in 1991.[2] In the first phase of ^{14}C measurements, some archaeological samples were measured with a modified Hiconex 834 reflection-type ion source at a sensitivity of 10^{-14} with an accuracy better than 1.7% for modern samples.[3]

---

① Coauthers: Chen Chia-erh, Guo Zhiyu, Yan Shengqing, Li Renxin and Xiao Min, Li Kun, Liu Hongtao, Liu Kexin, Wang Jianjun, Li Bin and Lu Xiangyang, Yuan Sixun, Chen Tiemei and Gao Shijun, Zheng Shuhui, Chen Chengye and Liu Yan. Reprinted from the Nuclear Instruments and Methods in Physics Research B92(1994) 47~50.
② supported by the National Natural Science Foundation of China (NSFC).

The Instrumental background was in the order of 10^{-15}. The dedicated ion source for AMS was installed and commissioned in autumn 1992. That is a high-intensity Cs sputtering negative ion source with spherical ionizer and multi-target mechanism.[4] Since then the ^{10}Be and ^{26}Al measurements had been carried out. The instrumental background was 10^{-14} but the blank samples gave a background level of about 10^{-13}. The routine radiocarbon analysis work was started at PKUAMS in spring 1993 and many application research projects are waiting for AMS measurements now. To meet the wider requirements, an upgrade plan was worked out and will be performed in the next three years under the support of NSFC.

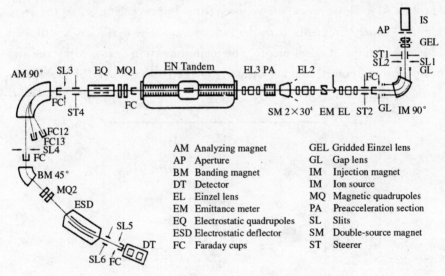

Fig. 1 Schematic layout of the PKUAMS facility

2. Measurements of ^{14}C, ^{10}Be and ^{26}Al

2.1 ^{14}C

The system set up and ^{14}C measuring procedure with fast cycling sequential injection have been described before.[2~3] The gas detector has two anodes for ΔE and E_{rest} measurements, respectively. An additional solid detector is mounted at the end of the detector chamber, which is helpful to judge the working state of the detector. Fig. 2 shows a typical ΔE-E_r spectrum of ^{14}C measurement, in which the isotopes ^{14}C, ^{13}C and ^{12}C are separated distinctly. Sometimes a ^{14}N peak may appear in the spectrum. It is believed that the ^{14}N comes from a slight leak on the acceleration tube at low energy side. But it has not been confirmed so far.

The dedicated high-intensity ion source has been used for the routine ^{14}C measurements. The sample tray of the source has 20 positions and the exchange of samples is performed by remote control. The throughput of PKUAMS improved remarkably with the new source comparing to the Hiconex 834. However, there is no essential improvement yet on the accuracy of ^{14}C measurements with the new source. The deviation is still a bit larger than the statistical error. The main limitation is caused by the stability of some power supplies, especially the power supply of Einzel lens at the tandem entrance.

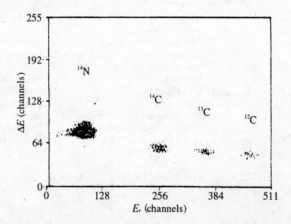

Fig. 2 Two-dimensional spectrum of ^{14}C measurements

2.2 ^{10}Be

The difficulties of ^{10}Be measurement come from three aspects. Firstly, the negative ion of beryllium has very low production rate so that the high-intensity ion source has to be used and the molecular ion BeO$^-$ is chosen. Secondly, stripping of BeO$^-$ by foil will introduce "Coulomb explosion" effect, that usually causes a bad transmission, while stripping by gas to break the molecular needs too much gas, which makes the vacuum worse and also causes the bad transmission. The solution is to use a combination of gas and foil[5], or, the best way, to use recirculated gas stripper.[6] Thirdly, the isobar ^{10}B is a serious interference. It should be stopped before the ^{10}Be detection by some medium, usually gas.

After the high-intensity ion source was put into operation, ^{10}Be measurements were carried out at PKUAMS with an ordinary gas stripper. It has been proved that the stripping yield of ^{10}Be ions is a function of terminal voltage for O_2 gas. The yield of 2+ reaches a maximum of 70% at about 4.5 MV and 3+, around 60% at 8~10 MV. And the yield of 3+ at 5~6 MV just lies in 20%~30%.[7] That means to select charge state 2+ at terminal voltage 4.5 MV may get higher counting rate than 3+ at 5~6 MV, which is the limitation of our EN tandem. But our experiment indicates that it is difficult to separate ^{10}Be from ^{10}B in detector effectively with 2+ at 4.5 MV due to the too small residual energy. And, as mentioned above, the ordinary gas stripper tend to a bad transmission, so the total transmission efficiency is only 3%~5% although the transmission efficiency is 70%~80% in the injection beamline and near 100% in the HE analysis beamline after the main 90° analyzing magnet.

The detector used for ^{10}Be measurements at PKUAMS is somewhat like the ANU one[8], which combines the roles of both ^{10}B absorber and ^{10}Be counter in a chamber so only a single kind of gas is needed. The cross section of the detector is shown in Fig. 3. The entrance window is a 2um Havar foil. The front part of the detector is used as a gas absorption chamber of ^{10}B, which consists of two parallel plates with the electric field directed perpendicular to the beam. The lower plate is biased at -300 V and the upper one is grounded through a microammeter, which measures the ^{10}B ionization current. That current is very helpful to the machine tuning and the transmission stability monitoring. The following ^{10}Be counting region of the detector has three anodes A_1, A_2 and A_3. The anodes and grid are biased to +600 V and +300 V respectively, and the cathode is maintained at ground potential. P-10 gas is used at the moment. The gas flow through the detector prevents the build up of oxygen and other contamination. The pressure used during ^{10}Be measurements is sufficient to stop ^{10}B in the absorption chamber and is available for ^{10}Be ions to pass through to the A_3 anode.

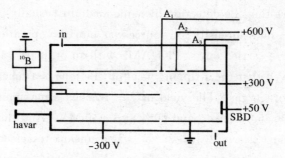

Fig. 3   Sketch of the ^{10}Be detector with ^{10}B absorption chamber

The signals from anode A_1 can be used to indicate whether ^{10}B ions have been stopped in the absorption chamber. Signals from anodes A_2 and A_3 correspond to the energy loss $-E$ and residual energy loss E_r of ^{10}Be ions which have passed through the region of ^{10}B absorber. The surface barrier silicon detector (SBD) is used to test whether all the ^{10}Be ions are stopped within the region of anode A_3 and if other higher energy ions exist. The ^{10}B ionization current varies with the different samples, typically around several hundred nA. Fig. 4 shows the two-dimensional spectra of ^{10}Be measurements; it is a spectrum from a sample with ratio ^{10}Be/^{9}Be= 6.7×10^{-9}. No ^{10}B ion is observed on the plot, which indicates all the ^{10}B ions were stopped in the absorption region. Fig. 4(b) is a spectrum from the measurement of a blank sample with a Mylar foil window on the detector entrance counting for about one hour, which presents a strong ^{7}Be background. The ^{7}Be arises from the well-known reaction H(^{10}B, ^{7}Be)^{4}He and the hydrogen in Mylar gives a great contribution to the background. By replacing the Mylar with Havar the background dropped down more than an order. It is expected to reduce the background further by using the hydrogen-free gas instead of P-10 gas.

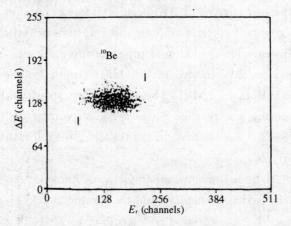

Fig. 4 Two-dimensional spectrum of ^{10}Be measurement from a sample with a ^{10}Be/^{9}Be ratio of 6.7×10^{-9}

Three samples with known ^{10}Be concentrations were measured to check the performance of PKUAMS. As shown in Fig. 5, the agreement between the measured and known ^{10}Be/^{9}Be ratio is satisfactory.

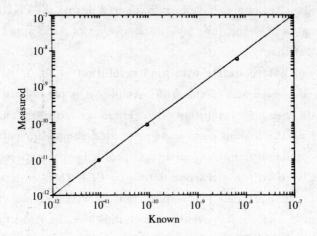

Fig. 5 Comparison of the measured ^{10}Be/^{9}Be ratio with the known value for three samples

2.3 ^{26}Al

Due to the low current of Al$^-$ beam extracted from ion source, it is important to get the highest stripping and transmission efficiency for the ^{26}Al measurement. The yield of stripped positive Al ions with various charge states were investigated at different terminal voltage within the range of 2.5 MV to 5.5 MV. The results indicate that the maximum stripping efficiency was reached at the terminal voltage of 3.0 MV with charge state 3+ for our facility. In that case the overall transmission efficiency is about 5%.

During ^{26}Al measurements a sample with known ^{26}Al/^{27}Al ratio of 1.114×10^{-10} is used as a standard and the commercial Al$_2$O$_3$ reagent sample, as a blank. The background is determined as 1×10^{-13} with the blank sample, and an accuracy of 5%~10% could be easily reached for meteorite sample with ^{26}Al/^{27}Al ratio ranged in $10^{-10} \sim 10^{-11}$.

3. Applications and Prospects

More than 70 samples have been measured during the PKUAMS routine operation in May and June 1993, and a series of AMS application projects will start this year.

3.1 Study of palaeoclimate with high resolution

Due to the small size of the AMS sample, it is possible to study the palaeoclimate with high resolution now. There are large arid and semiarid areas in the north of China, and loess is an ideal sample for palaeoclimate study. Six fossil soil samples from a loess profile in the north-west of China were analyzed with radiocarbon dating by PKUAMS to determine the time scale for that profile. The results coincide well with the expectation derived from known climate events, which indicates the AMS radiocarbon time scale is reliable. More analysis will be carried out in order to investigate the palaeoclimate evolution in some detail.

In the south of China the climate is damp and the carst developed thoroughly. PKUAMS will provide a hand in a project to rebuild the evolution process of palaeoenvirenment for the past thirty thousand years by using the carst information.

3.2 Archaeology

Many archaeological radiocarbon samples were measured by PKUAMS during the passed several months. Some of them have shown very interesting results. For example, pottery pieces unearthed from the Nan Zhuang Tou site, Hebei Province, China were dated as early as 10500 years B. P. It is the earliest pottery manufactures unearthed in China so far. The dating material is soot picked up from pottery surface and the amount is very little. So AMS is the only reliable means to date the pottery directly.

Another group of dating samples that measured by PKUAMS recently is the bone from a site in the north-east of China. There are metal earrings on the both side of one skull. Therefore, it is important to investigate the Chinese metallurgy history.

Some other samples excavated from different layers of a cave in the south of China were also analyzed with ^{14}C dating by PKUAMS, which will be helpful to understand the transition from the paleolithic culture to the neolithic culture in China.

3.3 ^{10}Be and ^{26}Al

The ^{10}Be concentrations of the sediments from Huang Hai (Yellow Sea) and Bo Hai have been measured and it is proved that all they are typical continental deposit. The ^{26}Al concentration in Jjlin meteorites from the north-east of China has also been measured. There are more ^{10}Be and ^{26}Al measurement projects on the study of sediment, meteorite, loess and soil to be started in the near future.

4. Upgrade Plan

The objectives of the upgrade plan are to reach an accuracy of 0.5% for modern carbon sample and a sensitivity of 10^{-15} level for ^{14}C and ^{10}Be measurements. For that purpose the ion source will be improved to include the sample scanning function; the recirculation gas stripper will be mounted at the terminal of EN tandem to improve the vacuum and transmission; the unstable power supplies will be replaced to raise the system stability; the ^{14}C sample preparation technique will be investigated further to reduce the contamination; the data acquisition and facility control system will be improved to gain its ability and to increase the throughput of PKUAMS.

Acknowledgements

The authors would like to thank Ms. Zhang Ruju, Mr. Qian Weishu, Ms. Yuan Jinglin, Ms. Yang Fengling and the staff of EN laboratory for their contribution to PKUAMS. Special appreciation will be given to Dr. K. van der Borg and Dr. J. R. Southon for providing Al standard sample and Havar foil and for their helpful discussions. The authors' many thanks also to Prof. A. E. Litherland, Prof. D. J. Donahue, Prof. W. Wölfli, Dr. G. M. Raisbeck, Dr. J. Klein, Dr. R. E. M. Hedges, Dr. M. Suter, Dr. G. Bonani, Dr. R. P. Beukens and Dr. Zhao Xiaolei for their beneficial comments and discussions during the past years.

References

[1] Chen Chia-erh, Li Kun, Guo Zhiyu, Yan Shengqing, Liu Hongtao et al., Nucl. Instr. and Meth. B52(1990) 306.

[2] Chen Chia-erh, Li Kun, Guo Zhiyu, Yan Shengqing, Liu Hongtao et al., Conf. Rec. of the 1991 IEEE Particle Accel. Conf., May 6~9, 1991, San Francisco, Vol. 4, p. 2612.

[3] Chen Chia-erh, Li Kun, Guo Zhiyu, Yan Shengqing, Liu Hongtao et al., Nucl. Instr. and Meth. B79(1993) 624.

[4] Si Houzhi, Zhang Weizhong, Zhu Jinhua, Du Guangtian and Zhang Tiaorong, Rev. Sci. Instr. 63(1992) 2472.

[5] J. Klein, R. Middleton and H. Tang, Nucl. Instr. and Meth. 193(1982) 601.

[6] G. Bonani, P. Eberhardt, H. J. Hofmann, Th. R. Niklaus, M. Suter, H. A. Synal, and W. Wölfli, Nucl. Instr. and Meth. B52(1990) 338.

[7] M. Suter, Nucl. Instr. and Meth. B52(1990) 211.

[8] L. K. Fifield, T. R. Ophel, G. L. Allan, J. R. bird and R. F. Davie, Nucl. Instr. and Meth. B52(1990) 233.

北京大学加速器质谱计研究与应用进展[①②]

摘　要

近年来,加速器质谱[1](Accelerator Mass Spectrometry,简称 AMS)在国际上的应用有较大的发展,每年测量的样品已达数万个.AMS 主要用于测量 ^{14}C, ^{10}Be, ^{26}Al 和 ^{36}Cl 等宇宙成因核素的同位素丰度,从而推断样品的年龄或进行示踪研究.与常规质谱相比,AMS 的灵敏度要高出 5~7 个数量级;与衰变计数法相比,AMS 具有样品量小、工效高的优点.因此,AMS 已被广泛应用于地球科学、考古学与古人类学、环境科学、生命科学、材料科学、核物理学与天体物理学等诸多领域[1,2].为满足我国相应学科基础研究的需要,北京大学加速器质谱计(PKUAMS)在国家自然科学基金的资助下,已顺利建成并投入运行,在多个学科领域中开展了应用研究工作.

1　PKUAMS 概况

北京大学加速器质普计 1988 年被列为国家自然科学基金重大项目,1992 年建成并进行了 ^{14}C 测量[3~5].至 1993 年 4 月,又先后完成了 ^{10}Be 和 ^{26}Al 测量方法的建立[6,7].1993 年 5 月 PKUAMS 正式投入运行并开始对外测量服务.此后,针对提高测量精度、降低本底、减小污染及提高工效等目标,我们又对设备和方法做了若干改进工作[8],使 PKUAMS 的性能得到了进一步提高.目前国际上先进的 AMS 实验室在日常运行中, ^{14}C 测量的精度一般在 0.5%~1%,测年上限一般在 45 ka 左右.北京大学加速器质谱计 ^{14}C 测量的精度可好于 1%(对于年龄在 10 ka 左右或更年轻的样品),样品测量的重复性亦可达到 1%.机器本底小于

---

① 合作者:郭之虞,李　坤,刘克新,鲁向阳,李　斌,汪建军,陈铁梅,原思训,高世君,袁敬琳,钱伟述.摘自《自然科学进展》,Vol.5,5(1995)513~516.
② 国家自然科学基金资助项目.

0.004 MC (Modern Carbon，1 MC 相应的 ^{14}C/^{12}C 值为 1.2×10^{-12})，样品的测量本底不大于 0.006 MC，相当于测年上限约 43 ka．样品量可小至 $0.4\sim0.5$ mg 碳（指样品中碳元素的含量），测样工效一般可达每天 $10\sim12$ 个样品．

2 运行情况

北京大学加速器质谱计自 1993 年 5 月至 1994 年 6 月测量了 ^{14}C 样品共 520 个，其中标准样品 107 个，本底样 44 个，已知年代样品 46 个，加速器质谱计性能改进研究用样品 38 个，为用户测量样品 285 个．被测样品来自 30 余个用户，其中有 6 个样品来自美国．被测样品中有地质样品 137 个，占 48%；考古样品 78 个，占 27.5%；环境科学样品 23 个，占 8%；生命科学样品 47 个，占 16.5%．这些样品的测量结果普遍得到用户的好评，其中许多结果具有重要的科学意义．

在所测样品中，有木炭、有孔虫、海洋沉积物（软泥）、烧骨或骨化石、古土壤、黄土、水样、贝壳（或螺壳）、石笋（或其他碳酸盐）、孢粉、湖相和陆相沉积物、考古文物、气溶胶、^{14}C 标记生物样品等．其中，地质和考古样品的年龄在 $3\sim30$ ka；环境与生命科学样品的 ^{14}C 含量在 $0.05\sim1500$ MC．

为了检验测量的可靠性，在测量样品的同时，我们都测量了已知年代样品（马王堆中的木炭）．根据 17 次运行的统计，每日测量误差（标准偏差）为 ±80 a．

3 应用研究

3.1 地球科学

为配合全球变化的研究，了解我国第四纪以来气候变化的趋势，需要掌握我国不同地区过去气候变化的高分辨率记录．"我国干旱半干旱区 150 ka 来环境演变的动态过程与发展趋势"这一国家自然科学基金重大项目，对我国北方干旱、半干旱区东西向地质大断面中的若干典型剖面进行了研究，以期系统总结该地区过去气候变化的规律，从而更好地预测未来环境发展的趋势,制定正确的环境政策和措施. 我们与中国科学院地质研究所合作,用 AMS ^{14}C 方法测定了渭南黄土地层剖面一系列样品的年龄[9]. 所测样品材料为古土壤或黄土,每个样品又分为胡敏酸与胡敏素两种组分对比测年. 17 个样品的测量结果与地层层次相符,从而提供了该剖面从晚更新世到全新世的高分辨率时间标尺. 该时间标尺与热释光方法得到的时间标尺吻合衔接良好,与深海沉积物 SPECMAP 曲线也相当吻合,为重现黄土中保存的古气候古环境记录所反映的地质历史情况提供了重要依据.

在中国南方的大片碳酸盐地区,可以采用岩溶沉积物作为获取古气候变化信息的载体. 我们与中国地质科学院岩溶地质研究所合作,对桂林盘龙洞采集的 1.22 m 高大型石笋按照沉积形成的微细纹层层序进行了分层取样 AMS ^{14}C 测年,并配合氧同位素、微量元素等方法对过去 30 ka 来的气候变化做了精细研究①. 30 多个样品的测量结果表明,石笋不同时期的沉积速率在每百年 0.12~28 cm 间变化. 在距今 12~32 ka 间是末次冰期中特别干燥的时期,石笋生长很慢,氧同位素数据表明其间气温很低. 在距今 10 ka 以来的全新世中,该地区气候总的具有潮湿和温暖的趋势,且显示出由冷到暖的多个气候旋回. 研究还表明在距今 10~11 ka 之间存在着一个短暂的寒冷时期,这可能是新仙女木事件在本地区的表现. 本工作的测年分辨率在暖湿期可在 100 a 之内,而在沉积速率较慢的干冷期也可达 300~500 a,从而为我国南方岩溶古环境重建提供了一个系统的剖面.

除了各种 ^{14}C 地质样品以外,我们还测定了黄海与渤海的沉积物和悬浮物中 ^{10}Be 的含量以及吉林陨石中 ^{26}Al 的含量.

3.2 考古学与古人类学

在研究人类从旧石器时代向新石器时代过渡的问题中,对有关遗址的年代测定是一项关键工作. 中国南方的石灰岩洞穴中有大量史前人类的文化遗存,其中广西柳州白莲洞遗址包含了从旧石器晚期至新石器早期的文化层,所对应的年代为距今 37~7 ka. 由于常规 ^{14}C 测量所需样品量很大,许多层位难于取得足够的样品量,因而所得年代数据一直不完

---

① 袁道先. 岩溶作用对环境变化的敏感性及其记录. 科学通报,1995,待发表.

整. AMS 小样品的特点为解决此问题提供了有力的手段. 1993～1994年我们测得了一批极有价值的年代数据,给出了若干关键层位的年代,证实了该遗址连续层位的完整性[10]. 这就为研究自玉木冰期盛冰期以来的环境与气候变化,以及旧石器晚期文化向新石器文化的过渡提供了可靠的依据.

由于具有样品耗量少这一优点,AMS ^{14}C 法可以解决许多其他测年方法无法解决的考古学中重要的疑难问题. 例如,古陶器的定年是一个复杂而困难的问题. 陶土中的基质碳只能给出陶土形成的地质年代,并不代表陶器的制作与使用年代[11],而陶器上的少量积碳(如烟炱、食物残渣等)往往因数量太少不能用衰变计数法测年. 故通常只能取同层位的木炭等样品测年,这样推测的年代可靠性较差. 我们用 AMS 测定了河北省南庄头遗址出土陶片上烟炱的年代,结果与遗址年代相符,从而证实了在新石器时代早期(约 10 ka 前),当地人已掌握了制陶术. 我们还用 AMS 研究了各类农作物的起源、早期制陶史、早期冶金史等一些有关我国新石器文化与文明起源的重要问题. 例如,我们曾测定了甘肃东灰山遗址出土的碳化小麦的年龄为距今 4 230 a. 这是目前所知我国最早的小麦.

此外,我们还在几乎无损的条件下,为用户测定了一些珍贵古代文物的年代,从而为确定其真伪提供了依据.

3.3 环境科学

大气气溶胶是重要的大气污染物,对其来源的研究是当前国际瞩目的重要课题. 传统上一般用多元统计方法研究其来源,但结论具有一定的局限性. 我们与北京大学环境科学中心合作,利用 ^{14}C 作为气溶胶来源的示踪剂(矿石源不含 ^{14}C,而生物源则含有 ^{14}C),用加速器质谱计测定了北京城区与郊区、山东、湖南等地的大气气溶胶中的 ^{14}C 含量①,据此计算了生物源与矿石源的相对贡献,并对采暖期与非采暖期的结果做了对比. 研究表明,^{14}C 示踪法可给出与多元统计法不同的结论. 例如对北京中关村地区采暖期气溶胶样品的分析,多元统计法的结论其主要来源为

---

① 邵敏. AMS 方法在大气气溶胶来源研究中的应用. 北京大学环境科学中心博士论文, 1994.

土壤,矿石源所占的比例很低,而^{14}C示踪法则给出其矿石源高达80%. 这是由于多元统计法是基于气溶胶中的微量元素组分,而^{14}C示踪法则是基于其碳组分,二者具有不同的来源特征. 故将^{14}C示踪法与多元统计方法相结合,可以加深人们对大气气溶胶来源的认识,这对于环境污染问题的研究有重要意义.

3.4 生命科学

1994年以来,我们与北京大学技术物理系应用化学专业有关课题组合作,用AMS研究了具有强烈致癌作用的烟碱亚硝基衍生物NNK与DNA的加合作用[12]. 在实验中将^{14}C标记的NNK灌入老鼠胃中,24 h后从其肝脏中提取DNA,再制备成石墨样品,用AMS测量. 测量的结果表明,DNA加合物与NNK的剂量呈较好的对数线性关系,实验所用的最低剂量相当于吸烟者吞咽20支卷烟的烟雾所吸取的NNK量,比放射免疫分析方法低4个数量级. 我们也用同样方法研究了环境剂量下烟碱(尼古丁)本身与DNA的加合作用[13]. 加合物含量与烟碱剂量亦成对数线性关系,但相同剂量下烟碱产生的DNA加合物较NNK所产生的明显偏少. 这就证实了NNK有较强的基因毒性. 在以上实验中,DNA加合物在正常DNA中含量的测量灵敏度达到了10^{-11},与目前使用的其他方法(如^{32}P后标记法、放射免疫分析法、荧光光谱法等)相比,其灵敏度要高3~6个数量级. 国际上类似工作是由美国劳伦斯利沃莫尔实验室首先开展的,其测量灵敏度亦为10^{-11},目前尚未见到其他实验室开展此类工作的报道. 本工作也为药物基因毒性的检验与筛选,建立了既经济又有效的方法.

4 展 望

在国家自然科学基金的资助下,我们正在对北京大学加速器质谱计做进一步改进,包括降低制样过程中引入的污染、提高离子源工作的可靠性、采用气体循环剥离技术、更新数据获取及测量控制系统,以使其性能达到一个新的水平. 同时,我们还将进一步发展样品的分组分测量技术,发展小样品测量技术和就地测量(in situ)技术,并进一步加强与各学科科学家的合作,拓展应用的领域,以期为我国科学事业做出更大的贡献.

致谢 作者在此谨向刘东生、袁道先、刘元方、唐孝炎、刘嘉麒、林玉石、覃嘉铭、周国兴、仇士华、李金龙、邵敏、李新松及其他合作者表示衷心的感谢.

参 考 文 献

[1] 郭之虞. 加速器质谱学. 见: "现代核分析技术及其在环境科学中的应用"项目组编著, 现代核分析技术及其在环境科学中的应用. 北京: 原子能出版社, 1994. 79~127

[2] Liu Yuanfang, Guo Zhiyu, Liu Xinqui et al. Applications of accelerator mass spectrometry in analysis of trace isotopes and elements. *Pure & Apple Chem*, 1994, 66 (2): 305~334

[3] Chen Chia-erh, Li Kun, Guo Zhiyu et al. Status of the tandem accelerator mass spectrometry facility at Peking University. *Nucl Instr and Meth*, 1990, B52: 306~309

[4] Chen Chia-erh, Li Kun, Guo Zhiyu et al. The test run of AMS system at Peking University, in: *Conf Rec of the* 1991 *IEEE Partical Accelerator Conf* (4), May 6~9, 1991. San Francisco: World Scientific, 1992, 2616

[5] Chen Chia-erh, Li Kun, Guo Zhiyu et al. Beam characteristics and preliminary applications of the tandem based AMS at Peking University. *Nucl Instr and Meth*, 1993, B79: 624~626

[6] 刘洪涛, 汪建军, 刘克新等. AMS-^{10}Be 粒子的探测和鉴别. 原子能科学技术, 1994, 28(6): 551~555

[7] Chen Chia-erh, Guo Zhiyu, Li Kun et al. Accelerator mass spectrometry at Peking University: experiments and progress. *Nucl Instr and Meth*, 1994, B92: 47~50

[8] Guo Zhiyu, Liu Kexin, Li Kun et al. Improvements and Applications of AMS Radiocabon Measurements at Peking University in: Proc of the 15th Inter Radiocarbon Conf, August 15~19, 1994, Glasgow, UK

[9] 刘嘉麒, 陈铁梅, 李坤等. 渭南黄土剖面的年龄测定及十五万年来高分辨时间序列的建立. 第四纪研究, 1994, (3): 193~202

[10] Yuan Sixun, Zhou Guoxing, Guo Zhiyu et al. Radiocarbon Dating the Transition from Paleolithic to the Neolithic in South China, in: Proc of the 15th Inter Radiocarbon Conf, August 15~19 1994, Glasgow, UK

[11] 陈铁梅, Hedges R E M. 湖南彭头山遗址陶片及我国最早水稻遗存的加速器质谱 ^{14}C 测年. 文物, 1994, (3): 88~94

[12] Li Xinsong, Liu Yuanfang, Liu Kexin *et al*. A study on DNA adduction with nicotine-derived nitrosamine by accelerator mass spectrometry. *Chinese Chemical Letters*, 1994, 5(10): 873~876

[13] Li Xinsong, Liu Yuanfang, Liu Kexin *et al*. Genotoxicity of nicotine studied by accelerator mass spectrometry. *Chinese Chemical Letters*, 1994, 5(12): 1043~1044

第三部分 束流物理与脉冲化技术研究

静电加速器脉冲化装置及有关特性[①]

摘　要

本文系统地讨论了射频扫描切割器、聚束器的工作原理与有关的束流特性,着重利用束流相图分析脉冲化束的波形、横向发散度及能散度与束流的初始品质及脉冲化器件工作参数间的关系,最后例举了 4.5 兆伏静电加速器上产生 100 微微秒脉冲束的组合聚束装置的设计参数.

一、前　　言

飞行时间法在核物理研究中的广泛应用,以及近年来超短脉冲束加速技术的提出[1~3],推动着束流脉冲化的技术迅速发展,许多单级的或串级的静电加速器设置脉冲化装置以获得毫微秒的乃至微微秒级的脉冲束.

脉冲化的器件通常包括设置在注入系统中的前扫描切割器、"速调管"式射频聚束器以及加速器出口端的后扫描切割器、后聚束器等,有关这些器件的设计和运行情况的文章发表得相当之多[4~12],尽管如此,有关脉冲化束的品质与入射束的初始条件以及器件的工作参数间的关系等问题尚缺乏系统的讨论. 为此本文自有关器件的工作原理出发系统分析脉冲化束的特性,最后例举了能提供 100 ps 脉冲束的组合聚束器的参数.

二、脉冲化装置的原理和工作性能

1. 射频扫描切割器

扫描切割器由一对扫描电极和一个切割狭缝所组成(图 1). 电极之

---

① 摘自《全国加速器技术交流论文选编》,1980,p.75~84.

间加有射频电压 $V=V_m\sin\omega t$. 在电场的作用下,离子束周期性地扫过切割缝,使一小部分相位合乎要求的离子穿过窄缝形成束团.

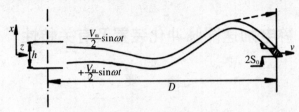

图 1　扫描切割器示意图

(1) 扫描束的瞬态包迹　Turner[4]给出了具有"硬边界"的电场中离子扫描运动的近似关系式,它适用于轴向速度高的轻离子,对于行动迟缓的重离子误差较大,Livingood[16]考虑了边缘电场的影响,用数字解和一系列曲线列出了各有关结果. 这样的处理比前者更切合实际,可惜其表述方式过于繁复,运用不便,且推演中遗弃了若干项次,造成了一些错误,为了改进这一不足,下面给出一个经过严格推导所得的简洁的解析式.

我们仍假定电场分布近似于梯形,梯形顶点伸入电极实际边界 $1.5h$[16](图 2),斜边延伸长 $7h$. 这样的分布下离子所见的电场为:

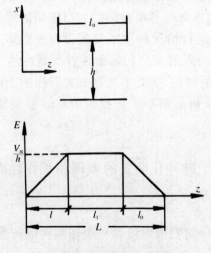

图 2　梯形电场示意图

$$\begin{cases} E_1 = \dfrac{V_m \cdot z}{hl}\sin\left(\varphi_0 + \dfrac{\omega z}{v}\right), \ 0 \leqslant z \leqslant l \\ E_2 = \dfrac{V_m}{h}\sin\left(\varphi_0 + \dfrac{\omega z}{v}\right), \ l \leqslant z \leqslant l + l_1 \\ E_3 = \dfrac{V_m}{h}\left[1 - \dfrac{z-(l+l_1)}{l}\right]\sin\left(\varphi_0 + \dfrac{\omega z}{v}\right), \ l + l_1 \leqslant z \leqslant L \end{cases} \quad (1)$$

其中 φ_0 是离子进入电场时的初相,ω 是射频电场的角频率,v 是离子的轴向速度,束流满足傍轴条件时 $v=$ 常数.

解横向运动方程 $\dfrac{d^2 x}{dz^2} = \dfrac{Q}{mv^2}E$($Q, m$ 分别是离子的电荷和质量),归并整理各项之后,我们所得离子在切割器之后飘移空间中运动状态的表述式为

$$\begin{cases} x = x_0 + x_0' z + A \cdot \left(z - \dfrac{L}{2}\right) \cdot K \cdot T_F \cdot \sin\left(\varphi_0 + \dfrac{\omega L}{2v} - \delta\right) \\ x' = x_0' + A \cdot T_F \cdot \sin\left(\varphi_0 + \dfrac{\omega L}{2v}\right) \end{cases} \quad (2)$$

式中 x_0、x_0' 分别是离子进入电场时的初始位置和散角;常数 $A = V_m(l+l_1)/2V_0 h$;QV_0 是离子的初始能量;渡越因子 $T_F = \left(\dfrac{\sin\varphi}{\varphi}\right) \cdot \left(\dfrac{\sin(\varphi+\varphi_1)}{\varphi+\varphi_1}\right)$,其中 $\varphi = \dfrac{\omega l}{2v}, \varphi_1 = \dfrac{\omega l_1}{2v}$;扫描相角差

$$\delta = \arctan\left\{\dfrac{v}{\omega(z-L/2)}[1 - (\varphi+\varphi_1)\cot(\varphi+\varphi_1) + 1 - \varphi\cot\varphi]\right\};$$

系数 $K = [1+\tan^2\delta]^{1/2}$

式(2)表明对于傍轴入射束,扫描电场的作用仅仅在于使束轴发生周期性的波动. 由此,如进一步假定入射束流是所谓"标准束",其初始特征长度为 λ_0[17],并注意到 t_0 时刻进入电场的离子行进至 z 处的时刻 $t = t_0 + z/v$,则离子束在飘移空间中的瞬态包迹为

$$x_\pm = \pm a\left[1 + \left(\dfrac{z-z_a}{\lambda_0}\right)^2\right]^{1/2} + A \cdot K \cdot \left(z - \dfrac{L}{2}\right) \cdot T_F$$
$$\cdot \sin\left[\omega\left(t - \dfrac{z}{v}\right) + \dfrac{\omega L}{2v} - \delta\right] \quad (3)$$

式中 a 和 z_a 分别是无扫描电场时束腰的半宽度和腰位;第一项取 "+" 号时,x_+ 表束流外廓的上边界,"−" 号时 x_- 表下边界.

(2) 通导时间 考察一个自下而上扫动的束通过狭缝 $2S_0$ 的过程: $t=t_i$ 时束流的上边界 $x_+ = -S_0$,束流开始通导;此后 $t=t_c$, $x_+ = a\left[1+\left(\dfrac{z-z_a}{\lambda_0}\right)^2\right]^{1/2}$,束心与几何轴线重合; $t=t_f$ 时束流下边界 $x_- = S_0$,束流截止.据此,由(3)我们有:

$$\sin\omega(t_f - t_c) = \sin\omega(t_c - t_i) = b/(A \cdot T_F \cdot D \cdot K),$$

其中 D 是狭缝至切割板中心间距, $b = S_0 + a\left[1+\dfrac{1}{\lambda_0^2}\left(D-\left(z_a-\dfrac{L}{2}\right)\right)^2\right]^{1/2}$, $t_c = \dfrac{D}{v} + \dfrac{\delta}{\omega}$,这样,经切割之后,脉冲化束的半宽度 $\tau_c = t_c - t_i = t_f - t_c$ 为

$$\tau_c = \frac{1}{\omega}\sin^{-1}\frac{b}{A \cdot K \cdot T_F \cdot D} \tag{4}$$

如果不考虑边缘电场,且离子速度较高使 $T_F \approx 1$, $K \approx 1$,则当 $\sin\omega\tau_c \approx \omega\tau_c$ 时

$$\tau_c \approx \left(\frac{m \cdot v \cdot b \cdot h}{QV_m}\right) \cdot \left(2 \cdot D \cdot \sin\frac{\omega L}{2v}\right)^{-1} \tag{5}$$

此即 Turner[4] 所给出的近似结果.

(3) 脉冲形状平均束强 假定离子在 (xt) 空间均匀分布, t 时刻的束强便应正比于束流在该时刻所占的微分相面积,于是由(3)脉冲化束的波形可由下式表述(参阅图 3):

$$\begin{cases} i_r = i_0 \cdot \dfrac{b}{2a}\left[1 + \dfrac{\sin\omega(t-t_c)}{\sin\omega\tau_c}\right], & t_i \leqslant t \leqslant t_c - \dfrac{1}{\omega}\sin^{-1}\dfrac{|a-S_0|}{b}\sin\omega\tau_c = t_r \\ i_m = \begin{cases} i_0 & (S_0 \geqslant a), \\ i_0 \dfrac{S_0}{a} & (S_0 \leqslant a), \end{cases} & t_r \leqslant t \leqslant t_c + \dfrac{1}{\omega}\sin^{-1}\dfrac{|a-S_0|}{b}\sin\omega\tau_c = t_d \\ i_d = i_0 \dfrac{b}{2a}\left[1 - \dfrac{\sin\omega(t-t_c)}{\sin\omega\tau_c}\right], & t_d \leqslant t \leqslant t_f \end{cases} \tag{6}$$

可见一般情况下脉冲近似于梯形.由此,每周期单次切割的脉冲所给出的平均束强,当 $\omega\tau_c < 1$ 时:

$$\bar{I} = f_r \cdot \left[\int_{t_i}^{t_r} i_r \mathrm{d}t + i_m(t_d - t_r) + \int_{t_d}^{t_f} i_d \mathrm{d}t\right] \approx i_0 \cdot f_r \cdot 2\tau_c \cdot \frac{S_0}{a+S_0} \tag{7}$$

其中 f_r 是脉冲重复频率.值得注意的是 $S_0 \gg a$ 时,脉冲接近于矩形, $\bar{I}$ 趋于极限值 $\bar{I}_{\max} = i_0 \cdot f_r \cdot 2\tau_c$;而 $S_0 = a$ 时, $\bar{I} = \dfrac{1}{2}\bar{I}_{\max}$.

(4) 横向品质 扫描切割使束流的横向特性发生显著的变化,由于

扫描中束流相图的参考中心连续地在(xx')平面中移动,这就使脉冲化束的有效相面积比入射时的初始值高出约$\frac{4S_0}{\pi a}$倍(图4),减小切割缝宽S_0虽可降低束的有效发散度,但由(7),束强随之下降.

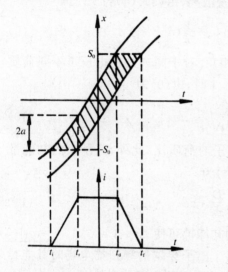

图 3　束流的(x,t)相图与脉冲波形

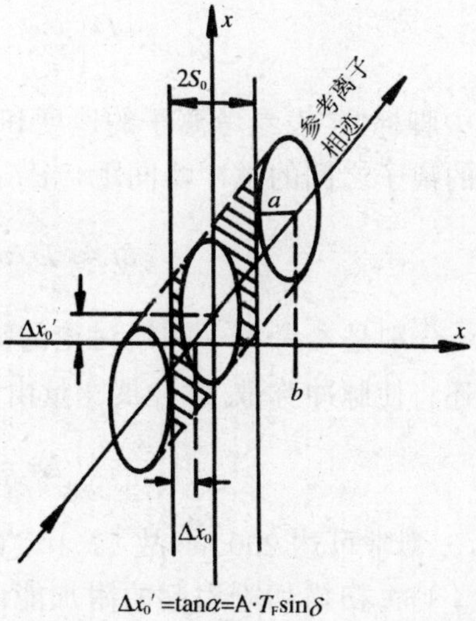

图 4　扫描束的横向相图

$\Delta x_0' = \tan\alpha = A \cdot T_F \sin\delta$

另一个值得注意的特点是,参考轨道的相迹一般不过原点.这意味着脉冲化束的束轴是倾斜的.其相对于几何轴的倾角$\alpha = \tan^{-1}[A \cdot T_F \cdot \sin(n\pi + \delta)]$,式中当束向$x > 0$的方向扫动时$n = 0, 2, 4\cdots$;反向扫动时$n = 1, 3, 5\cdots$.克服束轴倾斜的办法之一是使切割器的几何参数满足$\delta \equiv 0$.由式(2),这要求

$$\frac{\omega(l+l_1)}{2v}\cot\frac{\omega(l+l_1)}{2v} + \frac{\omega l}{2v}\cot\frac{\omega l}{2v} = 2 \qquad (8)$$

此即 Livingood 所提出的所谓"魔相"条件[16]. 不过在这里简明的解析式代替了[16]中的数组和曲线,用起来方便多了. 顺便指出[16]中遗弃了诸如$\frac{\omega l}{v}\sin\frac{\omega l}{v}$等项次,造成了某些差异.

通常"魔相"条件满足时,离子的渡越因子T_F很小,这意味着高频功耗将大大增加. 另一方面倘若离子在电场中的滤越角不大$\sin(\varphi+\varphi_1) \approx (\varphi+\varphi_1)$,则倾角$\alpha \approx \frac{b}{8v\tau_c} \times \frac{l^2 + (l+l_1)^2}{D^2}$. 可见只要距离$D$足够大,就可使

α 降到可允许的程度,有人曾将扫描电极放在离子源吸极之内[9],这是最大限度地增大 D 的一种方式.

(5) 色散效应 扫描切割器同时也是一种飞行时间速度分析器,由离子速度的岐离引起的脉冲中心位置散离,由(3)、(4)为

$$\delta t_c = \frac{D}{v_c}\left(\frac{v_c}{v} - 1\right) + \frac{1}{\omega}(\delta - \delta_c) \tag{9}$$

式中脚标"c"表参考离子的速度和相角. 对于同一加速电压下不同荷质比的离子之间的脉冲峰间距,当 $T_F \sim 1$ 时,由(9)为

$$\delta t \approx D/v_c\left(\sqrt{\frac{\varepsilon_c}{\varepsilon}} - 1\right) \tag{10}$$

$\varepsilon_c, \varepsilon$ 分别是参考离子和所讨论的离子的荷质比. 此外,离子束的初能散 ε_0 还将使脉冲弥散. 脉冲展宽量由(4)为

$$\Delta \tau \approx \frac{D}{v_c} \cdot \varepsilon_0 \tag{11}$$

D/v_c 常常可达 200 ns,故 5×10^{-3} 的初能散可使 $\Delta \tau \sim 1$ ns.

(6) 扫描切割引起的附加能散 离子穿越射频扫描电场时其能量受到调制,在"硬边界"近似下求得由此引起的能散的增长[19]:

$$\Delta \varepsilon \approx \left[1 + \frac{1}{2}\left(\frac{\tau_l}{\tau_c}\right)^2\right]\frac{\tau_l}{4\tau_c}\left(\frac{b}{KD}\right)^2 + 2\frac{\tau_l}{\tau_c} \cdot \frac{b}{KD} \cdot \left(\frac{x_0}{l} + \frac{x_0'}{2}\right)_m \tag{12}$$

这个式子偏保守,但与 Euge 的式子[15]相一致. 如入射束腰在狭缝处可证明:

$$\Delta \varepsilon \approx \left[1 + \frac{1}{2}\left(\frac{\tau_l}{\tau_c}\right)^2\right]\frac{\tau_l}{4\tau_c}\left(\frac{b}{KD}\right)^2 + \frac{2A_0}{v\tau_c}\left(1 + \frac{S_0}{a}\right) \tag{13}$$

其中 $\tau_l = \frac{l}{v_c}$,A_0 是束的初始发散度,通常上式中第二项比第一项大得多. 由式(13)可见只有小的 A_0,才能避免能散度的明显增长.

2. 头部聚束器

静电加速器头部通常设置速调管式的聚束器用以压缩束团的纵向尺寸. 实际上这是一种"时间透镜",它将束流的时间岐离转化成能量岐离或者反过来. 对于这种装置 J. H. Anderson[8]等曾在不考虑初始能散的条件下就其工作性能给出了有关的表述式,下面我们试图进一步考虑初能散对群聚脉冲波形等的影响,并讨论聚束电场对脉冲化束的横向品

质带来的损害.

(1)"时间焦距" 我们讨论常见的三电极聚束器(图5)的情况.这样的结构可以近似地等效为一个位于器中心的射频电隙,离子穿过时,其能量受到的调制为

$$\frac{\Delta W_m}{W} = \varepsilon_m \sin\varphi_b \tag{14}$$

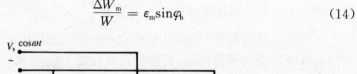

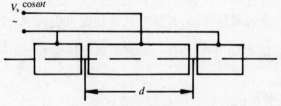

图5 三电极聚束器示意图

式中 ε_m 是调制系数,$\varepsilon_m = \frac{2QV_b}{W_c}\sin\frac{\omega d}{2v}$;$\varphi_b$ 是离子到达器中心时的射频相位.现在我们以零相时刻通过电隙的离子为参考,考察一束初能散为 ε_v 的束流,经聚束器调制后束流的能量岐离和相位分布情况.不难证明在距器心为 L 的位置上,束流满足

$$\begin{aligned}\varepsilon &= \varepsilon_0 + \varepsilon_m\sin\varphi_b \\ \varphi &= \varphi_b + \frac{\omega L}{v_c}[(1 + \varepsilon_0 + \varepsilon_m\sin\varphi_b)^{-1/2} - 1]\end{aligned} \tag{15}$$

通常 ε_0、$\varepsilon_m \ll 1$ 故

$$\varphi \approx \varphi_b - \frac{L}{F_z}\left(\sin\varphi_b + \frac{\varepsilon_0}{\varepsilon_m}\right) \tag{16}$$

其中 $F_z = \frac{v_c \cdot T}{\pi\varepsilon_m}$($T$ 是射频周期,v_c 参考离子的速度)称为聚束器的时间焦距.

为了说明 F_z 的含义,不妨令 $\sin\varphi_b \approx \varphi_b$,即讨论锯齿波聚束器的情况.在此条件下,离子在聚束器前后的滑相速度的变化 $\Delta\left(\frac{d\varphi}{dz}\right) = -\frac{\varphi_b}{F_z}$,其形式与普通薄透镜的作用式 $\Delta\left(\frac{dx}{dz}\right) = -\frac{x}{F}$ 完全相同,这表明齿波聚束器实际上是一个焦距为 F_z 的薄"时间透镜".有趣的是 $L = F_z$ 时,每周期内

的所有入射离子,都将被薄透镜压缩在 $\tau=\dfrac{T}{\pi}\dfrac{\varepsilon_0}{\varepsilon_m}$ 的区间之内,光学上这可以归结为薄透镜的腰-腰传输过程. 事实上如将入射束近似为腰位处在聚束器中心的标准束[17],其特征长度 $x_z=\dfrac{v_c T}{\varepsilon_0}$,那么腰-腰传输关系式[13]给出的象腰位置恰好在 F_z 处,放大倍数为 $\dfrac{1}{\pi}\dfrac{\varepsilon_0}{\varepsilon_m}$.

正弦波的透镜是非线性的,然而在 ±60° 的范围内 $\sin\varphi_0\approx\varphi_0$,故仍近似地具有薄透镜的那种性能,其线性区的焦距也仍为 F_z. 下面可以看到正弦波情况下每周期仍有 30%～60% 的离子压缩在 $\tau=\dfrac{T}{\pi}\dfrac{\varepsilon_0}{\varepsilon_m}$ 的间隔之内.

(2) 聚束脉冲的波形　假定离子在 (ε,φ) 空间均匀分布,φ 时刻的离子数 $N(\varphi)$ 便比例于 $(\varphi,\varphi+\mathrm{d}\varphi)$ 间离子所占的微分相面积,于是瞬时束强可写为

$$i(\varphi)=\frac{Q\mathrm{d}N(\varphi)}{\mathrm{d}\varphi}\cdot\frac{\mathrm{d}\varphi}{\mathrm{d}t}=\frac{i_0}{2\varepsilon_0}\sum\Delta\varepsilon_i(\varphi) \tag{17}$$

其中 $\sum\Delta\varepsilon_i(\varphi)$ 是 φ 时刻束流相图 ε 方向上诸边界间距之和,可由数值解或作图法求得. 群聚离子束的脉冲形状随飘移距离 L 变化的基本趋势 (参见图 6) 是: 脉冲高度先随 L 的增大逐渐升高至某一极大值 $i_{\max}=i_0\left(1+\dfrac{\varepsilon_m}{\varepsilon_0}\right)$ 然后下降,极大值位置当 $\dfrac{\varepsilon_0}{\varepsilon_m}\geqslant 0.14$ 时是 $L_m=\dfrac{F_z\pi}{2\left(1+\dfrac{\varepsilon_0}{\varepsilon_m}\right)}$; $\dfrac{\varepsilon_0}{\varepsilon_m}<0.14$ 时 $L_m\approx F_z\cdot\left[1+\left(\dfrac{\varepsilon_0}{1.06\varepsilon_m}\right)^{3/2}\right]$,脉冲宽度则先随 L 的增大而减小,$L=F_z$ 时脉宽减至极小,此时脉冲的半宽度 (FWHM) $\tau_{\min}\approx\dfrac{T}{\pi}\dfrac{\varepsilon_0}{\varepsilon_m}=\dfrac{\varepsilon_0 F_z}{v_c}$,该处的脉高 $i_p\approx i_0\left(1+\alpha\dfrac{\varepsilon_0}{\varepsilon_m}\right)$,其中 α 是经验系数,$0.86\leqslant\alpha\leqslant 1$,$L>F_z$ 后束流波形会出现双峰,按照(17),并注意到微分相面积的极值发生在 $\dfrac{\mathrm{d}\varphi}{\mathrm{d}\varepsilon}=0$ 处,则由(16)可推得双峰间距与 L 的关系为

$$\Delta\varphi_d=2\left\{\sqrt{\left(\frac{L}{F_z}\right)^2-1}-\cos^{-1}\frac{F_z}{L}-\frac{L}{F_z}\frac{\varepsilon_0}{\varepsilon_m}\right\} \tag{18}$$

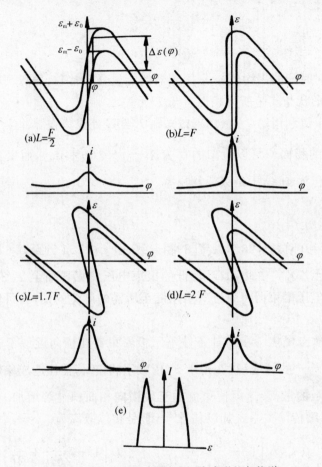

图 6　正弦波群聚束的相图,脉冲形状与能谱

上述各项讨论中如令 $\varepsilon_0=0$,则可得到与[8]相同的结果. 然而过去实验上观察到的与[8]的理论不符的一些现象诸如脉宽比理论值大得多,脉冲高度又小得多;出现双峰的位置明显地比理论值大等等都可以由上述诸式得到更好的解释.

(3) 群聚束的能谱　按照以上的假定,能量偏离处于 ε 与 $\varepsilon+\mathrm{d}\varepsilon$ 间的离子也应正比于该处的微分相面积,对于初能散为 ε_0 的束,经过聚束器之后,其能谱为一对称的曲线(图 6(e)):

$$\bar{i}(\varepsilon) = c \cdot \left(\frac{i_0}{2\pi}\right) \cdot \begin{cases} \left[\sin^{-1}\frac{|\varepsilon|+\varepsilon_0}{\varepsilon_m} - \sin^{-1}\frac{|\varepsilon|-\varepsilon_0}{\varepsilon_m}\right] & (0 \leqslant |\varepsilon| \leqslant \varepsilon_m - \varepsilon_0) \\ \left[\frac{\pi}{2} - \sin^{-1}\frac{|\varepsilon|-\varepsilon_0}{\varepsilon_m}\right] & (\varepsilon_m + \varepsilon_0 \geqslant |\varepsilon| \geqslant \varepsilon_m - \varepsilon_0) \end{cases} \quad (19)$$

式中 $\bar{i}$ 表对时间的平均值,c 是探测器的能量分辩常数,如所预期(19)给出的能谱形状与实验所得相似且与 L 无关.

(4) 束流利用率 由(16)(17),在 $L=F_z$ 处聚集在脉冲半宽度内的离子所占的相面积可以近似的写为 $2\left(\frac{\varepsilon_0}{\varepsilon_m}\right) \times 2(\varepsilon_0 + \alpha\varepsilon_m)$,而聚束之前束流的相面积为 $4\pi\varepsilon_0$,于是束流利用率

$$\eta \approx \frac{1}{\pi}\left(\alpha + \frac{\varepsilon_0}{\varepsilon_m}\right) \quad (20)$$

由上式,$\varepsilon_0=0$ 时,$\eta \approx 28.6\%$ 与[8]相一致,但 $\frac{\varepsilon_0}{\varepsilon_m}$ 由 0.1 升至 0.6 时 η 可由 30% 上升至 50%. 由此,如以较低的聚束电压将束压缩至 $T/3$ 以下,之后再用高电压聚束器将束进一步压缩至所需的脉宽的话,当可使束流利用率提高.

(5) 横向聚焦 在傍轴条件下三极聚束器在横向近似于一个焦距为 $F_r \approx -\frac{2F_z}{\cos\varphi_b}$ 的薄透镜[19]. 由于 F_r 依赖于相位 φ_b,往往造成象腰的位置与放大倍数的散离. 这就使束流的有效横向相面积显著增加. 对此可用常见的腰-腰传输关系式加以估计[13],于是腰位散离为

$$\Delta L_m = \frac{L_0\left(1-\frac{L_0}{2F_z}\right) - \frac{x_0^2}{2F_z}}{\left(1-\frac{L_0}{2F_z}\right)^2 + \left(\frac{x_0}{2F_z}\right)^2} - \frac{L_0\left(1+\frac{L_0}{2F_z}\right) + \frac{x_0^2}{2F_z}}{\left(1+\frac{L_0}{2F_z}\right)^2 + \left(\frac{x_0}{2F_z}\right)^2} \quad (21)$$

放大倍数散离为

$$\Delta\mu = 4F_z^2[(x_0^2 + (2F_z - L_0)^2)^{-1} - (x_0^2 + (2F_z + L_0)^2)^{-1}] \quad (22)$$

式中 L_0: 物腰位,x_0: 横向特征长度,当 $F_z \gg L_0, X_0$ 时,由(21)(22),$\Delta L \approx \frac{x_0}{F} \cdot x_0 < x_0$,$\Delta\mu \to 0$. 因此若令入射束的物腰位于聚束器中心,且 x_0 足够小,当可减少聚束器对束流横向品质带来的损害. Dandy[7]等人曾在聚束电压由 0 增至 1.68 kV 时,观察到束斑由 ⌀4 mm 猛增到 21 mm 的现象. 这是横向品质因聚束而变坏的例证. 他们曾用屏蔽丝网限制横向

散焦作用使束斑控制在 $\varnothing 6.5$ mm 左右.

3. 后聚束器及组合聚束器

后聚束器的工作原理与头部聚束器相同,只是这里入射离子的能量比头部高出百倍以上,故所需的聚束电压高达几百千伏乃至兆伏量级. 为此,后聚束器常常采用高 Q 值的谐振腔制成.

近年来发展起来的螺旋波一类加速腔[1,14,15],结构简单、尺寸小,工作频率和相速都较低,且在相当宽的能量范围内都有较高的腔效率,故适宜于用作重离子束的后聚束器,国外曾用这类腔获得 40~70 ps 的重离子脉冲束,我们也曾用柱形螺旋线腔方便地得到 2 ns 后聚束脉冲[20].

螺旋线波导一类聚束腔内的驻波电场可近似地表述为

$$E(z) = E_{sm} \sin\frac{2\pi z}{\beta_p \lambda} \cdot \sin(\omega t + \varphi_0) \qquad (23)$$

离子穿过腔时其能量受到的调制可象头部聚束器一样写为 $\frac{\Delta W_m}{W} = \varepsilon_m \sin\varphi_0$,其中调制系数

$$\varepsilon_m = \frac{1}{W}\left[2E_{sm}l_c \frac{\sin\frac{\pi l_c}{\beta_p \lambda}\left(\frac{\beta_p}{\beta} - 1\right)}{\frac{\pi l_c}{\beta_p \lambda}\left(\left(\frac{\beta_p}{\beta}\right)^2 - 1\right)}\right] \qquad (24)$$

式中 β_p 是腔内行波的相对论速度,E_{sm} 是驻波电场振幅,β 是参考离子的速度 v_c/c,l_c 是腔的有效电长度,通常 $l_0 < 3\beta_p\lambda$ 的短腔可等效为一个位于腔心的射频电隙,此时 φ_0 即为离子到达该处时的相位,聚束腔的工作原理与头部聚束器同,只要令式(15)~(20)中的 ε_m 项由式(24)所给定,各关系式都继续有效.

后聚束器较前聚束器更易于得到 1 ns 以下的脉冲,一方面这是由于静电加速器输出的束能散度小,另一方面后聚束器的结构一般不受空间限制,可获得较高的聚束电压,后聚束的缺点是终能散较大,所需的飘移距离较长,这些不足之处可通过与头部聚束器组合运行的方式来克服,在组合运行中,头部聚束器先以较低的频率和电压将束流压缩到 1~10 ns,接着后聚束以较高的频率和聚束电压将束压缩至 0.1~1 ns 以下. 根据前面的讨论可以预期,组合聚束可在脉冲化束的品质和束流利用率方面优于单级聚束器.

事实上经头部切割、聚束后,脉冲化束的相面积 $\tau_1\varepsilon_1 \approx \frac{T_1}{\pi}\varepsilon_0\left(1+\frac{\varepsilon_0}{\varepsilon_m}\right)$,经静电加速之后这一面积将收缩 100 倍以上(收缩倍数 $K \approx W_f/W_0$,即静电的能量增益). 此时,如在脉宽小于后聚束电压周期的 1/3 的位置上,即 $\tau_1 < T_2/3$ 处,设置后聚束器,那么由刘维定理,终脉宽 $\tau_2 = \frac{\tau_1\varepsilon_1}{K\varepsilon_2}$,这就是说,如允许终能散 ε_2 等于头部的能散 ε_1 的话,$\tau_2 < \frac{1}{100}\tau_1$! 实际上往往要求 $\varepsilon_2 \leqslant 0.1\varepsilon_1$ 再加上静电加速电压波动的影响,τ_2 做不到这样短,尽管如此,τ_2 仍可小于 $0.1 \sim 0.2\tau_1$. 表一列出了 4.5 MV 静电加速器上组合聚束器的设计参数,这个装置最短可给出 0.1 ns,串级静电加速器的能量增益倍数 K 比单级静电高得多,故如采用组合聚束器可得到更短的脉冲.

三、结束语

1. 扫描切割器

解析式(2)(3)简洁地描述了有关扫描离子的运动状态和束的瞬时包迹,改进了[4][16]给出的结果,可方便地用来确定切割器的束流通导时间、脉冲波形、"魔相条件"及色散等性能. 为了减低脉冲化束的束轴倾角及附加能散,高的注入能量、长的飘移距离和好的初始束流品质是重要的.

2. 射频聚束器

式(15)~(20)给出了束流初能散对群聚束的脉冲宽度、脉冲高度、双峰间距等的影响,推广了[8]等给出的结果. 由 $\tau \approx \frac{\varepsilon_0 F_z}{v_c}$ 可见,高的注入能量、低的初能散和短焦距的聚束器是获得短脉冲的重要条件. 另一方面组合聚束器在获得超短脉冲化束,及提高束流利用率方面将优于单级聚束器. 此外,为了减少聚束电场对束流横向品质带来的损害,控制入射束的腰位和特征长度是必要的.

表1　组合聚束器工作参数与束流性能

终能量 $W_f = 4.5 \cdot \dfrac{Q}{A} \text{MeV}(K=100)$　　　电压纹波 $\dfrac{\delta V}{V}=5\times 10^{-4}$

初能散 $\varepsilon_0 = 5\times 10^{-3}$　　　终能散 $\varepsilon_2 = 0.01$

头部聚束	聚束频率/MHz	27	27	18
	调制系数 ε_m	0.03	0.1	0.1
	初始脉宽 $2\tau_c$/ns	12	12	18.5
	终脉宽 τ_1/ns	1.95	0.59	0.89
后聚束	聚束频率/MHz	27	54	54
	调制系数 ε_m	5.5×10^{-2}	8.6×10^{-2}	5.7×10^{-2}
	入射束能散 ε_0'	8.5×10^{-4}	1.45×10^{-3}	1.45×10^{-3}
	终脉宽 τ_2/ns	0.18	0.1	0.15
	焦距 F_2/m	6.3	2	3

参 考 文 献

[1] L. M. Bollinger et. al., IEEE. Trans., NS-22, 3(1975) 114.
[2] L. M. Bollinger et. al., IEEE. Trans., NS-24, 3(1977) 107.
[3] P. Arndt et al., IEEE. Trans., NS-24, 3(1977) 116.
[4] C. M. Turner, R. S. I. Vol. 29, 6(1956) 480.
[5] H. W. Lewis, R. S. I. Vol. 30, 10(1959) 923.
[6] C. D. Moak, R. S. I. 35(1964) 672.
[7] D. Dandy et al., N. I. M. 30(1964) 1.
[8] J. H. Anderson, N. I. M. 41(1966) 30.
[9] J. J. Kritzingen, N. I. M. 101(1972) 573.
[10] K. Tswkada, N. I. M. 39(1966) 249.
[11] H. Naylor, IEEE. Trans., NS-11 3(1965) 6.
[12] K. H. Puser, IEEE. Trans., NS-14 3(1967) 174.
[13] A. V. Luccio, N. I. M. 80(1970) 197.
[14] K. W. Shepard, IEEE. Trans., NS-24, 3(1977) 147.
[15] H. Ingwersen IEEE. Trans., NS-24, 3(1977) 1107.
[16] J. J. Livingood, Particle Accelerators, 7(1976) 243.
[17] A. P. Banford, "The Transport of Charged Beams" §2.3.2.
[18] L. Carneberg N. I. M. (1961) 335.
[19] 陈佳洱, 北京大学内部报告, BDJ-B-1/1978.

[20] 陈佳洱,螺旋波导聚束器初步设计,《加速器通讯》,第二期(1973)(内部资料).

BEAM CHARACTERISTICS OF RF PULSING DEVICE FOR VAN DE GRAFFS

Abstract

The principles of RF chopper and bunchers are discussed. The wave form, effective emittance and energy disperssion of the pulsed beam are analyzed. The design parameters of a buncher combination which might provide a pulsed beam of 100 ps are given as an example.

THE BEAM DYNAMICS OF A
SINE WAVE SWEEPER[①]

I. Introduction

The pre-tandem double-drift hormonic buncher of the Stony Brook tandem-LINAC accelerator (SUNYLAC) bunches 60%~70% of the DC beam into ~1 ns bunches. The remaining 40%~30% of the beam must be removed before injection into the LINAC. The conven-tional method of accomplishing this is to switch off the beam before the buncher during those intervals where bunching does not occur. At SUNYLAC floor space and injection energy considerations render this approach impractical. The practical solution in this case is to employ a sine wave beam sweeper and a DC coupled beam deflector (switch) after the tandem and before the LINAC.

Upon first inspection it might appear that the difficulty of the post-tandem approach would be increased in proportion to the rigidity of the beam after the tandem. However several factors work to actually make the post-tandem location very convenient. 1) The bunch is already completely formed. Changing voltages due to finite rise time in deflector or fringe fields in the sweeper are of no consequence except when the bunch is present (i.e.: 1 ns out of 100 ns). 2) The super buncher, the first 90° magnet, and slits of the 180° turn on the LINAC injection path constitute an excellent time (phase) filter. Any errrant beam within a 5.5 ns window around the bunch will be rejected by the slits. The sweeper therefore need

---

① Coather: Joseph M. Brennan. Reprinted from Stoney Brook Tandem LINAC research report. Jan. 1983, Stoney Brook.

only chop the beam down to 5.5 ns bursts. 3) The floor plan is such that a very long (7 m) lever arm can be used for the sweeper and the deflector.

Each of these three considerations works to mitigate the inevitable degradation of the beam emittance in both transverse and longitudinal phase spaces. It is the purpose of this report to show by direct calculation that within the present design specifications this phase space degradation can be made negligible.

Conceptual design[1] and some calculations[2] based on a "hard edge" field approximation have been reported previously. In this note, the effect of the fringing field of the sweeping plates is included and analytical expressions convenient for a practical design are given.

A program named SWEEP was written for calculating beam emittance. Finally, parameters for the sweeper are specified and a small low voltage compensating deflector, the corrector, is suggested for correcting the coherent divergence of the chopped beam bunch with respect to the beam axis.

II. Equations of Motion

Assume the field across the sweeping electrodes is uniform, except at both ends, where the fringing field is approximated by a linear ramp as illustrated in Fig. 1, and can be expressed as following

$$\begin{cases} E_y = \dfrac{z \cdot V_m}{l \cdot g}\sin\varphi & (0 \leqslant z \leqslant l) \\ E_y = \dfrac{V_m}{g}\sin\varphi & (l \leqslant z \leqslant l+l_0) \\ E_y = \left\{1 - \dfrac{z-(l+l_0)}{l}\right\}\dfrac{V_m}{g}\sin\varphi & (l+l_0 \leqslant z \leqslant L) \end{cases} \quad (1)$$

Where L is the electric length of the sweeping field, $L = 2l + l_0$ as shown in Fig. 1.; g is the gap height between the plates and φ is the RF phase seen by an ion, $\varphi = \varphi_0 + \omega\int\dfrac{dz}{v}$. Referring to {3}{4}, we assume here $l = 1.5g$ and $l_0 = l_s - g/2$ for paraxial rays.

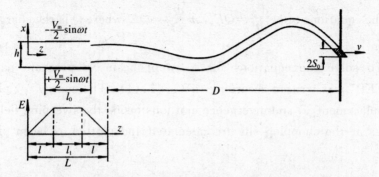

Fig. 1 Sweeping plates and the form of their fringing field

In the approximation that the field is quasi stable while an ion is passing through, the axial component of the fringing field near the axis may be written as

$$E_z = \frac{\partial E_z}{\partial y} \cdot y + \cdots\cdots \approx \frac{\partial E_y}{\partial z} \cdot y \qquad (2)$$

Consequently

$$\begin{cases} E_z = \dfrac{y \cdot V_m}{l \cdot g}\sin\varphi & (0 \leqslant z \leqslant l) \\ E_z = 0 & (l \leqslant z \leqslant l + l_0) \\ E_z = -\dfrac{y \cdot V_m}{l \cdot g}\sin\varphi & (l + l_0 \leqslant z \leqslant L) \end{cases} \qquad (3)$$

For the convenience of computation, we put the equations of motion as

$$\begin{cases} \dfrac{dw}{dz} = QE_z \\ \dfrac{d\phi}{dz} = \omega/v \\ \dfrac{dy}{dz} = \alpha \\ \dfrac{d\alpha}{dz} = \dfrac{Q}{2w}(E_y - \alpha E_z) \end{cases} \qquad (4)$$

by substituting $\dfrac{d^2y}{dt^2}=v^2\dfrac{d^2y}{dz^2}+v\dfrac{dv}{dz}\dfrac{dy}{dz}$, $\dfrac{d^2z}{dt^2}=v\dfrac{dv}{dz}$, $v=\dfrac{dz}{dt}$ and $w=\dfrac{1}{2}mv^2$

into the equations of $m \frac{d^2y}{dt^2} = QE_y$, $m \frac{d^2z}{dt^2} = QE_z$ where Q is the charge of the ion.

To solve the equations for a swept beam, a program named "SWEEP" was written according to (1)(3)(4). The energy w, phase ϕ, displacement y, and divergence α of ions inside the sweeping field as well as at the chopping slit are calculated and plotted as beam phase plots.

III. The Sweeping Voltage and Geometrical Parameters

In practice, for a post tandem sweeper, the fractional energy change caused by the fringing field will be less than 10^{-3}, hence can be neglected while discussing the transverse motion. Consequently analytical expressions for the motion of a swept ion at the plane of slit can be solved analytically and expressed according to Chen[5] as

$$y = y_0 + (D+L)\alpha_0 + (D+L/2) \cdot K \cdot A \cdot T_F \cdot \sin(\varphi_0 + \omega L/2v - \delta)$$

$$\alpha = \alpha_0 + A \cdot T_F \cdot \sin\left(\varphi_0 + \frac{\omega L}{2v}\right) \quad (5)$$

where

$$A = Q \cdot \overline{V}_m (l + l_0)/2 \cdot w \cdot g$$

$$T_F = \left\{ \sin\left(\frac{\omega l}{2v}\right) \Big/ \left(\frac{\omega l}{2v}\right) \right\} \cdot \left\{ \sin\left(\frac{\omega(l+l_0)}{2v}\right) \Big/ \left(\frac{\omega(l+l_0)}{2v}\right) \right\}$$

$$K = (1 + \tan^2\delta)^{1/2}$$

$$\tan\delta = \frac{v}{\omega(D+L/2)} \left\{ \left(1 - \frac{\omega(l+l_0)}{2v} \cot\frac{\omega(l+l_0)}{2v}\right) + \left(1 - \frac{\omega l}{2v} \cot\frac{\omega l}{2v}\right) \right\}$$

The above expression is accurate enough for a practical design and was verified by results from SWEEP, although it is much simpler than those given by Livingood[3]. Expression (5) implies that the sweeping field never changes the original shape of the beam ellipse in (y, α) space, but does displace it coherently across the chopping slit. This is clearly illustrated in Fig. 2 where the swept beam ellipse at various instants were plotted by SWEEP. The sweeping voltage necessary for providing a time window τ_c can be deduced from (5) as:

$$V_m = W_0 \cdot G \cdot B \bigg/ \left(Q \cdot K \cdot T_F \cdot (l+l_0) \cdot \left(D + \frac{L}{2}\right) \cdot \sin \frac{\omega \tau_c}{2} \right) \quad (6)$$

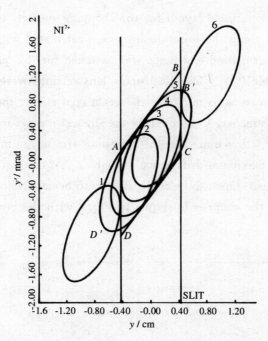

Fig. 2 Swept beam ellipses in (y, y') space at various instances:
1) $t_0-2.75$ ns, 2) t_0-1 ns, 3) $t_0-0.5$ ns, 4) t_0, 5) $t_0+0.5$ ns,
6) $t_0+2.75$. The effective emittance is indicated by rhomb ABCD (for
5.5 ns injected bunch or AB'CD' (1 ns injected bunch)

where B is the vertical size of the beam plus the slit width.

As the product $T_F \cdot (l+l_0)$ in denominator of (6) varies with the ion velocity, V_m will be minimal when $\frac{\omega(l+l_0)}{v} = \pi$. For a sweeper optimized for $\beta = 0.055$ at a frequency of $f = \frac{1}{32}(150) = 4.6875$ MHz[2], this gives an estimated length of $(l_0 + l) = 176$ cm. Unfortunately long sweeping plates have the undesirable effect of big coherent divergence and energy spread (see next sections), hence short plate length (~ 50 cm) is preferred. On the other hand, it is desirable to have a long sweeping arm which helps by reducing both RF power consumption and coherent divergence.

To get an optimized layout for the chopping system, beam envelopes of typical ions along the injection line were calculated with the program Optic II. A normalized emittance was assumed for this purpose as 20π mm·mrad·MeV$^{1/2}$.[8] The quadrupole lens settings were then optimize by the program so as to minimize the beam spot size at the chopping slit and following buncher. Consequently the slit width is set to the minimized beam size (i.e., 8.6 mm) and the sweeping arm is 7.4 m, which corresponds to the maximum drift space available. A 2 cm gap height for the sweeper was also shown to be compatible with beam envelopes computed. The layout of the chopper is shown in Fig. 3 with the beam envelope of O^{7+}.

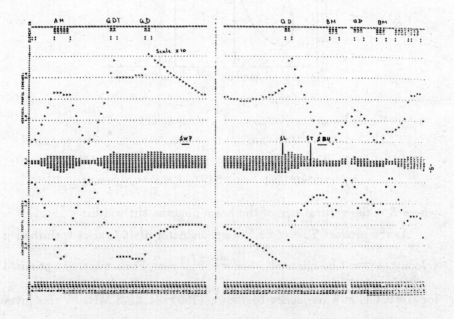

Fig. 3 Typical beam envelope along injection path with beam pulsing elements. H-horizontal beam profile, V-vertical profile (outer H, V -1×10 expanded plot) SWP-sweeper, SL-chopping slit, SBU-super buncher, ST-stripper

IV. The Effective Emittance and Coherent Divergence of the Chopped Beam

The movement of the beam ellipse in (y, α) space may effectively increase the phase area of the chopped beam, as can be seen in Fig. 2. When the slit width equals the vertical beam size without sweeping, the increase in emittance with respect to the incident beam may be up to $15\% \sim 27\%$, depending on the original shape of the beam ellipse and the duration of beam pulse. The effective emittance of chopped beam may be reduced at a cost of beam intensity; i.e., by closing in the chopping slit, which is sometime undesirable.

As the ion trajectories are bent by the RF field inside the sweeper, an originally central ray with $y_c = 0$ and $\alpha_c = 0$ will deviate from the axis by Δy_c if it emerges parallel to the axis at the exit. Alternatively, a central ray which is on axis at the slit will emerge with a residual divergence $\Delta \alpha_c$. This is seen in (5) by the phase difference δ between the variations of transverse displacement and divergence, and can be expressed on the plane of chopping slit as

$$\begin{cases} \Delta y_c = \left(D + \dfrac{L}{2}\right) \cdot T_F \cdot K \cdot A\sin(n\pi - \delta) \\ \alpha_c = 0 \end{cases}$$
$$\begin{cases} y_c = 0 \\ \Delta \alpha_c = A \cdot T_F \sin(n\pi + \alpha) \end{cases} \tag{7}$$

where $n = 0, 2, 4 \cdots$ for up sweep and $n = 1, 3, 5 \cdots$ for down sweep. Values of Δy_c, $\Delta \alpha_c$ are listed in table 1 for typical ion species; in addition, the situation is illustrated in Fig. 4 by the locus of the centre of swept beam ellipse.

Table 1

$f_0 = 4.6875$ MHz $l = 3$ cm $g = 2$ cm $B/2 = 0.86$ cm $D = 740$ cm

(a) $l_s = 50$ cm

Ion	W/meV	V_m/kV	$\Delta\alpha_c$/mrad	Δy_c/mm
C^{4+}	40	11.03	4.62×10^{-2}	0.354
O^{7+}	64	10.06	4.21×10^{-2}	0.323
Ni^{7+}	64	10.40	8.11×10^{-2}	0.622
Mo^{8+}	72	10.47	10.02×10^{-2}	0.769

(b) $l_s = 100$ cm

Ion	W/meV	V_m/kV	$\Delta\alpha_c$/mrad	Δy_c/mm
C^{4+}	40	5.64	0.172	1.37
O^{7+}	64	5.10	0.157	1.24
Ni^{7+}	64	5.82	0.315	2.49
Mo^{8+}	72	6.31	0.402	3.18

Actually $\Delta\alpha_c$ is the coherent divergence of the chopped beam bunch as can be seen in Fig. 4, and the change of its sign in up and down sweep may cause some undesirable timing effects in experiments. The straightforward way to get rid of this trouble is to make $\delta = 0$, that is to use either an extremely short sweeper so that $\dfrac{\omega(l+l_0)}{v} \approx 0$ or the so-called 'Magic phase' sweeper suggested by Livingood[3]. Unfortunately both result in high RF power consumption. However the coherent divergence can be compensated in an easy way if $\Delta\alpha_c$ is a small fraction of $B/2D$ and τ_c is small compared to $2\pi/\omega$ as it usually is. In practice $\Delta\alpha_c$ as small as one tenth of $B/2D$ can be approached by using a long sweeping arm (as is the case in table 1), since $\Delta\alpha_c \approx \dfrac{B}{4v\tau_c} \left[\dfrac{l+l_0}{D+L/2}\right]^2$ which is obtained from (5) when $\sin\delta \approx \delta$, $l_0 \gg l$. Suppose now we have a short sinusoidal compensating deflector working under above conditions. As the amount of angle to be deflected is a small fraction of that from the sweeper and a flat field near the peak rather than a rapidly changing field near zero phase can be

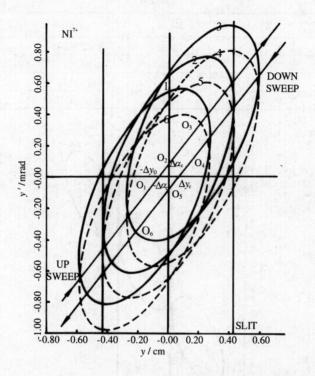

Fig. 4 The locus of the centre of the swept beam ellipse for both up sweep and down sweep ($l_s=50$ cm, $D=7.4$ m, $g=2$ cm)

used, the product of $V_m(l+l_0)$ for the 'corrector' could be two orders of magnitude less than that of the sweeper. Consequently, a peak value of a few hundred volts is enough for the divergence compensation, but it would not be high enough to cause noticeable displacement at it's exit. The practical length and voltage of a corrector can be determined by (5). As an example, voltages for a 10 cm. long corrector for various ion species are listed in Table 2, where $\delta\alpha_c$ is the divergence to be compensated and δy_c is the extra displacement caused by the compensation. (See also Fig. 5).

Table 2

Ion	W/meV	$\overline{V}_C$/V	$\delta\alpha_c$/mrad	δy_c/mm
C^{4+}	40	168	4.62×10^{-2}	3.23×10^{-3}
O^{7+}	64	140	4.21×10^{-2}	2.95×10^{-3}
Ni^{7+}	64	270	8.11×10^{-2}	5.68×10^{-3}
Mo^{8+}	72	329	10.02×10^{-2}	7.02×10^{-3}

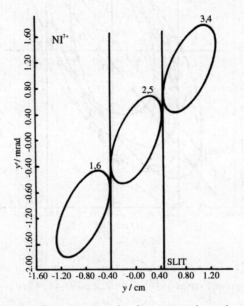

Fig. 5 The emittance plot for a swept beam bunch after being compensated by corrector

V. Induced Energy Spread

As the ions approach or leave the sweeper they are accelerated or deaccelerated by the axial component of the fringing field. The amount of energy which an ion gains or looses depends on its transverse position and phase excursions between the entrance and exit. The extra energy spread thus introduced can be estimated by the expression given by Chen[6] and Enge[7] based on a hard edge approximation

$$\Delta\varepsilon \approx \frac{2B}{D} \cdot \left(\frac{l_0}{v\tau_c}\right) \cdot \sqrt{\left(\frac{y_0}{l_0}\right)^2 + \left(\frac{\alpha_0}{2}\right)^2} \tag{8}$$

To be more accurate, integration of the equations of motion over appropriate fringing fields has been carried out. For a field of linear ramp, the distribution of the axial component has a simple rectangular profile whereas a more realistic profile can be expressed according to Ben-Zvi[4] as

$$E_z = -\frac{\overline{V}_m}{g} \frac{0.5y/g}{\cosh\left\{\frac{2|z - L/2| - l_s/2}{g} - 0.6\right\}} \tag{9}$$

Typical results for both fields calculated by SWEEP are listed in Table 3. The expansion of longitudinal phase area due to this is shown in Fig. 6. It is seen that the amount of extra energy spread introduced in the present design is well within 10^{-4} for all cases, though it increases with decreasing velocity as indicated by the formula (8).

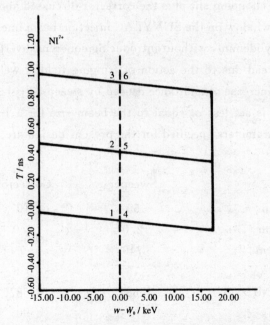

Fig. 6 Logitudinal phase plot for a chopped beam which is injected monoenergetically and then expanded to an area as shown

Table 3

Ion	$\Delta\epsilon_1$	$\Delta\epsilon_2$	$\Delta\epsilon_3$
O^{7+}	1.4×10^{-4}	1.37×10^{-4}	1×10^{-4}
Ni^{7+}	2.7×10^{-4}	2.71×10^{-4}	2×10^{-4}
Mo^{8+}	3.3×10^{-4}	3.38×10^{-4}	2.4×10^{-4}

($\Delta\epsilon_1$ is calculated by (8), $\Delta\epsilon_2$ by linear ramp, $\Delta\epsilon_3$ by (9))

Ⅵ. Conclusion

A post-tandem beam sweeper is shown to produce negligible beam quality degradation when used with a bunched beam. The coherent divergence and the induced energy spread decrease with increasing ion velocity, and the transverse emittance degradation is small for a bunched beam. The sweeper, chopping slit plus the corrector discussed above can provide a 5.5 ns time window on the SUNYLAC injection beam line, for ions from carbon to molybdenum, without introducing coherent divergence. The extra energy spread due to the action of fringing field is well within 10^{-4}. The effective increase of emittance caused by sweeping is less than 20% if the slit width is set less or equal to the beam size for a 1 ns prebunched beam. The parameters specified for the present design are summarized as follows:

	Sweeper	Corrector
l_s/cm	50	10
g/cm	2.0	2.0
D/cm	740	0
V_m/kV	15	0.5
f_0/MHz	4.6875	4.6875
φ_c	0	90

The authors are indebted to A. Scholldorf for his help with the program Optic Ⅱ and for many fruitful discussions. Thanks are also due to

Professor P. Paul, Professor G. Sprouse and Dr. J. W. Noe for their kind support and encouragement. One of us (Chen) was supported by a Fong Shu-Chuen Fellowship through the Committee for Educational Exchange with China.

Reference

[1] J. M. Brennan, "Bunching System for the SUNYLAC" Internal Report and Seminar for the Dept. of Physics, Nuclear Phys. Group SUNY at Stony Brook (1982).
[2] John Noe, T. Gentile, P. Paul, "Some Simple Discussions of the High Energy Sweeper" Internal Report Dept. of Physics SUNY at Stony Brook.
[3] J. J. Livingood, Particle Accelerators 7(1976) 243.
[4] I. Ben-Zvi, Z. Segalov, Particle Accelerators, 10(1979) 31.
[5] Chen, C. E., Physica Energiae Fortis et Physica Nuclearis, 4(1980) 401.
[6] Chen, C. E., Proc. of the National Particle Accelerator Conference May 1979, Chengdu China (1980) 75.
[7] Enge, Private Communication referred by L. Carneberg NIM (1961) 335.
[8] J. W. Noe, A note on Beam Emittance 3/5/77.

BUNCHING CHARACTERISTICS OF A BUNCHER USING HELICAL RESONATORS[①]

Abstract

A RF Buncher consisting of two coupled Helices tuned to 28.4 MHz was tested on the beam line of a 400 kV DC injector. Bunched waveforms versus RF power input were measured and analysed. With 8.4 W power input the proton beam was compressed to 1 ns (FWHM) bursts. The beam pulse splitted into double peaks at a higher power level depending on the initial energy dispersion. Both peak separation and the pulse height changed remarkably with increasing power input, and the overlapping of peaks in neighbouring cycles occured alternately. By making use of phase diagram in time-energy space a simple formalism is presented for the analysis of the bunching characteristics. Results obtained are well consistnet with the experiment. The effect of the quadrupole lens on bunched waveform was also observed and discussed.

Introduction

Interest in beam bunching technique has been increased considerably with the development of the time-of-flight method in heavy ion experiments and the installation of post accelerators in many laboratories. Sine wave bunchers have been used and studied extensively for various purposes for a long time[1~4]. Since discrepancies existed between one dimensional theory and the experimental waveforms, the dynamics of bunching process remained to be studied further, so as to get better performance. Dur-

---

① Coauther: Guo Zhi-yu, Zhao Kui. Reprinted from the IEEE. Trans. NS-30, 2(1983) 1254.

ing the design studies of our 4.5 MV heavy ion electrostatic accelerator project and its RF booster, a RF buncher consisting of He lices was developed and tested on the beam line of a 400 kV DC injector. Waveforms of the bunched beam were measured versus increasing RF power input under different beam energies and different bunching lengths. Beam bursts of 1 ns was obtained with comparatively low power consumption. In order to get better understanding of the buncher performance, efforts have been made to analyse the experiment in time-energy space[5]. It turned out that the results calculated in this way were well consistent with the reality.

It is the purpose of this paper to examine carefully the bunching characteristics by systematic measurements and to present a simple formalism developed in the time-energy space for the analysis of the performance of a single sine wave buncher. Effect of the matching between the quadrupole and the cavity on the bunched waveform was also observed and discussed. In the following sections the experimental layout and the bunched waveforms obtained will be described first and then an analysis will be presented.

Experimental Layout

The vertical beam from the 400 kV injector was bended into horizontal direction by a 90 degree bending magnet and then focused by a quadrupole doublet to a waist at about 3.5 m downstream close to the center of the cavity. With an emittance of 30 mm-mrad., the size of the waist in both X and Y directions was less then 1 cm. After being modulated by the RF field of the bunching cavity, the beam was focussed again by a pair of QD doublets and then drifted downwards to a fast response target, which was set 4.129 m and 1.904 m away from the center of the cavity in two cases. A 90 degree double focusing analyzing magset with a radius of curvature of 250 mm was set before the target to calibrate the amplitude of energy modulation by the buncher or to measure the energy distribution of the CW incident beam when the buncher was off. The energy dispersion of the incident beam depended upon the performance of the ion source, typically about 2×10^{-3}. The layout of the experiment is shown in Fig. 1.

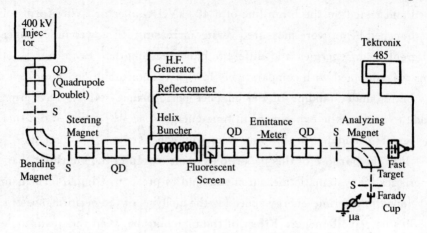

Fig. 1   The layout of the experiment

The bunching cavity consists of two strongly coupled Helices tuned to 28.4 MHz, each has a radius of 40 mm, a pitch of 5.65 mm and a length of 170 mm[6]. With a resonably high bare impedence of 22 Meg-Ohm/m, the cavity can be run in a wide range of field levels at a comparatively low power consumption. The amplitude of the energy modulation is measured in terms of ε_m, the fractional energy change $\Delta E/E_i = \varepsilon_m \cdot \sin\varphi_c$, ($E_i$ mean energy of the incident beam, φ_c the RF phase seen by the particle at the centre of the cavity.) As the operating frequency of the cavity was rather low, the dispersion in modulation caused by the radial inhomogeneity of the E_z component would be less than 2×10^{-4} for a beam having a diameter less than 1 cm.

The structure of a fast response target[7] is shown schematically in Fig. 2. A biased tungsten grid is set infront of the collector to schield off the displacement current induced by the ions before entering the target. With a positive bias of 250 V, secondary electrons emitted at the collector are collected by the grid and results in amplified signals of the ion pulse current. An effective 2~3 fold increase in primary beam current of protons is obtained under such condition. The rise time of the target-cable assembly measured by a TDR is 1.2 ns and the output impedence of it is 50 Ohm. The output signals were displayed on a Tektronix 485 oscilloscope with a rise time of 1 ns.

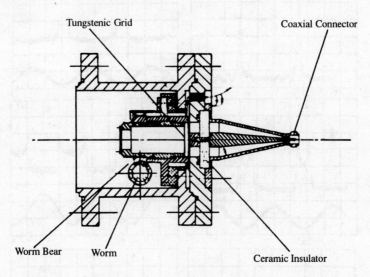

Fig. 2 The fast target assembly

Bunched Waveforms

A family of bunched waveform photos illustrating the change of the beam pulse shape with successive increase in RF power input P to the Helices loaded cavity is given in Fig. 3. The experiment was carried out with a 250 keV proton beam having a mean target current of approximately 50 μa. The power input to the cavity was indicated by a signal from a probe. It was calibrated by the temperature rise of the cavity cooling water. The bunching length in first case was 4.129 m.

In order to estimate the change of the bunched waveforms with the power input, half width (FWHM) and pulse height are plotted against the longitudinal focusing power L/F in Fig. 4, where the time focus F (see next section) is inversely proportional to the amplitude of modulation ε_m,

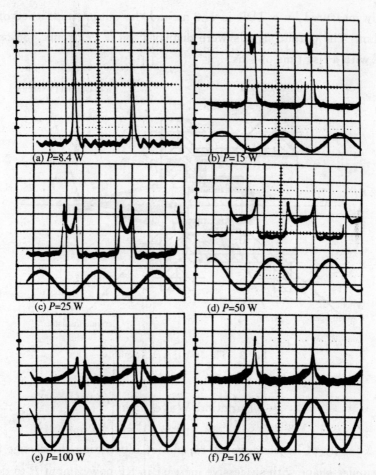

Fig. 3 Photos of bunched waveforms

and 1, the fixed bunching length. It is shown that the half width drops first with L/F to a broad minimum of 1.4 ns at $L/F=1-1.22$ ($P=5.6\sim 8.4$ W). Taking the effect of finite rise time of the measuring system $\tau_r=1.02$ ns in-to account, it is estimated that the incident beam was compressed to bursts having a half width of 1 ns. The maximum of the pulse height i_p appears near $L/F=1.26$, and the maximum peak to incident DC current ratio is about 14.5 : 1. The beam appears to be over bunched immediate-ly after the maximum and the peak starts splitting into double peak forms, which refer to dashed lines on the figure. The separation of peaks increses with L/F, while the height of the peaks drops with it. On the other hand, the valley to peak ratio R increases with L/F first and then tends to saturation. It is very interesting to notice that at $L/F=4.73$ ($P=126$ W) the peak separation equals to the period of the RF field, and overlapping of peaks between neighbouring periods takes place, consequently instead of double peak a combined single peak appears. As the power input increases further this kind of combined peak appears periodically.

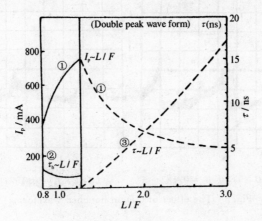

Fig. 4 Waveform parameters versus L/F

Families of bunched waveforms similar to above were obtained for three other cases, i.e. $E_i = 250$ keV; $L = 1.904$ m; $E_i = 350$ keV, $L=4.129$ m; $E_i=350$ keV, $L=1.904$ m. In each case a curve similar to that of figure 4 was obtained. Shortest half width of $\tau_m = 1.06$ ns was recorded at $P=66.8$ W in the last case. This means we might have got bursts having a half width less than 0.5 ns, unfortunately we were not able to confirm this without a wider bandwidth oscilloscope at that time.

The effect of the quadrupole on the bunched waveform was observed during the experiment. For an over bunched beam, the height of one peak could be altered with respect to the other by adjusting the focusing power of the matching quadrupole (Fig. 5). In an extreme case of high power input, it was even possible to cut off one of the peaks by deliberately mismatching the quadrupole lens. Actually, due to the coupling between transverse and longitudinal motions inside the cavity, the transverse emittance of the beam becomes phase dependent after bunching, and this might explain the phenomena described above. It might be possible to make use of this effect to limit the phase range in a bunched beam. This would be similar to the function of a Dee gap and associated vertical phase selective slits assembly in the central region of a Cyclotron.

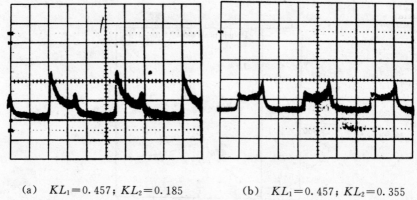

(a) $KL_1=0.457$; $KL_2=0.185$ (b) $KL_1=0.457$; $KL_2=0.355$

Fig. 5 The effect of QD on bunched waveforms

Analysis of Waveforms

An ideal linear bunching system, characterized by the time focus $F = v_i/\pi \cdot \varepsilon_m \cdot f_0$ (v_i is the mean velocity of the incident beam, f_0 is the operating frequency), compresses a CW beam to a longitudinal waist at its focal point, where the bunched pulse is in a rectangular form having a full width of

$$\tau = \varepsilon_{0m} \cdot F/v_i \tag{1}$$

ε_{0m} in above is the initial energy dispersion of the beam.

For a pratical bunching system using sine wave, however, the nonlinear effect must be taken into account. It is for this purpose a theory, which deduces the bunched waveform directly from the phase plot in the time-energy space, has been developed.

In case of weak modulation, the bunching cavity can be taken effectively as a single gap buncher at the center of the cavity[8]. Those particles which see a RF phase of φ_c at the gap will arrive at the target at a relative phase of φ (with respect to those with $\varphi_c=0$) after drifting a distance of L.

$$\varphi = \varphi_c + \frac{\omega L}{v_i}(1+\varepsilon)^{-1/2} \qquad (2)$$

where ε is the fractional energy modulation, $\varepsilon=\varepsilon_0+\varepsilon_m \cdot \sin \varphi_c$ ($\varepsilon_0=\Delta E_i/E_i$ initial deviation to the mean energy).

The area occupied by the bunched particles is bounded by two curves $\varepsilon_1(\varphi)$, $\varepsilon_2(\varphi)$, which can be written as following if considering $\varepsilon \ll 1$, $\varepsilon_0=\pm\varepsilon_{0m}$:

$$\begin{aligned}\varepsilon_1(\varphi) - \varepsilon_m\sin\left(\varphi + \frac{L}{F}\frac{\varepsilon_1(\varphi)}{\varepsilon_m}\right) - \varepsilon_{0m} = 0 \\ \varepsilon_2(\varphi) - \varepsilon_m\sin\left(\varphi + \frac{L}{F}\frac{\varepsilon_2(\varphi)}{\varepsilon_m}\right) + \varepsilon_{0m} = 0\end{aligned} \qquad (3)$$

Suppose the ions distribute uniformly in the (φ,ε) space with a constant density ρ, then the number of those ions lying in $(\varphi,\varphi+d\varphi)$ is proportinal to the differential area occupied. Hence $dN=\rho \cdot d\varphi \cdot \sum_i \Delta\varepsilon_i(\varphi)$, where $\sum_i \Delta\varepsilon_i(\varphi)$ is the sum of the height of all area elements along ε direction (refer to figure 6), and thus the instaneous intensity of the beam will be:

$$i(\varphi) = e\frac{dN}{dt} = e\omega\rho\sum_i \Delta\varepsilon_i(\varphi)$$

For an incident beam with an intensity of i_0, obviously $\sum_i \Delta\varepsilon_i(\varphi)=2\varepsilon_{0m}$, so

$$i(\varphi) = \frac{i_0}{2\varepsilon_{0m}}\sum_i \Delta\varepsilon_i(\varphi) \qquad (4)$$

Theoretical bunched waveforms versus L/F were calculated on a computer according to (2)~(4). The results are presented in a graphical form in Fig. 7a. The waveforms thus calculated are well consistent with the ex-

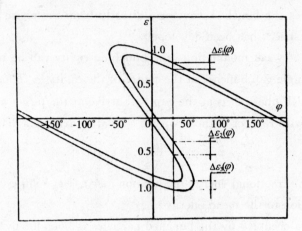

Fig. 6 Phase plot in (φ, ε) space

periment though they are quite different from those calculated from the existing one dimensional theory, which predicts a infinite peak of zero width at $L=F$ and infinite double peaks for any $L>F$ (see Fig. 7b). There are three more points worth mentioning:

1) The calculated half width τ_b(FWHM) depends on both time focusing power and initial energy dispersion. It gets down to a minimum at $L/F=1$, where it can be expressed effectively by $\tau_b = \varepsilon_{0m} \cdot L/v_i$. It appears that a great number of particles concentrated in a small interval of $|\varphi| \leqslant \varepsilon_{0m}/\varepsilon_m$, in the linea region, so that the effect of nonlinearity on τ_b is much less than that estimated by (9).

2) The existence of initial dispersion makes the maximum of the pulse height occur at a higher power level than that need for obtaining a minimum half width, and meanwhile it makes the maximum peak a finite height pulse. The maximum peak to initial DC current ratio can be estimated by:

$$\frac{i_{pm}}{i_0} = 1 + \frac{\varepsilon_m}{\varepsilon_{0m}} \sin\left[\left(\frac{L}{F}\right)_m \cdot \frac{\varepsilon_1(0)}{\varepsilon_m}\right] \approx 1 + \frac{0.85}{\pi} \frac{T}{\tau_b} \left(\frac{L}{F}\right)_m \tag{5}$$

where the term $(L/F)_m$ satisfies

$$\left[\left(\frac{L}{F}\right)_m^2 - 1\right]^{1/2} - \cos^{-1}\left(\frac{F}{L}\right)_m = \pi \frac{\tau_b}{T} \tag{6}$$

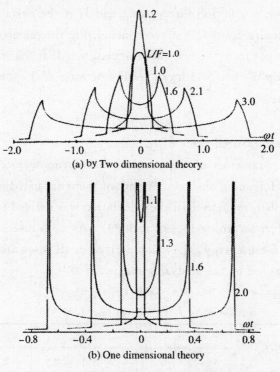

Fig. 7 Calculated waveforms

which means two boundary curves are symmetrically tangential to the ε axis in the (φ, ε) space.

3) The double peak form occurs immediately after the peak maximum, and the separation of the peaks also depends upon both L/F and ε_{0m}. Actually it can be given by:

$$\tau_d = \frac{T}{\pi}\left[\sqrt{\left(\frac{L}{F}\right)^2 - 1} - \cos^{-1}\left(\frac{F}{L}\right)\right] - \frac{\varepsilon_{0m} \cdot L}{v_i} > 0 \qquad (7)$$

τ_d increases with L/F, and any time if $\tau_d = nT$ holds (n integer, T period of the RF field), overlapping of peaks across n periods takes place.

To compare the theoretical waveforms with the experimental results, the frequency response of the measuring system was taken into account. According to (3) the frequency response G may be written as:

$$G = \sin(H \cdot \pi \cdot f_0 \cdot \tau_r)/(H \cdot \pi \cdot f_0 \cdot \tau_r) \qquad (8)$$

where $\pi \cdot f_0 \cdot \tau_r = 0.09$ for our system, and H is the harmonic number. Actually harmonics up to $H=30$ were included in the calculation. The interesting point got from the response calculation is that the effect of the finite rise time on the half width can readily be estimated approximately by the expression:

$$\tau_m = (\tau_b^2 + \tau_r^2)^{1/2} \tag{9}$$

where τ_m is the measured value of τ_b.

A set of parameters of the bunched waveforms were calculated according to the foregoing analysis. A part of them are listed in table 1 and 2. The initial dispersion used in the calculation was derived from the measured value of τ_b according to (2) and (9). As τ_b is insensitive to slow change of the mean energy, the value derived in all cases are slightly less than that measured by the analyzing magnet.

Table 1

$E_i = 250$ keV, $L = 4.129$ m, $\varepsilon_{0m} = 1.64 \times 10^{-3}$

P/W	L/F	$\tau_{b,d}$/ns		i_p/i_{pm}		R	
		a	b	a	b	a	b
8.4	1.22	1.40	1.40	1.00	1.00		
15	1.64	3.40	3.31	0.58	0.54	0.30	0.36
25	2.11	7.80	7.81	0.42	0.43	0.56	0.56
49	2.96	16.5	16.5	0.32	0.33	0.65	0.68
100	4.22	30.7	30.1	0.25	0.24	0.74	0.73
126	4.74	35.2	35.7				

Table 2

$E_i = 350$ keV, $L = 1.904$ m, $\varepsilon_{0m} = 1.4 \times 10^{-3}$

P/W	L/F	$\tau_{b,d}$/ns		i_p/i_{pm}	
		a	b	a	b
66.8	1.10	1.06	1.07	1.00	1.00
178.9	1.80	5.20	5.10	0.47	0.46
285.7	2.27	10.0	10.0	0.40	0.38
417.5	2.75	15.0	14.5	0.35	0.32

a-measured data, b-calculated data.

Conclusions

It has been shown that short bursts of 1 ns (FWHM) can be produced conveniently by Helices loaded cavity. The characteristics of bunched waveforms versus RF field level of the cavity have been presented with beam energy and bunching length as variables. In comparing with the experimental waveforms, the theory developed in (φ, ε) space seems to give much better results than that given by one dimensional theory.

The authors wish to thank Professor F. C. Yu for his encourangement and advice. Thanks are also due to Professor J. M. Hu for his valuable suggestions. We are also indebted to Mr. X. J. Zhang, Mr. Z. Z. Han, and Mr. P. L. He for their help during the experiment.

References

[1] F. E. Whiteway, A. W. R. E. Report No. 0~12/61 (1961).
[2] J. H. Anderson, N. I. M., 41(1966) 30.
[3] L. R. Evans, D. J. Warner, Proc. Proton Linear Accelerators conference (1972) 349.
[4] L. M. Bollinger et al., IEEE. Trans. NS-22, 3(1975) 114.
[5] Chen Chia-erh, Proc. of the Particle Accelerators Conference, May 1979, Chengdu, China, (1980) 75.
[6] Chen Chia-erh et al, "Status Report of a Helix Booster" Paper submitted to the 1980 National Particle Accelerators Conference, Shanghai, China, October 20 1980.
[7] Guo Zhi-yu, M. A. Thesis of Peking Univ., October 1981.
[8] Chen Chia-erh, Internal Memo, June 1976.
[9] S. J. Skorka, Proc. of the 3ed International Conference on Electrostatic Accelerator Technology, (1981) 130.

BUNCHING SYSTEM FOR THE STONY BROOK TANDEM LINAC HEAVY-ION ACCELERATOR[①②]

Summary

A versatile high-efficiency bunching system which is used to couple the Stony Brook 9 MV tandem Van de Graaff to the recently completed superconducting heavy-ion LINAC is described. The major elements of the system are: a pre-tandem double-drift harmonic buncher working at 9.4 and 18.8 MHz, a post-tandem sine-wave beam chopper at 4.7 MHz, a pre-LINAC super-conducting double-drift harmonic buncher at 150 and 300 MHz and a high voltage DC-coupled single pulse selecting beam switch.

I. Overview

The superconducting LINAC is an attractive choice for an energy booster for existing tandem Van de Graaff accelerators. An essential requirement is that a large fraction of the continuous beam from the tandem be bunched into the 100 ps phase acceptance window of the LINAC. Harmonic bunchers[1,2] are capable of compressing 60%~70% of the DC beam from the ion source into approximately 1 ns bunches. Further compression is then effected immediately before injection into the LINAC by a high voltage rebuncher. Here, the rebuncher is a high frequency (150 MHz) superconducting resonator. The phase acceptance range of the rebuncher can be extended, just as for the first buncher, by employing harmonics.

---

① Coauthers: J. M. Brennan, J. W. Noe, P. Paul, G. D. Sprouse and A. Scholldorf. Reprinted from the IEEE Transactions on Nuclear Science, Vol. NS-30, No. 4, (1983) 2798.

② Supported in part by the National Science Foundation.

This relaxes greatly the requirements of the tandem buncher and allows an even greater fraction of DC beam utilization. Furthermore, if the rebuncher operates on the same fundamental frequency as the LINAC, DC beam can be injected directly into the harmonic rebuncher to fill every RF cycle of the LINAC. Debunching after the LINAC then produces an essentially continuous beam.

We have chosen double-drift harmonic bunchers, since they are optimal for the desired low operating frequency (9.4 MHz) of the pre-tandem buncher and are most readily adapted to the superconducting technology of the rebuncher. A 300 MHz Quarter Wave Resonator has been developed here for use in this application.[3] The presence of both harmonic bunchers provides the flexibility of producing either beams with pulse spacing of 106 ns or quasi-continuous beams.

Experiments using the pulsed beams would generally require high purity, that is, low dark current between pulses. Some of the 30% of the beam not bunched by the pre-tandem buncher would enter the LINAC at accelerating phases and result in full energy small satellite pulses in the output beam. To eliminate this, a sine-wave beam chopper is placed between the tandem and the rebuncher and operates at one half the pro-landem buncher frequency (the chopper passes beam twice per cycle). Beam phase space degradation introduced by the chopper is minimal ($<20\%$ for transverse, $<10\%$ for longitudinal) because the beam is already bunched at the chopper location. In addition, the beam injection path here involves two 90° bends after the rebuncher. Beam phase at the rebuncher translates onto momentum dispersion and displacement after the first 90° bend. The final clean up of the time structure is done with a vertical slit after the 90° bend. The chopper then has only to eliminate beam that would arrive outside $\pm 150°$ (5.5 ns) with respect to the bunching phase. The 5.5 ns pulse width is a rather modest requirement of the chopper.

For those experiments that require pulse spacing greater than 106 ns a high voltage beam switch is installed adjacent to the chopper and deflects in the parpendioular plane. The fact that the beam is already bunched is also advantageous here since the voltage pulse rise time can be as long as the beam pulse separation, 106 ns. The system is operated in the normally on mode so that the beam is normally rejected. The voltage is pulsed off to accept a single pulse and in this way the fringe field of the deflecting plates has no effect on the selected pulse.

II. Pre-Tandem Buncher

The fundamental frequency of the bunching system is a design parameter dictated by the requirements of the experimental program carried out with the accelerator. In this laboratory a frequency of 10 MHz was chosen. The LINAC frequency being 150.4 MHz implies that the pre-tandem buncher should work on the 16th, subharmonic of the LINAC, 9.40 MHz. At this frequency double-drift harmonic bunchers have been shown to give good bunching performance with high capture efficiency.[1]

Double-Drift Harmonic Buncher

The double-drift scheme uses two simple sine-wave bunchers, operating at the fundamental and 1st. harmonic frequencies and separated by a drift distance of $\sim 15\%$ of the system focal length. The first buncher overbunches strongly to focus particles at $\pm 120°$ of RF phase; the second buncher, operating 180° out of phase from the first, then removes the excess modulation from the particles at low phases and will have little effect on the particles at the extreme phases. The drift space between the two allows the bunch to begin to coalesce before the second buncher and greatly improves the overall system efficiency.

The optimum values for the amplitudes of the two bunchers and the length of drift space between them were obtained with a least squares minimization computer program. The optimized values for the buncher strengths and separation are:

$$d_1/d_f = 0.52, \quad d_2/d_f = 1.67, \quad d/d_f = 0.15$$

where: d_f is the system focal length, measured from the first buncher, $d_{1(2)}$ is the focal length of buncher 1(2), and d is the separation between the two. These values correspond to a capture efficiency of 60 and yield a compression factor of ~ 200. They are essentially universal, being independent of beam velocity, beam energy and system frequency and focal length.

Buncher Tube Structures

Skorka[4] has shown that the minimum longitudinal phase space product is obtained by the proper choice of the injected beam energy into the tandem. One strikes a balance between low injection energy where ion source noise dominates the time factor and very high injection energy where buncher modulation dominates the energy factor. This follows from the observation that the time factor is bounded from below by the buncher aberrations and the energy factor is bounded by the tandem stripper energy straggling.

The optimum injector energy as a function of ion mass is plotted in Fig. 1. For moderate heavy ions this energy E is quite high, ~ 300 keV. Since the RF power required to drive a buncher increases as E^3 one must take care that impractically high power levels are not called for. This means high Q tank circuits and low capacitance structures. The common practice of using multiple drift tube lengths and some switching mechanism is quite costly in terms of capacitance. The structure used here, shown in Fig. 2 uses only one drift tube length for each frequency and is capable of spanning the mass range 4~100 if the beam energy follows approximately the optimum value.

Also plotted in Fig. 1 are the transit time factors $T(\beta)$, which relate the actual voltage drop seen by the the beam to the RF voltage on the buncher drift tube. The transit time factor is well represented by the analytical form

$$T(\beta) = \sin\left(\frac{\pi}{2} \cdot \frac{\beta_0}{\beta}\right) \cosh^{-1}\left(\frac{\pi g'}{\beta \lambda}\right)$$

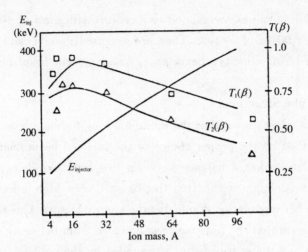

Fig. 1 Optimum injection energy and transit time factors for buncher 1 (fundamental) and 2 (1st harmonic) as a function of mass. The points are measured values, those for 1 are normalized to $T_1(\beta)$ at $m=32$

Fig. 2 Buncher high voltage structures and standard 6″ Tee vacuum chamber. Buncher spacing is 27 cm

where β = the beam velocity, v/c,

λ = the RF free-space wave length

g' = effective gap = $\sqrt{(\text{true gap})^2 + (\text{tube bore})^2}$

$$=\sqrt{5^2+15^2}=16 \text{ nm}$$

β_0 = optimum velocity, 2(tube lengty)/λ

The first factor reflects the phasing of the beam in drifting from the first to the second gap and the second factor accounts for the changing RF voltages as the beam traverses the extended region (g') of electric field. $T(\beta)$ also depends on the radius of the beam from the buncher tube center line, being higher at the outer radii where the extent of the field distrioution is less. This effect leads to a multiplicative factor which is independent of the tube bore and is given approximately by $1+(\pi r/\beta\lambda)^2$. A 9 mm collimator between the bunchers ensures that this factor is always less than 7%.

Figure 1 shows some measured values for $T(\beta)$ obtained with $^{12}C^-$, $^{16}O^-$, $^{24}C_2^-$, and $^{32}S^-$ ions injected at various energies. Bunched beams were detected at the 0° port of the tandem analyzing magnet via a fast secondary-electron emission detector which uses a micro-channel plate electron multiplier (rise time < 0.5 ns). Secondary electrons are emitted when the beam impinges on a thin (0.002″) tungsten wire which is held at -2 kV to quickly accelerate the electrons to the detector and swamp out the distribution of initial electron energies. Typical time spectra are shown in Fig. 3.

The buncher voltages that truly correspond to the measured focal lengths are deduced by the double peak method proposed by Chen[5]. This method takes into account that the minimum pulse width occurs when the beam is actually somewhat overbunched. Measuring the double peak separation when the beam is strongly overbunched gives a reliable means of adjusting the buncher focal length precisely to the target distance.

Effective target distances were calculated in a computer program that sums the effective drift lengths of all the accelerating sections of the tandem, the dead spaces within the tandem and the drift length to the target according to standard formulas. The physical distance to the entrance of accelerating tube is 205 cm and the total effective distances ranged from 213 cm to 240 cm, depending on beam injection velocity and tandem terminal voltage.

The optimum velocity β_0 was chosen at a somewhat high value, 0.00425, in order to extend the mass range to $m=4$. The fall-off of $T(\beta)$ at high mass is mitigated by the fact that heavy ions need less voltage to bunch than light ions at a given energy since they are slower. The drift tube length for the second buncher is $\sim 15\%$ less than half that of the first one even though the β_0 is essentially the same. This was done in anticipation of the asymmetrically distributed fields about the gaps that occur when the length to bore ratio is not large[1].

The two insulators supporting each drift tube are Alumina, Al_2O_3 (Alsimag 614 made by 3 M Technical Ceramics Division). They are very strong when used in compression, have negligible dielectric losses, and are thermally stable.

Tank Circuits

The tank circuits for the bunchers are conventional lumped-element LC parallel-resonance circuits. The inductors are wound from 3/8 inch copper tubing on a 5 inch diameter tefion coil form with a pitch of 0.6 inch/turn. Tuning is done with $5 \sim 25$ pf vacuum-variable high voltage (20 kV) capacitors. The circuits are enclosed in 12 inch cubical aluminum boxes equipped with fans for forced air cooling. Relevant parameters are listed in Table 1.

Table 1

f/MHz	Turns	L/MH	C/pf	Q	Shunt Impedance/MΩ
9.40	10.5	10.0	29	800	3.7
18.80	4.0	2.86	25	1300	3.5

Direct coupling is used to match the tank circuits to the 50 Ω output impedance of the driver power amplifiers. This is done by tapping onto the inductor very near the grounded end. The precise location of the tap for critical coupling is found empirically by observing the reflected power wave form when the drive power is pulsed.

Control Circuit

Successful operation of harmonic bunchers relies on accurate adjustment of the amplitudes and relative phase of the constituent harmonics. The high Q tank circuits that must be used are susceptible to significant phase fluctuations between the driving power source and the resonating voltage as the resonant frequency varies over a range comparable to the band-width. An attractive feature of the double-drift scheme is that the harmonics are completely de-coupled electrically. Nevertheless, if long term stable operation is to be achieved active feedback stabilization of the amplitude and phase of the bunchers must be employed.

Phase control is achieved with a phase locked loop scheme where the phase reference signal derives from the LINAC reference oscillator divided by 16 or 8. Phase adjustment is provided by a 360° voltage controlled phase shifter between the tank circuit pick-up signal and the loop phase comparator. A high quality temperature compensated VCO keeps loop stress to a minimum and gives essentially instantaneous lock-up upon turn on.

The amplitude feedback circuit compares the detected value of the pick-up voltage, after it passes through an adjustable attenuator, with a standard level. Amplitude adjustment is done by varying the adjustable attenuator. The amplified error signal controls the drive power to the tank circuit in a closed-loop fashion to keep the detected value equal to the standard level. This inverse programming approach has the advantage that the RF levels inside the control circuit are always the same. Linear response to the amplitude control knob (or remote programming voltage) is provided by an analog voltage divider before the variable attenuator.

A specialized feature relevant to the double-drift buncher control is that the standard level can be switched between two values having the correct ratio between double-drift and simple buncher amplitudes independently of the specific amplitude. This feature is very useful in tuning the double-drift buncher (see below).

Tuning

Figure 3 illustrates the straightforward tuning of the system. The amplitude of each buncher is first separately adjusted for a time focus at the target. With the first buncher off, the phase of the second buncher is set such that the bunches arrive symmetrically about the arrival time from the first buncher. The two amplitudes are then adjusted by the above mentioned factors, 1/0.52 and 1/1.67, and both bunchers are turn-turned on.

The measured capture efficiency from these data is that 66% of the beam falls within the range bounded by the full width at 1/10 maximum limits of the peak (~3 ns). Full width at half maximum is 1.4 ns. No discernable change in transmission through the tandem was observed when the bunchers were in operation.

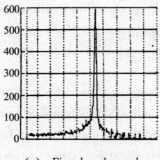

(a) First buncher only

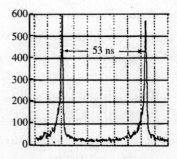

(b) Second buncher only phase adjusted for symmetry about a

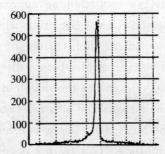

(c) Both bunchers on with amplitudes adjusted for double-drift bunching

Fig. 3 Timing spectra and tuning procedure for double-drift buncher

Acknowledgement

The authors are indebted to the technical staff of this laboratory for skillful implementation of these designs. In particular the work of Mr. C. Pancake and Mr. A. Morrongiello was vital to the success of this project. One of us, Pr. C. E. Chen would like to acknowledge the support of a Fong Shu-Chnen Fellowship through the Committee for Educational Exchange with China.

References

[1] W. T. Milner, IEEE Trans. Nucl. Sci., NS-26 1(1979) 1455.
[2] F. L. Lynch, R. N. Lewis, L. M. Bollinger, W. Henning and O. D. Despe, Nucl. Instr. and Meth. **120**(1974) 245.
[3] I. Ben-Zvi and M. Brennan, NIM, in press.
[4] S. J. Skorka, Proc. of the 3rd. Int. Conf. on Electrostatic Accelerator Technology, Oak Ridge, TN, USA, 13~16 April 1981, IEEE, p. 130~8.
[5] C. E. Chen et al., Proc. of the 7th Conf. on the Application of Accelerators in Research and Industry, November 1982, Denton, Texas.

600 kV 强流纳秒脉冲加速器中的前切割与后聚束系统[①]

摘　要

介绍了 600 kV 强流纳秒脉冲加速器中束流脉冲化系统的物理设计、工艺结构特点、粒子动力学计算、聚束腔的静态特性调整与测量、高频功率发射系统以及载束实验结果.

关键词　切割器，聚束器，粒子的纵向运动，束流脉冲

为了在 14 MeV 能域利用飞行时间法开展脉冲中子束双微分截面的测量，中国原子能科学研究院建造了一台 600 kV 强流纳秒脉冲高压倍加器. 我们承担研制束流脉冲化系统的任务. 此系统是该加速器产生纳秒短脉冲束的核心装置，可以为中子物理双微分截面测量实验提供 1~1.5 ns 的短脉冲束. 这台加速器是我国研制的第一台强流（直流 D^+ 束 3~5 mA）脉冲束高压倍加器，达到了当前国际上同类加速器的先进指标. 脉冲化系统的主要设计指标是：离子种类为 D^+；聚束器最佳调制能量为 300 keV；聚束频率为 6 MHz；脉冲重复频率为 1.5 MHz；脉宽（FWHM）≤1.5 ns；脉冲束平均流强为 30~50 μA.

根据加速器的总体设计指标和特点，我们曾作过头部单波聚束器、头部双漂移谐波聚束器、尾部单波聚束器和尾部双漂移谐波聚束器等设计，对这些方案作了计算和反复比较[1]后决定采用前切割和后聚束方案. 这个方案的优点是，它比头部聚束较为容易达到对脉宽的要求，而且对离子源能散的要求可适当放宽. 采用单波聚束是为了减小高频功率和造价，并且便于实验调整与操作，但是后聚束方案所要求的高频聚束功率比头部聚束要大得多，因而造价也高，同时给聚束腔的工艺设计提出

---

① 合作者：吕建钦，谢大林，全胜文. 摘自《核技术》，Vol. 20, 9(1997) 555~560.

了更严格的要求.产生能量为 300 keV 的强流短脉冲束,在国内尚属首次.由于被聚束离子的能量高、流强大,并且聚束器工作频率较低,要实现≤1.5 ns 的短脉冲有相当的难度.

1 束流脉冲化系统概述[2]

1.1 总体布局

束流脉冲化系统的配置如图 1 所示.从离子源引出的束经过单透镜

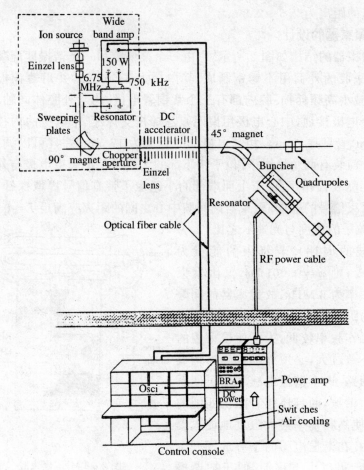

图 1 束流脉冲化系统布局

Fig. 1 Layout of the beam pulsing system

聚焦后在头部 90°分析磁铁的物点处成腰,此腰位就是扫描切割板所在的位置.切割孔安放在这个磁铁的像腰处,为了避免 Mobley 磁聚束效应,扫描方向必须位于磁铁的垂直平面内.切割器的工作频率为 0.75 MHz,可以产生时间长度为 20～55 ns、重复频率为 1.5 MHz 的脉冲束团.切割后的脉冲束再经过加速,使能量达到 300 keV,然后再经过一个 45°偏转磁铁,使束流偏转到中子实验线上,聚束器的中心就在 45°磁铁的像腰位置附近.能量受到聚束器射频场调制后的束穿过一个四极透镜对,使其横向像腰投射到靶点上,此处也形成纵向束腰.聚束器出口到靶点的距离为 2.5～2.8 m.

1.2 聚束器的设计

聚束器的结构如图 2 所示.它由一个聚束腔和一个谐振匹配装置组成.聚束腔的外筒用不锈钢制成,其内径为 $\phi350$ mm,并用薄铜板作内衬,以减小高频耗损.腔内部有三个紫铜管电极,均与外腔体同轴.端部的两个电极接地,中心电极用聚四氟乙烯杆支撑.三个电极的内径均为 $\phi30$ mm,管壁厚 5 mm.相邻电极间的距离为 17 mm,并且可以调节.聚束腔外筒装有两个绝缘子.位于中下部的大绝缘子把聚束电极与外部的谐振匹配装置相连;位于上面端部的小绝缘子带有信号提取探针,用作腔内聚束信号的监测.聚束器两间隙中心之间的距离应满足 $L=0.5\beta cT$(β 为粒子的速度与光速 c 之比,T 为调制波的周期).若氘束的能量为 300 keV,则 $\beta=0.001794$.又因聚束器的频率为 6 MHz,故聚束器的间隙之间的距离为 $L=44.8$ cm.由于聚束器的工作频率较低,为了使聚束腔的体积不致过大,必须外加电感以构成谐振回路.这个谐振电感不仅可使系统产生谐振,而且还起着阻抗匹配的作用,使高频功率最大限度地馈入聚束腔内.在谐振电感的下方放置一个调谐板,可以上下移动,便于把谐振频率准确地调到 6 MHz.在其侧面还

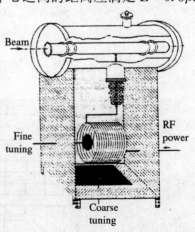

图 2 6 MHz 聚束腔
Fig. 2 The 6 MHz buncher cavity

有微调谐板.因为聚束功率较大,还须考虑系统的散热冷却问题,因而谐振电感用双层紫铜管绕制而成,冷却水从内管流入,从内外管壁之间流出.

1.3 切割器的设计

由于中子物理实验所需要的脉冲重复频率为 1.5 MHz,故切割器的工作频率取 0.75 MHz. 一般正弦波聚束器只能把相位处在±60°之内的粒子群聚到一起,所以切割器产生的束流脉宽应为 50～56 ns. 切割板的几何尺寸为:长 80 cm、宽 40 mm、间隙 30 mm. 切割缝放置在磁铁的像腰处,缝宽为 5～6 mm. 为了使束流受到均匀的扫描,切割板形状为燕尾槽式,其几何形状是利用 POISSON 程序[3]进行静电场的数值计算后确定的.

2 粒子动力学计算

2.1 粒子的纵向运动

我们设计了一个聚束系统计算程序 LMOVE[4]来模拟粒子的纵向运动在相空间中的演变.此程序可以计算大量单个粒子在由速调管聚束器、单狭缝谐波聚束器、双漂移谐波聚束器、漂浮空间和加速空间任意组合而成的聚束系统中的纵向运动.所输入的纵向相图可以是矩形、椭圆或直线段(相应于单能束).程序可以显示出每个元件之后的纵向相空间图形.这个程序用 MS FORTRAN 语言写成,适用于任何 IBM-PC 及其兼容机.

2.1.1 脉冲宽度.我们利用 LMOVE 程序对离子的纵向运动作了大量的计算.结果表明,如果加速器引出束的能散在±500 eV 以内,则脉冲束的宽度可以达到 1.5 ns (FWHM)以内.图 3 为聚束后的纵向相图和脉冲波形.

2.1.2 能散对脉宽的影响.确定了聚束腔的几何尺寸、工作频率和聚束距离后,影响脉宽的关键因素就是束流的能散度.在后聚束的情况下,这种能散度来自三个方面的贡献:(1)离子源引出束的能散;(2)由于切割器而引起的附加能散;(3)头部电源和加速电压不稳定(高压的摆动与纹波等)而引起的能散.由于聚束脉宽与束流的初始能散成正比,要想获

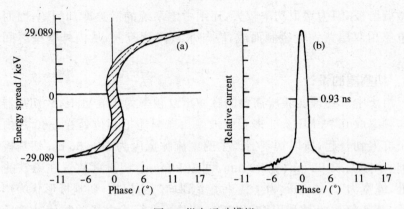

图 3 纵向运动模拟
(a) 纵向相图,(b) 束流脉冲
Fig. 3 Simulation of longitudinal motions
(a) Longitudinal phase space diagram, (b) Beam pulse

得满足要求的短脉冲,必须对束流的能散度加以限制.因此,对能散的限制主要是对离子源的能散和对加速器各电压稳定性的限制.表1给出了聚束距离为 2.50 m 时,在不同氘核能量 E_0 和不同能散 ΔE 下所能获得的脉宽.

表 1 不同能散下的束流脉宽 τ(FWHM)
Table 1 Beam pulse widths (FWHM) under different energy spreads

E_0/keV	ΔE/eV				聚束电压 Bunching voltage/kV
	300	400	500	600	
250	0.87	1.08	1.30	1.54	12.08
300	0.69	0.87	1.07	1.29	16.00
350	0.54	0.77	0.80	1.01	20.00
400	0.46	0.64	0.67	0.86	24.60
450	0.31	0.46	0.50	0.70	30.00

在以上的计算中,我们假定加速器引出束流的总能散度为 ±500 eV,其中,离子源和切割器对能散的贡献假定为 200 eV.由于扫描切割产生的能散为~87 eV,则离子源的能散不能大于 180 eV.

2.2 粒子的横向运动

束流脉冲化系统的各部件在加速器的束流线上需要合理安放,因此我们用 LEADS 程序[5]对粒子的横向运动作了许多计算. 单透镜把从离子源发出的束会聚在 90°分析磁铁的物点处并形成束腰,束斑直径约 6 mm,扫描切割板就位于此处. 在这个磁铁的像腰处放置一个直径为 5~6 mm 的切割孔. 在加速管的入口处有一个由三膜片组成的单透镜,它把加速后的束腰投射到 45°偏转磁铁的物点处. 聚束腔则位于此偏转磁铁的像腰处. 其后的磁四极透镜对把束流会聚到最终靶点上,整个系统的束包络见图 4.

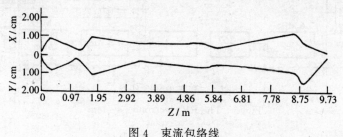

图 4 束流包络线

Fig. 4 The beam envelopes

3 聚束腔的静态特性测量[6]

聚束腔的静态特性测量主要包括三个方面. (1)谐振频率及调谐范围:大范围内的频率调节主要靠调节螺线管的圈数,在谐振点附近通过粗调和细调两个调谐板把频率调准,调谐板的调谐范围为 ±70 kHz. (2)输入阻抗调整:主要测量仪器为 QF 1052 信号发生器和 DT-1 矢量电压表,把阻抗调至 50 Ω. (3)品质因数 Q 值:无载品质因数 $Q_0 \approx 1000$.

4 高频功率发射系统

高频功率发射系统包含两个高频放大线路,其频率分别为 6 MHz 和 0.75 MHz. 它们是基于同一个主振荡器,经过不同的倍频、移相和放大线路,再经过不同的调谐匹配回路馈送到各自的负载. 前者为聚束器供电,后者为切割器供电. 由于切割系统处于 600 kV 高电位,激励信号处在低电压,所以从激励源产生的 0.75 MHz 信号需经过光缆传送到处在高压端的宽度放大器. 为了实现良好的聚束,两信号源之间的相位要能够调节,以便使切割后的脉冲束团处在聚束相位之内,因此,在 6 MHz 放大线路内设置一个 0~360°的移相器. 束流脉冲化系统的高频电子学线路示于图 5.

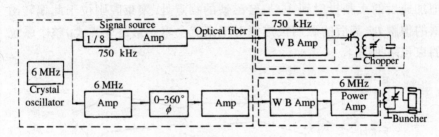

图 5　高频电子学线路

Fig. 5　The RF electronics system

5　载束实验

在 300 kV 加速电压下,对 H_2^+ 束进行了聚束实验. 当直流束强度为 1~1.2 mA 时,在示波器上观察到的束流脉宽为 1~1.2 ns (FWHM),脉冲峰值流强为~20 mA. 切割器可以把直流束强切割为原来的 1/25~1/30,这就意味着切割后的束流脉宽可达到 20~30 ns,图 6 为聚束器与切割器联合运行时所得到的束流脉冲.

致谢　张胜群同志在项目的初始阶段做了许多部件制作与调整等工作,中国原子能科学研究院 600 kV 倍加器组的同志在脉冲化系统载束实验中给予了大力支持和帮助,在此表示衷心的感谢.

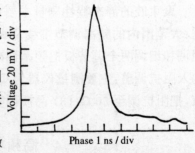

图 6　聚束实验结果

Fig. 6　Experimental result of beam bunching obtained with the chopper-buncher system

参 考 文 献

[1] 吕建钦,谢大林,张胜群. 600 kV 纳秒脉冲中子发生器束流脉冲化系统设计,中国原子能科学研究院加速器方案论证会,北京,1990.
[2] Lü Jianqin, Xie Dalin, Zhang Shengqun, et al. Proc 3rd European particle accelerator conference, 1992; 2:1244.
[3] Menzel M T, Stokes H K. User's guide for the POISSON/SUPERFISH Group of Codes LA-UR-87-115, 1987.
[4] 吕建钦. 原子能科学技术,1992;26(6):36.
[5] Lü Jianqin. Nucl. Instr. Meth., 1994; A355:31.
[6] 吕建钦,张胜群,谢大林. 中国粒子加速器学会第四次代表大会第五次年会论文集,1992.
[7] Lü Jianqin, Quan Shengwen. Nucl. Instr. Meth., 1995; A346:253.

A PRE-CHOPPER AND POST-BUNCHER SYSTEM FOR THE 600 kV INTENSE NANOSECOND PULSED BEAM ACCELERATOR

Abstract

This paper briefly describes the design of a beam pulsing system for the 600 kV intense pulsed beam accelerator, the structure of the buncher cavity and the measurements of its static features, the beam dynamics calculations, the RF power system and some experimental results.

Key words Chopper, Buncher, Particle longitudinal motion, Beam pulse

HIGH-EFFICIENCY TWO-HARMONICS BEAM CHOPPER[①]

Abstract

A novel beam chopper with two-harmonics (f and kf) sweeping voltages is described. It is shown that beam utilization efficiency and the beam quality of the harmonic chopper are much better than that of the conventional sinewave chopper. The beam utilization efficiency might be even higher than the square-wave chopper for low-velocity ions. In practice, a chopper with 1.5 and 13.5 MHz ($k=9$) was constructed with great convenience for the beampulsing system of a 4.5 MV Van de Graaff in Peking University. The bench tests indicated that the beam efficiency of the harmonic chopper was higher than a single-frequency chopper by about 47%, which agrees with the theory and the design.

Introduction

rf beam choppers are widely used in time-of-flight equipment for neutrons and heavy ions as well as in the injection line of linear accelerators. A good chopper should be able to pulse the beam with minimum deterioration of beam quality and high beam utilization efficiency (BUE). Among the commonly used choppers, the square-wave chopper[1] is capable of reaching high BUE when the ion speed is high. However, it is quite difficult to get the amplitude high enough to fit the need of the chopper. Actually, the rise time of the beam pulse is interrelated to the beam spot size

---

① Coauthers: G. S. Xu, Z. B. Qian. Reprinted from the Rev. Sci. Instrum. Vol. 57, 5 (1986) 795~797.

and ion velocity. Under the same sweeping voltage amplitude, the lower the ion velocity, the longer the pulsed beam rise time. Consequently, the BUE is low, and in some cases, even lower than that of a sine-wave chopper. Meanwhile, the beam quality gets worse. Using traveling wave deflection, the rise time can be shortened but the system will be rather complicated, and the beam efficiency will be limited by the fringing field effect. The sine-wave chopper is simple in construction and is able to make short beam pulses, yet the efficiency is low. When the chopping slit width equals the beam spot diameter, $\eta=50\%$. In addition, the coherent divergence caused by sweeping is a problem.[2~4]

An ion beam pulsing system for the 4.5 MV Van de Graaff was developed in Peking University. The dc beam is to be chopped into 67 ns (FW) bursts first and then bunched to $1\sim2$ ns (FWHM) by the system inside the high-voltage terminal. A new beam chopper with two-harmonics sweeping voltage has been designed, constructed, and tested. It turns out to have high BUE and good beam quality and can be constructed with case. The working principle, layout, and the performance of the chopper will be described as follows.

I. The Principle of the Harmonic Chopper

The sweeping voltages applied to the harmonic chopper ran be expressed by

$$V(t) = V_1[\sin\omega t + B_k \sin(k\omega t + \phi)], \qquad (1)$$

where ωt is the phase of the fundamental wave when a particle reaches the center of the sweeping field and k is the harmonic number. The static distribution of the sweeping field can be expressed by a linear ramp approximation as illustrated in Fig. 1, where $l=1.15g$, $l_1=l_g-0.5g$, g is the gap height, and l_g is the plate length.

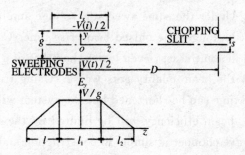

Fig. 1 Sweeping plates and the form of their fringing field

A. Beam Utilization Efficiency (BUE)

Referring to the results for a single sine-wave chopper[2] and the principle of superposition, the displacement of the central ray $X_0(t)$ at the chopping slit is

$$X_0(t) = C[\sin\omega t' + b_k \sin(k\omega t' + \psi)], \qquad (2)$$

where

$$t' = t - D/v - \delta(\omega)/\omega, \quad C = A_1 T_F(\omega) K(\omega) D,$$
$$b_k = [T_F(k\omega)/T_F(\omega)][K(k\omega)/K(\omega)] B_k,$$
$$\psi = \phi - \delta(k\omega) + k\delta(\omega),$$
$$T_F(\omega) = (\sin\phi/\phi)[\sin(\phi + \phi_1)/(\phi + \phi_1)],$$
$$\delta(\omega) = \tan^{-1}\{(v/\omega D)[2 - (\phi + \phi_1)\cot(\phi + \phi_1) - \phi\cot\phi]\},$$
$$K(\omega) = [1 + \tan^2\delta(\omega)]^{1/2},$$
$$A_1 = V_1(l + l_1)/2(V_0 g). \qquad (3)$$

QV_0 and v are ion energy and speed, respectively, $\phi = \omega l/2v$ and $\phi_1 = \omega l_1/2v$. By selecting suitable ϕ, b_k, and C, so that $\psi = \pi$, $X_0(t)$ can be illustrated as in Fig. 2 for $k = 9$ ($k = 3, 5, 7, \cdots\cdots$). It is obvious that the pulsed beam has the same repetition rate as a single sine-wave chopper of frequency ω.

Assume the beam distribution at the slit with diameter "a" is uniform in (X, t) space. The instantaneous pulsed beam current $i(t)$ and BUE are given by

$$\frac{i(t)}{i_0} = \begin{cases} \min(1, s/a), & |X_0(t)| < |a - s|, \\ (s + a - |X_0(t)|)/(2a), & |a - s| < |X_0(t)| < a + s, \\ 0, & |X_0(t)| > a + s, \end{cases} \qquad (4)$$

$$\eta = \frac{1}{\tau}\int_0^T \frac{i(t)}{i_0}\mathrm{d}t, \tag{5}$$

where i_0 is dc beam current, τ full width of the chopped beam, and T the repetition period. Figure 2 is an example of pulse waveform when the sweeping voltage consists of 1st and 9th harmonics and $b_9 = 0.14$, $\omega\tau/2 = 0.1\pi$. $\eta(b_k)$ and $V_1(b_k)/V_{10}$ are shown in Fig. 3, where V_1 is the voltage of the fundamental wave, V_{10} is that of single sine-wave chopper for the same s and τ.

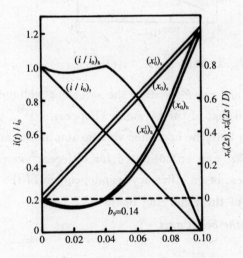

Fig. 2 The waveform of pulsed beam (The subscript "s" denotes sine chopper; "h" harmonic chopper)

It is seen in Fig. 3 that when $s = a$, the BUE of the harmonic chopper with 9th harmonic is higher than the single-frequency chopper by about 52%. It can be easily show that for any τ and ω value, the beam efficiency can be enhanced by 50%~70%, provided that V_1, k, and B_k are suitably selected. Meanwhile, the value of V_1 is fairly close to that of V_{10} and $V_k = V_1 B_k$ is much smaller than V_1.

In order to make the beam efficiency of the single-frequency chopper as high as the maximum beam efficiency (75.9%) of 1st +9th harmonic chopper for $s = a$, the chopping slit s has to be opened as wide as $[\eta/(1-$

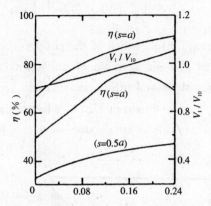

Fig. 3　The η and V_1/V_{10} of 9th harmonic chopper

$\eta)] \ a = 3.15a$, and consequently, the sweeping voltage has to be 1.75 times higher than that of the harmonic chopper.

By adjusting both the harmonic voltage amplitudes and the phase, the waveform of $X_0(t)$ can be optimized for different ion velocities. Consequently the highest beam efficiency is independent of the ion velocity.

B. The Quality of the Emerging Beam

1. *The slope of the beam axis*

Let
$$V(t) = V_1[\sin(\omega t + \delta_1) - B_k \sin(k\omega t + \delta_k)], \qquad (6)$$
thus,
$$X_0(t) = C(\sin\omega t' - b_k \sin k\omega t'), \qquad (7)$$
$$X_0'(t) = A_1 T_F(\omega)[\sin(\omega t' + \delta_1) - b_k \sin(k\omega t' + \delta_k)], \qquad (8)$$
where X_0' is the divergence of the central ray, $t' = t + D/v$, $\delta_k = \delta(k\omega)$. Assume $X_0 = X_0' = 0$ when $V(t) = 0$. For sine-wave chopping, when $X_0(t) = 0$,
$$X_0'(t) = (-1)^n A_1 T_F(\omega) \sin\delta_1 \neq 0 \quad (n = 0, 1, 2, \cdots),$$
so there is a coherent divergence of the emerging beam. But for harmonic chopping, if
$$b_k = \sin\delta_1 / \sin\delta_k, \qquad (9)$$

when $X_0(t)=0$, $X_0'(t)=0$, i.e., the coherent divergence can be eliminated. A new possible way of eliminating or reducing the coherent divergence is found. It is very significant that the value of b_k from Eq. (9) is very close to that at the highest η. So with the harmonic chopper, one can get high BUE with small coherent divergence.

2. Transverse emittance

Since the transverse emittance ellipse moves parallel with (X_0, X_0') at the slit, the transverse emittance of emerging beam can be obtained from $X_0(t)$, $X_0'(t)$. Calculation shows that the harmonic chopper has better time-average transverse emittance than that of single sine chopper as shown in Fig. 2.

3. Induced energy spread

Assuming the movement of the ions between the sweeping plates is small and the average position can be expressed as $\overline{Y}$, the induced energy spread can be estimated from Eq. (6) as follows:

$$\Delta E = -\frac{2QV_1\overline{Y}}{g}T_F(\omega)\frac{\omega(l+l_1)}{2v} \times \cos\omega t\left(1 - kb_k\frac{\cos k\omega t}{\cos\omega t}\right)$$

$$= \Delta E_{10}\frac{V_1}{V_{10}}\left(1 - kb_k\frac{\cos k\omega t}{\cos\omega t}\right), \tag{10}$$

where ΔE_{10} is the induced energy spread of a single sine-wave chopper for the same s and ω. The result of calculation shows the normalized energy spectrum of a "1st+9th" harmonic chopper ($b_9=0.14$) better than that of a sine-wave chopper.

II. The Layout and the Beam Test of the Harmonic Chopper

A. The Layout

Considering the condition in the terminal of the 4.5 MV electrostatic accelerator, the geometrical parameters of chopper are chosen as $l_g = 3$ cm, $g=4$ cm, $D=55$ cm (see Fig. 1). As the ion source extraction voltage might be as high as $V_0=30$ kV and the beam spot size $a=s=0.25$ cm, the sweeping voltages are determined accordingly and two rf power supplies (1.5 and 13.5 MHz) have been constructed. The voltages measured between sweeping plates can be as high as 3000 and 900 V, respectively, which are more than enough for light ions. Both amplitude and phase of the supplies can be adjusted.

When $V_1 = 1500$ V, $QV_0 = 30$ keV for different atomic number $A=1$, 2, 3 ($V_g = 0.28$, 0.43, 0.60 kV, respectively) particles, the highest BUE calculated of this harmonic chopper is 0.759.

B. The Beam Tests

In the bench beam tests, H_2^+ beam was used to simulate the deuteron ions. The beam was tuned first and accordingly the chopping slit was set to $s = a = 0.265$ cm. The beam bursts were measured by a 50 Ω coaxial fast target and were displayed on a TEK 485 oscilloscope.

The effect of the 9th harmonic superimposed onto the fundamental sweeping voltage is shown in Fig. 4. It can be seen that the harmonic component really changes the beam pulse shape from triangular into trapezoidal type and thus enhances the beam utilization efficiency. This result agrees well with theory and design.

 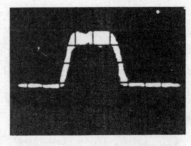

Fig. 4 The waveform of emerging beam burst of sine-wave and harmonic chopping, upper: sine-wave chopping $V_1 = 648$ V, $I_0 = 40$ μA, bottom: harmonic chopping $V_1 = 648$ V, $V_9 = 149$ V, $I_1 = 56$ μA ($V_0 = 21$ kV, dc beam at target $i_0 = 46$ μA, transverse 20 ns/div, and vertical 50 mV/div)

The measured (V_9/V_1) is shown in Table 1, where τ is the full time width at one tenth maximum of the beam burst. I is the output current at the target. The increase of beam efficiency of the harmonic chopper compared with the single sine-wave chopper is given by

$$\Delta\eta/\eta_0 = (I_1 - I'_0)/I'_0;\ I'_0 = (\tau_1/\tau_0)I_0, \tag{11}$$

where the subscript "0" denotes sine-wave chopper, whereas "1" denotes harmonic chopper and $\Delta\eta = \eta_1 - \eta_0$. The maximum $\Delta\eta/\eta_0$ measured is 47%, which agrees with the theoretical maximum.

Table 1 The results of beam tests of the harmonic chopper
($V_0 = 18$ kV dc beam current at target $I_d = 45$ μA)

Number	1	2	3	4
V_9/V	112	149	197	316
V_1/V	844	730	798	821
τ_0/ns	57.5	65	65	60
τ_1/ns	63.8	66.2	66.2	68.8
$I_0/\mu A$	18.8	21.6	21.2	20.0
$I_1/\mu A$	26.4	31.6	31.8	30.0
$\Delta\eta/\eta_0$	0.32	0.46	0.47	0.25

III. Conclusion

The new harmonic chopper has been developed and described. Both theoretical analysis and experimental results show that it has the merit of high-beam efficiency, good beam quality, and simple construction. In light of the preliminary success, it is being developed further so that it can be a new practical and widely used chopper.

References

[1] Anderson et al., Nucl. Instrum. Methods 20. (1964).
[2] C. F. Chen, Proceedings of the National Particle Accelerator Conference. May 1979, Chengdu China.
[3] Livingood. Part. Accel. 7, 243(1976).
[4] Ben-Zvi. Part. Accel. 10, 31(1979).

A DOUBLE-DRIFT HARMONIC BUNCHER FOR 4.5 MV VAN DE GRAAFF ACCELERATOR[①]

Introduction

RF beam bunching system is widely used in the time-of-flight experiment for neutrons and heavy ions as well as in the injection of linear accelerator or cyclotron. A good buncher should be able to bunch the beam into a narrow phase width ($3° \sim 5°$)[1] with high beam utilization efficiency (BUE), consequently with high beam intensity. Since 1970's, the double-drift harmonic buncher has been used on Tandem and got about twice BUE than that of the single drift buncher.[1~5] As the space and high voltage are very much limited in the terminal of the single ended Van de Graaff, the double-drift harmonic bunching system has not yet been fully adopted. However the beam intensity of the single ended Van de Graaff is higher than that of the Tandem, it suits many experiments which needs high beam intensity, such as neutron data measurement. Hence it is highly desirable to develop the double-drift harmonic buncher for single ended Van de Graaff. The design layout and the bench test of the buncher for the 4.5 MV Van de Graaff are discussed in the following text.

The Principle and Design of the Double-Drift Harmonic Buncher

For uniform D.C. beam the ideal modulation voltage with 100% BUE

---

① Coauthers: Jiang Xiao-Ping, Qian Zu-Bao. Reprinted from the Proc. 3rd Japan-China Joint Symposium on Accelerators for Nuclear Science & their Applications, Riken, Siatama, Japan, 1987: p. 212.

should be as follow[6]

$$V(t) = V_0((1 - t/\tau)^{-2} - 1) \qquad (1)$$

where V_0 is the source extraction voltage, $\tau = L/v_0$ is the drift time and L is the drift distance from the buncher to target, v_0 initial velocity of particles. As seen from Fig. 1 the ideal modulation waveform is a unsymmetrical curve to the origin of coordinates. Usually, the repetition rate of the modulation $T \ll \tau$, so Eq. (1) can be approximate to $V(t) = (2V_0/\tau)t$ and the waveform approximate to a straight line. However in the terminal of the single ended Van de Graaff $T \sim \tau$. It means the buncher has a short focus and the linear approximation is not suitable. In the double-drift harmonic buncher, the drifted fundamental voltage waveform superposes onto the 2nd harmonic voltage to approch the ideal waveform.

The double-drift harmonic buncher, as shown in Fig. 2, consists of the fundamental buncher 1 and second harmonic buncher 2. Generally each buncher can be equivalent to a RF gap located at its center.[7] The phase different between them is $\pi + \Delta\varphi$ ($\Delta\varphi \ll 1$).

Consider that a bunch of particles with initial relative energy deviation ε_0 ($\varepsilon_0 = \Delta E/E_0$) passing through the bunching system. Let the particle which passes through the first gap at zero phase as a reference. For ions injected at phase φ_1, the energy deviation ε and phase φ at the target are given as following

$$\varepsilon = \varepsilon_0 + \varepsilon_1 \sin\varphi_1 + \varepsilon_2 \sin(2\varphi_2 + \Delta\varphi) \qquad (2)$$

$$\varphi = \varphi_2 + \frac{\omega(L_T - L_0)}{v_0}((1 + \varepsilon)^{-1/2} - 1) \qquad (3)$$

where

$$\varphi_2 = \varphi_1 + \frac{\omega\tau_0}{v_0}((1 + \varepsilon_0 + \varepsilon_1 \sin\varphi_1)^{-1/2} - 1) \qquad (4)$$

$$\varphi_1 = \omega t_1, \Delta\varphi = \Delta\phi + \frac{2\omega L_0}{v_0} + \pi$$

$\varepsilon_1 \approx \dfrac{2QV_{1m}}{E_0}\sin\dfrac{\omega d_1}{2v_0}$, $\varepsilon_2 = \dfrac{2QV_{2m}}{E_0}\sin\dfrac{2\omega d_2}{2v_0}$ are energy modulation coefficient at the centers of buncher, d_1, d_2 are the central electrode lengths V_{1m}, V_{2m} are amplitudes of the fundamental and harmonic voltage. ω is fundamental frequency. Q is the charge of injected particles. E_0 and v_0 are initial energy and velocity of injected partlcles. It is seen from Eq. (2), (3) and (4) that the effective waveform of the modulation seen by the particles is a function of L_0, $\Delta\varphi$, ε_1, and ε_2. In the design, the parameters of the buncher were optimized by means of a special computer programe to make the voltage modulation waveform approching the ideal waveform as far as possible, so that the BUE can be maximized. Fig. 3 showes a few computed curves between BUE and the working parameters of the double-drift harmonic buncher. It appears that the curves vary in a similar way as in the case of small signal. Every curve has a peak value, but the peak height drops down with the decreasing phase acceptance.

When initial energy dispersion of injected beam $\varepsilon_0 \neq 0$ the position of the peak will move and the BUE will reduce considerably. It is significant to optimize L_0 and to reduce initial encrgy dispersion.

The pulse beam energy spectrum and phase diagram computed for uniform distributed beam in phase space in the case of large signal are different from the case of the small signal and are shown in Fig. 4. The energy spectrum is unsymmetrical and the center of the phase waveform is not at $\varphi = 0$ but at $\varphi = \varphi_c$. In the case of the short focus, the drift time τ is closed to bunching period T, the ideal modulation voltage waveform is not a symmetrical straight line, so it could not obtained by superimposing two harmonics with phase difference of π (e. g. $\Delta\varphi = 0$). Normally, when the modulation voltage is rather small, ε_0, ε_1, $\varepsilon_2 \ll 1$. Eq. (3), (4) are as follow

$$\varphi = \varphi_2 = \left(1 - \frac{L_1}{L_T}\right)\left(\frac{\varepsilon_0}{\varepsilon_{11}} + \frac{\varepsilon_1}{\varepsilon_{11}}\sin\varphi_1 - \frac{\varepsilon_2}{\varepsilon_{11}}\sin(2\varphi_2 + \Delta\varphi)\right) \quad (5)$$

$$\varphi_2 = \varphi_1 - \frac{L_1}{L_T}\left(\frac{\varepsilon_0}{\varepsilon_{11}} + \frac{\varepsilon_1}{\varepsilon_{11}}\sin\varphi_1\right) \quad (6)$$

where ε_{11} is a modulation coefficiency of buncher 1 when its linear focus equals to the distance from buncher 1 to target (e. g. $L_T = F_{z1}$) the phase diagram is symmetrical to origin.

The computed $\Delta\varphi$ and φ_c are given in table 1 when $\varepsilon_0 = 6.36\%$, $\varepsilon_{11} =$

0.12, burst width=2 ns. It indicated that $\Delta\phi$ and φ_k tend to be zero gradually along with the decrease of modulation voltage.

Table 1

L_1/L_T	0.11	0.12	0.13	0.14	0.15	0.16	0.17	0.18
$\Delta\varphi$	0.18	0.17	0.17	0.17	0.16	0.15	0.15	0.14
$\varphi_k^{(0)}$	5.4	5.6	5.7	5.7	6.0	6.2	6.4	6.5

Fig. 5 is a set of pulse beam waveforms of the double-drift harmonic buncher calculated by means of reference[8] for pulse length and energy deviation at injection. It indicates that when the pulse length≤66 ns clear beam pulse can be obtained without background and the BUE is about 60%, the pulsed beam intensity will increase with the increasing of injected beam length but more and more background appears. In addition, the pulsed beam will be lower and wider with initial energy deviation increase, so the initial energy dispersion must be reduced as much as possible. For comparision the beam pulse waveform of the single buncher is given simultaneously.

The Layout and Experimental Studies

The double-drift harmonic buncher has been set up on the test bench with a layout, shown in Fig. 6, which is the same as in the terminal of 4.5 MV Van de Graaff. The beam extracted from an ion source goes through einzel lense, chopper, $E \times B$ analyser and then enters the double-drift buncher. Finaly, it reaches the ns pulse measurement system.

Since the terminal space of Van de Graaff and RF power are very much limited, 9 MHz is chosen as the fundamental frequency of the buncher. But the repetition rate reduces to 3 MHz by the chopper to meet the requirements of time-of-flight experiments. When the initial energy E_0 is 20~30 keV, accelerating voltage is 2.5~4.5 MV the effective drift distance (L_T) of a deuteron beam is about 460~500 mm, $\tau=295$ ns, $T/\tau=1/3$. As mentioned above, optimal parameter $L_1/L_T=0.14$ the distance L_1

between the centers of two bunchers is 64～71 mm. In order to get enough transit time factor,[7] the length of two central electrodes are 80 mm and 40 mm respectively. The parameters of the buncher are given in table 2. This is really a compacture. (Fig. 7)

Table 2   Parameters of Buncher

I. D. of electrode	25 mm
gap	2 mm
length of 9 MHz central electrode	80 mm
length of 18 MHz central electrode	40 mm
central ground electrode	10 mm
others	16 mm
distance between two centers	70 mm
effective drift distance	508 mm
L_1/L_T	0.14

The RF signal originates from a 1.5 MHz crystal master osillator and feeds through frequency doubler, amplifier booster to the chopper and buncher. Each RF source has a mechnical phase shifter, the phase can be adjusted smoothly in 0～360° for these to match each other. In the test H_2^+ was used to replace deuteron so as to limit the radiation. The pulse beam measurement system consists of the 50 Ω fast target and TEK 485 oscilloscope.

By operating the buncher and chopper simultaneously, the BUE of buncher can be obtained by means of measuring injected beam width. Along with the increase of chopping voltage, the tail of pulse will get less and less till a clean beam pulse without "tail" displayong on the oscilloscope, it means all the beam through the chopper has been compressed into the pulse. The BUE equals to the average intensity i divided by initial D.C. beam intensity i_0. As bunching frequency is 3 times chopping frequency, the BUE $F_M = 3i/i_0$. The results with different injection energy and phase width (FWHM) are shown in table 3. When pulse width is 2 ns, the BUE is about 60%. The results of single 9 MHz buncher is given in talbe 4, the typical BUE measured by same method is about 30%.

Typical beam pulse are shown in Fig. 8. The BUE of the double-drift harmonic buncher is twice that of the single buncher. It agrees well with the theory.

The limited rise time (1.1ns) of measuring system and error in observation lead to errors in the measurement of the beam pulse width.

Conclusion

In order to enhance the pulsed beam intensity, it is profitable to install the double-drift harmonic buncher in the terminal of single ended Van de Graaff so as to get 1~2ns pulsed beam. Both the results of theory and experiments indicated that the BUE is about 60%. This paper discussed the performance of double-drift bunching with short focus and high modulation voltage as well as effect of initial energy deviation, consequently, it extended the results of reference [1]. The experiment agrees well with the theory.

Table 3 The measured BUE of double-drift harmonic buncher

	$E_0=19$keV				$E_0=25$keV		
τ/ns	$i_0/\mu A$	$i/\mu A$	$F_M/\%$	τ/ns	$i_0/\mu A$	$i/\mu A$	$F_M/\%$
1.8	82	16	58.5	1.6	30	5.3	53
2.6	82	18	65.9	2.0	30.5	6.1	60
4.0	80	20	75	2.6	29	6.7	69

Table 4 The measured BUE of single 9MHz buncher

	$E_0=20$keV				$E_0=25.2$keV		
τ/ns	$i_0/\mu A$	$i/\mu A$	$F_M/\%$	τ/ns	$i_0/\mu A$	$i/\mu A$	$F_M/\%$
1.7	36.2	3.6	29.8	1.8	68	6.8	30
2.0	36	3.8	31.7	2.1	72	8.2	34.2
2.8	36	4.3	35.8	4.0	73	8.7	35.8
5.6	36	5.2	43.3				

Reference

[1] W. T. Milner. IEEE Trans. NS-26 (1979) 1445.
[2] J. M. Brennan, C. E. chen et al. Bunching system for the Stony Brook Tandem Linac Heavy Ion accelerator, IEEE Trans. NS-30, 4(1983) 2798.
[3] S. A. Wender IEEE Trans. NS-28(1981) 1465.
[4] V. Ratzinger et al. N. I. M. 205(1983) 381.
[5] C. Goldstein et al. N. I. M. 61(1968) 229.
[6] J. L. Xie. The group bunching theory of klystron.
[7] C. E. Chen. 1979 National Accelerator Conference Collected Works p75.
[8] C. E. Chen et al. IEEE Trans. NS-30 2(1983) 1254.

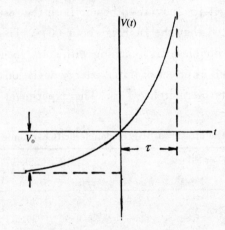

Fig. 1   The ideal modulation waveform

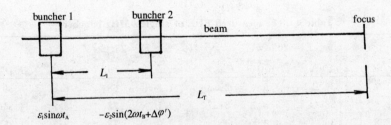

Fig. 2   The principle diagram of the double-drift harmonic bunching system

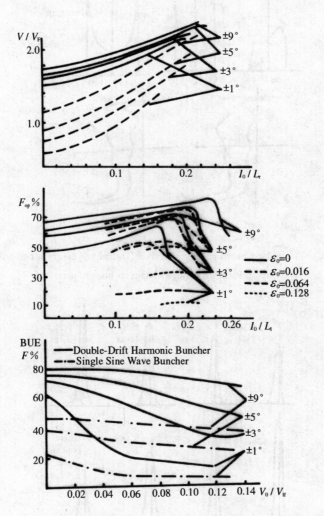

Fig. 3 The relative curves between BUE and the working parameters of the double-drift harmonic buncher

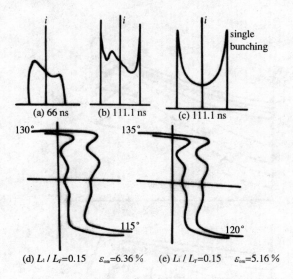

Fig. 4 The pulsed beam energy spectrum and phase diagram computed for uniform distributed beam in phase space

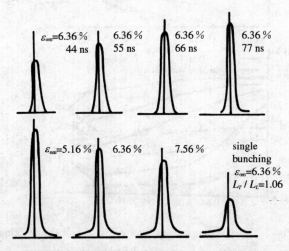

Fig. 5 The beam pulse waveform of the double-drift harmonic buncher

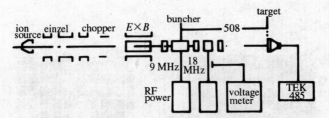

Fig. 6 The layout of the test bench of the double-drift harmonic buncher

1. flange I
2. ground electrode
3. link pole I
4. 9 MHz central electrode
5. central ground electrode
6. 18 MHz centyal electrode
7. link pole II
8. ceramic isolator
9. shield
10. flange II

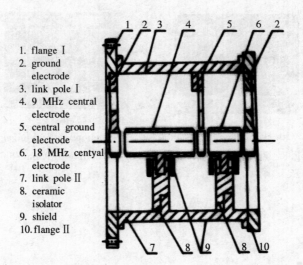

Fig. 7 The structure of the double-drift harmonic buncher

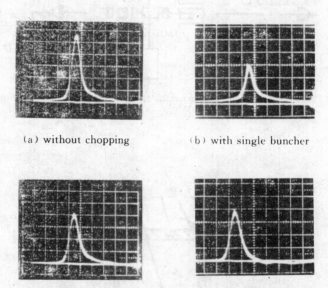

(a) without chopping   (b) with single buncher

(c) with chopping and gained   (d) with chopping and ungained

Fig. 8 The beam pulse waveform of the double-drift bunching

离子束 ns 脉冲测量用可变栅距同轴靶[①]

摘　要

本文讨论了影响同轴靶响应的诸因素以及在离子束 ns 脉冲测量用同轴靶的设计中的主要问题. 在此基础上制成了一个具有可变栅距结构的同轴靶. 该靶适用于速度在宽范围内变化的各种离子,曾成功地用于测量 $\beta=v/c$ 值低至 6.5×10^{-4} 的 Ar^+ 束脉冲. 该靶同轴结构的上升时间约为 0.2 ns,曾用其观测到半高宽约为 1 ns 的质子束脉冲.

关键词　可变栅距同轴靶,离子束 ns 脉冲,质子束脉冲.

一、引　言

同轴靶[1~4]在质子束流脉冲的测量中已得到了广泛应用. 为了克服位移电流效应和二次电子发射效应对同轴靶响应速度的影响,可以在靶前设置栅网[1]并加以适当的偏压[2]. 对于低 β 离子束流脉冲,位移电流效应的影响尤为严重. 本文下面的讨论表明,此时栅网参数的恰当选择对于正确的实时测量是非常重要的.

本同轴靶是为测试 4.5 MV 静电加速器[5]束流脉冲化装置的性能而研制的. 该束流脉冲化装置由头部与尾部的聚束器和斩束器组成,可产生多种轻离子与重离子束流脉冲. 离子的速度有很宽的变动范围. 满能量质子的 β 值接近 10^{-1}. 而在头部,重离子的 β 值可低达 10^{-4}. 本文基于对离子在栅靶间渡越效应的分析,提出了同轴靶的可变栅距结构,使检测不同种类、不同能量离子群聚脉冲的不同要求得以兼顾,从而提高了同轴靶的通用性.

同轴靶的基本结构示于图 1. 这是一个端入射自然冷却同轴靶. 用光刻工艺制作的钨栅网固定在可移动栅筒上,通过蜗轮蜗杆可以从外部

[①]　合作者:郭之虞,张征芳,赵夔. 摘自《原子能科学技术》,Vol. 20, No. 6 (1986) 710~715.

移动栅筒来调节栅距.

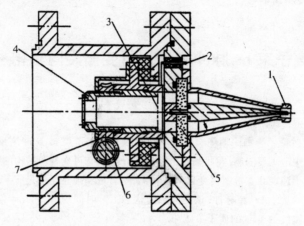

图 1 可变栅距同轴靶结构示意图
1——同轴插座； 2——V_g； 3——栅网； 4——栅筒；
5——绝缘子； 6——蜗杆； 7——蜗轮

二、栅网设计

影响同轴靶响应的因素除了靶体本身的同轴结构外,还有栅外位移电流效应、栅内位移电流效应、二次电子发射效应和靶端失配电容.这些因素都与栅网有关.

1. 栅外位移电流屏蔽失效率

我们采用平行线栅网屏蔽栅外束流运动引起的位移电流.屏蔽效果可用屏蔽失效率[6]

$$\sigma = \left(\frac{\mathrm{d}E_{\mathrm{in}}}{\mathrm{d}E_{\mathrm{out}}}\right)_{V_g} = \mathrm{const.}$$

来衡量.其中 E_{in} 与 E_{out} 分别为栅内外的电场强度,V_g 为栅偏压.σ 的大小取决于栅网几何参数 s/d 和 r/d,如图2所示.这里 s 为栅距,d 为栅线中心距,r 为栅线半径.选用较大的 s/d 和 r/d 可使 σ 下降,但大的 r/d 同时使束流损失率上升.实验表明,为使栅外位移电流对脉冲波形无明显影响,一般 σ 取 10%～15% 即可.

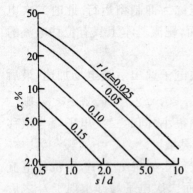
图 2 σ 与栅网几何参数的关系

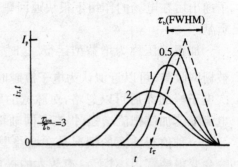
图 3 三角形束脉冲的位移电流波形

2. 栅内位移电流引起的脉冲波形畸变

设有一电量为 Q 的无限薄束流层以速度 v 在栅靶之间渡越,其渡越时间 $\tau_{ion}=s/v$. 假定栅线足够密,栅网可视为一导电平面,且 s 足够小,边缘效应可被忽略. 此时靶平面上的感应电荷量 $q_T=-Qvt/s$, 其中 t 为薄层束过栅以后的飞行时间. 故靶流在渡越时间内恒为

$$i_T = -\frac{dq_T}{dt} = \frac{Q}{\tau_{ion}}.$$

$Q=1$ 的薄层束在数学上对应于 δ 函数. 由上式可知,栅靶系统对于 δ 函数的脉冲响应 $h(t)$ 是一个时宽为 τ_{ion}, 高度为 $1/\tau_{ion}$ 的方波. 于是对于任意的输入束流脉冲 $i_b(t)$, 栅靶系统的响应可由下列卷积求得:

$$i_T(t) = \int i_b(t) h(t-\tau) d\tau.$$

当 $i_b(t)$ 为三角形脉冲时,我们可以得到如图 3 所示的位移电流波形,其中 t_T 为束流脉冲前峰到达靶表面的时刻. 由图 3 可知,位移电流峰的形状强烈地依赖于离子渡越时间 τ_{ion} 与束流脉冲时宽(FWHM) τ_b 的比值. 当 $\tau_{ion}/\tau_b \ll 1$ 时,靶流的畸变很小,位移电流峰的半高宽基本上可反映入射束脉冲的半高宽. 但当 $\tau_{ion}/\tau_b > 1$ 时,位移电流峰的半高宽成为 τ_{ion}. 故对于低 β 束的窄脉冲而言,栅内位移电流引起的波形畸变是个重要问题.

3. 抑制型运行与增益型运动

在栅网上加以足够高的负或正偏压可以有效地抑制二次电子发射

效应造成的脉冲波形拖尾现象. 加负偏压称为抑制型运行. 此时二次电子逸出后在电场的作用下很快返回靶表面, 靶流基本上只有位移电流的贡献.

加正偏压称为增益型运行. 这时二次电子逸出后被电场加速, 最后被栅网收集. 可以近似认为电子渡越时间 $\tau_e \leqslant s\sqrt{2m_e/eV_g}$ 若 $\tau_e \ll \tau_b$, 则二次电子峰的滞后可以忽略, 对脉宽的影响很小. 同时二次电子峰与位移电流峰迭加使靶流增强, 提高了同轴靶的灵敏度.

一种值得注意的情况是在增益型运行下, 如果 $\tau_{ion} > \tau_b$, 则位移电流峰会变得矮宽, 且对于 t_T 有较大的提前, 而二次电子峰却出现于 t_T 之后成为脉冲波形的主峰. 二者叠加的结果会出现"位移电流平台", 即在主峰上升之前脉冲基部有一低平台. 此时只要 V_g 足够高使 τ_e 足够小, 主峰的宽度可反映入射束脉冲的宽度.

4. 靶端失配电容的影响

设置栅网以后在同轴靶端部引入一个失配电容 C_g, 其大小取决于 b/a 与 s/b[7], 其中 a、b 分别为靶端内外导体直径. 此失配电容产生一个附加的上升时间 $\tau_g = Z_0 C_g$, 其中 Z_0 为同轴线特性阻抗. 靶直径较大时, 此电容也较大.

5. 栅网参数的选择

栅网参数中最重要的是 s, 以上所讨论的 σ、τ_{ion}、τ_e、τ_g 均与其有关. 图 4 给出了 s 对 τ_{ion}、τ_e 和 τ_g 的影响. 本同轴靶内导体端面直径 $a=15$ mm. 由图 4 可知在 s 小至 0.5 mm 时, τ_g 仅 0.2 ns. 故对本同轴靶, τ_g 不是响应速度的主要限制. 实际上 s 的选取应当主要考虑其对 τ_{ion} 的影响. 在离子种类、入射能量 W_0、束流脉宽等入射条件给定以后, 或取较小的 s 使 $\tau_{ion} \ll \tau_b$, 或取较大的 s 使 $\tau_{ion} \gg \tau_b$. 必须避免 $\tau_{ion} \approx \tau_b$ 的情况, 那将使同轴靶的响应显著变坏. 在 s 选定后, 可选择 V_g 使 $\tau_e \ll \tau_b$, 选择 d 与 r 使 σ 取合适的值且束流损失率尽可能小. 表 1 中给出了选择栅网参数的几个具体例子.

由于 4.5 MV 静电加速器使用的离子种类很多, 而且对于头部聚束器而言各种离子又有其最佳入射能量[8], 这就造成了离子 β 值的宽范围分布, 最佳栅距也彼此不同. 因此本同轴靶采用了可变栅距结构, 这对于

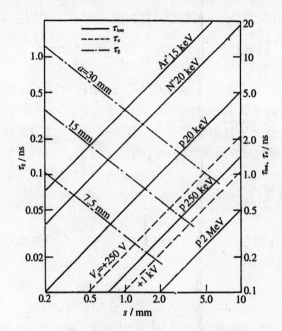

图 4 栅距 s 对 τ_{ion}、τ_e、τ_g 的影响

表 1 选择栅网参数的例子

入射条件			栅网参数				栅网性能			
离子种类	W_0/keV	τ_b/ns	s/mm	d/mm	r/mm	V_g/V	τ_{ion}/ns	τ_e/ns	τ_g/ns	σ/%
P	250	1~2	1.0	1.0	0.06	+300	0.15	0.2	0.1	14
N^+	20	2~3	5.0	1.0	0.06	+1000	10	0.5	0.03	3
Ar^+	15	3~4	3.0	1.0	0.06	+1000	11	0.3	0.045	5

各种不同 β 值离子束脉冲的正确检测显然是有益的.

三、同轴结构设计

1. 介质绝缘子

根据传输线理论,介质绝缘子可等效为一个 π 节低通滤波器,而其界面处内外导体阶跃所产生的不连续则可等效为并联在该处的分路电容 C_d,其大小取决于绝缘子的几何尺寸与材料. 为补偿 C_d 引起的失配,

可取绝缘子段的阻抗 $Z_{01}=KZ_0(K>1)$. 此时绝缘子的视在阻抗为[9]

$$Z_I = \frac{Z_{01}}{\sqrt{1+2\omega C_d Z_{01}\cot\beta l-(\omega C_d Z_{01})^2}},$$

其中 β 为介质的相位常数，l 为绝缘子厚度.适当选取 K、l 和介质材料可使 Z_I 在足够大的带宽内接近于 Z_0.

本同轴靶绝缘子材料使用陶瓷，因为其导热系数大，有利于在自然冷却条件下承受较大的入射束功率.绝缘子设置在靶的前部靠近端面是为了尽量缩短传热通道，进一步改善靶的热性能.据估算，靶的端面温升不大于 2℃/W.

2. 锥形过渡节

当内外导体表面为一对共顶点圆锥时，空气介质锥形传输线的特性阻抗为[10]

$$Z_c = \frac{1}{2\pi}\sqrt{\frac{\mu_0}{\varepsilon_0}}\ln\left[\frac{\tan(\theta_0/2)}{\tan(\theta_i/2)}\right],$$

其中 θ_i 与 θ_0 分别为内外锥的半顶角.本同轴靶 $\theta_i=5°$，此时 $Z_c=49.5\ \Omega$.

3. 同轴结构上升时间的测量

同轴结构的上升时间可在时域反射计上用开路回波法进行测量.先测出不接靶时开路回波的上升时间 t_{r1}，然后再测出接入靶以后开路回波的上升时间 t_{r2}. 此时信号在靶中传输两次，若开路完全则有

$$t_r = \sqrt{(t_{r2}^2-t_{r1}^2)/2}.$$

所使用的时域反射计可提供上升时间为 40 ps 的阶跃信号.实际测量的结果，本同轴靶的同轴结构的上升时间 $t_r \leqslant 0.2$ ns.

四、实验结果

为检验同轴靶的性能，先后在螺旋波导腔试验线和重离子源试验台上用该靶进行了质子和重离子聚束脉冲波形的实时测量.图 5 至图 8 为其中部分波形照片.图 5 和图 6 分别为在抑制型运行下与增益型运行下的 250 keV 质子聚束脉冲波形.由于二次电子的放大作用，增益型运行的峰高约为抑制型的 2.5 倍.但因位移电流峰超前于 t_T，而二次电子峰

滞后于 t_i,二者叠加的结果使在增益型运行下脉冲波形的半高宽比在抑制型运行下有所增加.本例中抑制型运行的半高宽为 1.2 ns,而增益型运行的半高宽则为 1.5 ns.可见在 $\tau_{ion} < \tau_b$ 的情况下,一般应采用抑制型运行,除非有必要提高同轴靶的灵敏度.图 7 为 20 keV 的 N^+ 聚束脉冲波形,半高宽约为 2.2 ns.图 8 为 15 keV 的 Ar^+ 聚束脉冲波形,半高宽约为 3.7 ns.二者均为大栅距增益型运行,在主峰前可以看到位移电流平台,相应的离子渡越时间分别为 5.7 ns 和 11 ns.对于 $\tau_{ion} \gg \tau_b$ 的情况,采用增益型运行是必要的.照片中脉冲尾部的起伏脉动是检测系统的高

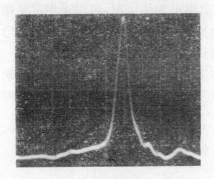

2 ns/div

图 5 质子聚束脉冲波形
$W_0 = 250$ keV, $V_g = -250$ V

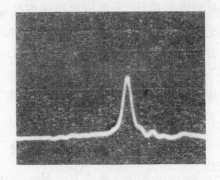

2 ns/div

图 6 质子聚束脉冲波形
$W_0 = 250$ keV, $V_g = +250$ V

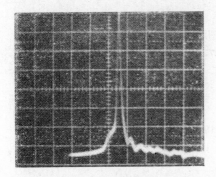

10 ns/div

图 7 N^+ 聚束脉冲波形
$W_0 = 20$ keV, $V_g = +250$ V

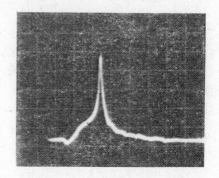

10 ns/div

图 8 Ar^+ 聚束脉冲波形
$W_0 = 15$ keV, $V_g = +250$ V

频截止特性所造成的.

实验中还进行了聚束双峰的观测,双峰时宽的实验测量值与根据纵向 $\varepsilon\text{-}\phi$ 相空间二维聚束理论计算所得的理论值符合良好[11]. 这也说明本同轴靶的性能是令人满意的.

五、结 论

对用于测量低 β 离子束 ns 脉冲的同轴靶而言,设计中的关键问题是栅网的设计. 为适应离子速度变动范围很大的情况,可以采用可变栅距结构. 在使用时,或者采用 $\tau_{ion} \ll \tau_b$ 的抑制型运行,或者采用 $\tau_{ion} \gg \tau_b$ 的增益型运行,应当避免采用 $\tau_{ion} \approx \tau_b$ 的运行状态.

作者衷心地感谢虞福春教授对本工作所给予的指导,感谢北京大学技术物理系加速器教研室和北京师范大学低核所的许多同志在同轴靶实验工作中所给予的协助.

参 考 文 献

[1] Mobley, R. C., *Rev. Sci. Instr.*, **34**, 256(1963).
[2] Moak C. D. et al., *Rev. Sci. Instr.*, **35** 672(1964).
[3] Anderson, J. H., *AWRE*-O-53/66 (1966).
[4] Evans, L. R., Warner, D. J., Proc. of Proton Linear Accelerators Conf. 1972, p. 349.
[5] 北京大学静电加速器小组,1979 年全国加速器技术交流会论文选集,p. 43.
[6] Bunemann, O. et al., *Can. J. Res.*, **27**, sec. A, 191(1949).
[7] Green, H. E., IEEE. Trans. *MTT*-13, (5), 676(1965).
[8] 赵夔,北京大学硕士论文(1981).
[9] Cornes, R. W., *P IRE*, **37**, 94(1949).
[10] Harris, I. A., *P IEE*, **103**, Pt. C 1(1956).
[11] 陈佳洱等,IEEE, Trans. *NS*-30, (2), 1524(1983).

A NANOSECOND COAXIAL TARGET WITH ADJUSTABLE GRID FOR PULSED PROTON AND HEAVY ION BEAMS

Abstract

The waveform of the beam pulses measured by a coaxial target is carefully studied versus the ratio of the ion transit time to the duration of the beam burst. Other factors involved in the design of a fast response target are also discussed. A coaxial target with adjustable grid-target gap is then designed and constructed to fulfil the pulse measurements for various ions. The rise time of the target is determined to be 0.2 ns. Beam bursts of H^+, N^+, N_2^+, and Ar^+ with velocities ranging from $V/C = 6.5 \times 10^{-4}$ to 2.5×10^{-2} are measured successfully with the target. The results are well consistent with the design values.

Key words　Nanosecond coaxial target, Adjustable grid-target gap, Fast response target.

束流相图仪的标定[①]

摘　要

　　测量束流相图是束流输运线设计最佳化和束流控制的重要环节.因此,多年来人们广泛地把它用于离子源试验台和束流输运线上,用以测定和诊断束流的品质[1~4].几年前,我们根据缝-针法研制了一种束流相图仪[5],并在螺旋波导增能器的束流线上进行了一系列的实验.为了给出用相图仪进行测量的准确度,我们首先用相空间的概念对测量过程进行了分析,并给出了不同束特性情况下的数据处理关系式;其次,研究了影响测量精度的多种误差因素,采取了一定的改善措施;最后,在 400 keV 高压倍加器输运线上,用设置标准接受相图的方法,对相图进行了实验标定,初步得出了测量相图斜率和相角宽度的误差限.

一、测量过程的相空间描述

　　用缝-针法测量束流相图的装置示意于图 1. 探针相对缝宽为 a 的光栅在实空间里作扫描平移时,即表现为缝针系统接受相平行四边形在 (X, X') 相空间中沿 X' 轴的连续移动[6],见图 2(a). 假定在相域中密度为均匀分布,由图 2(b)所示的简单几何关系可知,探针上的收集电流 i 随 X 变化的扫描波形为图 2(c). 图 2(b)和 2(c)中的序号一一对应.扫描波形的 12 段和 34 段按抛物线规律变化. 5 到 8 正好是 1 到 4 的反过程.实际测量中由于密度分布的不均匀性,一般观察不到图 2(c)中 4~5 的平顶,而仅出现一个突起的峰.

　　对于发散束和相图边界斜率 $|F| < 1/l$(单位为 mrad/mm)的会聚束,扫描波形起止点 1、8 所对应的相坐标应为图 2(d)中的 C、A 点:

[①]　合作者:严声清,沈定予,赵渭江,傅惠清.摘自《核技术》,1(1983) p.39~42.

C 点相坐标：

$[r_a, (r'_{b1} - r_a)/l]$ 即 $[r_a, (r_{X1}/k - r_a - b/2)/l]$,

A 点相坐标：

$[r'_a, (r_{b8} - r'_a)/l]$ 即 $[r'_a, (r_{X8}/k - r'_a + b/2)/l]$, $\quad$ (1)

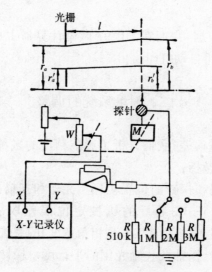

图 1　测量装置示意图

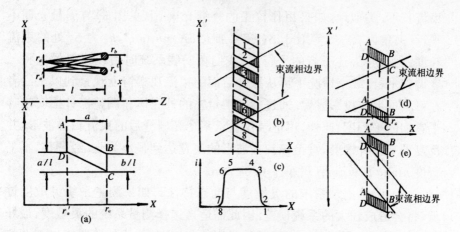

图 2　测量过程的相空间描述

（a、b 分别表示光栅缝宽和探针直径）

对 $|F|>1/l$ 的会聚束(见图 2(e))则为 D、B 点,其坐标分别为:
$$D[r_a', (r_{X1/K} - r_a' - b/2)/l],$$
$$B[r_a, (r_{X8/K} - r_a + b/2)/l], \qquad (2)$$

(1)、(2)式中的 K 是记录仪 X 轴的放大倍数,r_X 是探针的几何位置在记录仪上的指示值.

对于测得的数据,我们在 HP-97 微型计算器上用专用程序进行了处理,相边界坐标可自动打印给出.

二、测量系统的调整

为了减小测量误差,我们分析了影响精度的多种因素,并对测量系统作了调整,其中包括:

(1) 确定光栅和探针的位置坐标　光栅和扫描探针部件用读数显微镜预先作了校正;安装后的机械定位误差取决于加工公差(约 ± 0.1 mm). 它将造成相角 X'(不是相角宽度)的系统误差.

(2) 调整光栅、探针及狭缝光阑的平行度　探针相对光栅缝倾斜 α 角时,将造成束流角宽度增大 $\delta(\Delta\theta) \approx \dfrac{h}{l}\tan\alpha$,$h$ 为探针处通过光栅缝的束流长度. 类似地,如果用作校正的狭缝光阑 S_1(见图 3)与测量光栅不平行(夹角为 β),则等效于 S_1 缝宽加大 $\delta X \approx h_1\tan\beta$,$h_1$ 为 S_1 处的束流长度. 为改善上述三者的平行度,我们在安装系统时用经纬仪尽量将三者调整平行;载束情况下可从探针后面的束流观察窗上观察到以上三者在石英屏上的投影成像. 通过旋转测量室的外壳体和调整探针摆套改善了它们的平行度;此外,用记录仪显示某个光栅缝后的束流扫描波形,并反复调节探针摆套,直至该扫描波形的底宽达到最小为止. 经调整,不平行度 $h\tan\alpha$ 和 $h_1\tan\beta$ 可小于 0.04 mm.

(3) 调整记录系统的灵敏度与动态特性　如果测量系统灵敏度过低,将会造成较大的系统误差. 因此需适当选择测量系统的灵敏度,最好使扫描波形最高幅度达到记录仪量程的 2/3 左右. 此外还要求仔细调整记录仪的阻尼和增益,使仪器处于最佳动态特性. 理想情况下波形起止点附近应为抛物线段,实际上由于记录仪存在启动灵敏阈,记录的波形总是起点尖锐,终点略呈圆弧形.

(4) 其他 测定了探针传动螺杆、齿轮和电位器系统的非线性度,约为 0.5% 左右,它将会带来一定的非线性误差.

三、相图仪的标定

1. 方法

图 3 中的光阑缝 S_1 和 S_2 构成了一个束流接受系统,通过这一系统

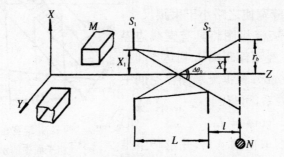

S_1——缝宽可调光阑 S_2——相图仪取样光栅 N——扫描探针 M——导向磁铁

图 3 标定装置示意图

的最大束角宽度为:

$$\Delta\theta_0 = [2(X_1 + X_2)]/l. \tag{3}$$

把这里的 $\Delta\theta_0$ 作为标准值,并用以与测量值 $\Delta\theta$ 比较,就可实现测量标定. 系统地改变设置量 X_1 和 X_2,经过统计平均可得到测量相角宽度的误差值:$(\Delta\theta - \Delta\theta_0) \pm \delta\theta$;其中 $\Delta\theta - \Delta\theta_0$ 为系统误差,$\pm\delta\theta$ 为统计误差. 相角宽度的实测值 $\Delta\theta$ 是通过测量束边界与 Z 轴的距离 r_b 得到的,当束流充满 S_1-S_2 接受系统时,这一距离为

$$r_b = (1 + l/L)X_2 + X_1 l/L. \tag{4}$$

我们可以采用以下两种方法来标定相图仪,一是使束流通过光栅 S_2 上的某一栅缝,然后依次改变 S_1 的缝宽,测出相应的 r_b;二是保持 S_1 不变而用调节导向器磁场的办法使束流依次通过 S_2 的各缝,测出相应的 r_b. 图 4 和图 5 是这两种方法的相空间表示法,图 5 中的阴影部分是各 S_1-S_2 所构成的接受相图. 这两种方法分别改变 X_1 和 X_2,可得到直线 r_b—

X_1 和 r_b-X_2 的斜率,分别为

$$dr_b/dX_1 = l/L,$$
$$dr_b/dX_2 = 1 + l/L, \quad (5)$$

其中 l 和 L 为预先设置值.

由此可见,上述方法不仅可以标定相图角宽度,还可通过线性拟合与理论值比较,检验测量的可信度.

如果使束流腰位接近 S_1(见图 3),并调节 S_1 的缝宽使之略小于束腰尺寸,则通过 S_1 的束流相图将具有直线边界(见图 6(a)). 经漂浮距离 L 到达 S_2 时,相图变为图 6(b). 根据无场漂浮空间中的线性变换关系,可得到:

$$\cot\phi = L. \quad (6)$$

测出表示相图取向的 ϕ 角,并进一步推算漂浮空间腰位,与设置的 L 值比较,便可确定腰位的测量误差.

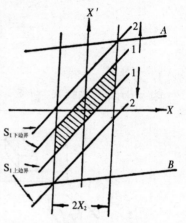

图 4 调节 S_1 缝宽在(光栅 S_2 处)相平面中的示意图
(A、B 是束流相边界)

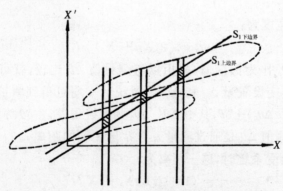

图 5 各 S_1-S_2 所构成的接受相图
(虚线部分表示束流边界;调节导向磁场,束流相域随之移动)

2. 实验结果

(1) 线性度 实验中设置距离 L 和 l 分别为 420 mm 和 160 mm. 直线斜率的相应计算值和实验值均列于表 1,图 7 和图 8 分别是 $r_b \sim x_1$ 和

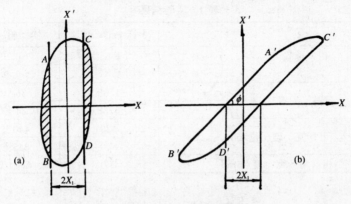

图 6 相图取向标定的相空间表示

$r_b \sim x_2$ 典型的实验曲线. 可见, 测量结果具有较高的可信度.

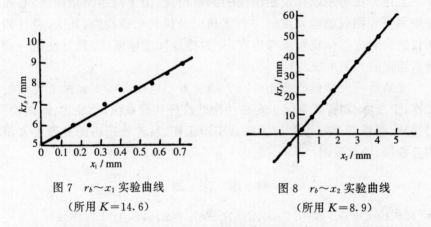

图 7 $r_b \sim x_1$ 实验曲线
（所用 $K=14.6$）

图 8 $r_b \sim x_2$ 实验曲线
（所用 $K=8.9$）

(2) 相图取向标定 把 $r_b \sim x_2$ 曲线变换成 $X' \sim X$ 曲线, 利用(6)式计算了束流光腰的算术平均值及标准偏差, 结果为 440 ± 11(mm).

在图 6 所述方法中设置 $L=840$ mm, 标定结果为 845 ± 52 (mm). 上述结果表明, 由所测相图的取向推算漂浮空间腰位的相对误差将小于 10%.

(3) 相角宽度标定 实验结果系统误差 $\Delta\theta - \Delta\theta_0 = +0.35$ mrad. 多次测量中标准偏差的最大值 $\sigma_{max} = \pm 0.3$ mrad, 此即统计误差的误差限 $\pm\delta\theta$ 值. 由于倍加器束流条件的限制, 标定实验所用的束角宽度相当窄, 大致位于 $0.6 \sim 2.6$ mrad 范围内. 以上实验结果均列于表.

表 1 实验结果

			标 准 值	实 验 值
线性度	相关系数[6] r	$r_b \sim x_2$	1	>0.9999
		$r_b \sim x_1$	1	>0.90
	斜 率	$r_b \sim x_2$	1.38	1.36 ± 0.01
		$r_b \sim x_1$	0.38	0.34 ± 0.05
束 流 腰 位 /mm			420	440 ± 11
			840	845 ± 52
相 角 宽 度 /mrad			$\Delta\theta_0$	$\Delta\theta_0 \pm 0.3$

四、结 束 语

上述方法可用以标定相图仪的测量精度. 为了减小测量误差, 必须仔细调整相图仪缝丝系统的平行度和记录仪的动态特性; 其次, 设计相图仪的同时应把标定要求考虑在内. 为提高标定精度, S_1 最好使用一套缝宽固定的光阑系统.

北京师范大学低能所 400 keV 倍加器小组配合本实验做了大量的工作, 于茂林、刘智辉等同志安装并帮助改进实验系统, 刘经之、何佩伦、吕建钦、张桂筠等同志参加了部分实验工作, 吕建钦还协助处理了大量实验数据, 特此致谢.

参 考 文 献

[1] R. S. Lord et al., *IEEE Trans. NNcl. Sci.*, NS-14, 3, 1151 (1967).
[2] W. Kuhlmann et al., *Nucl. Instr. Meth.*, 80, 89 (1970).
[3] Robert W. Goodwin et al., Proc. 1970 proton linear accelerator conf., Batavia, Vol. 1, 1970, p. 107.
[4] J. H. Billen, *Rev. Sci. Instr.*, 46, 33 (1975).
[5] 束流相图测量室的设计及其使用, 北京大学技术物理系资料, 1976 年.
[6] 数学手册, 人民教育出版社出版, 1979 年, 第 836 页.

HIGH-POWER MULTI-ELECTRON-BEAM CHERENKOV FREE-ELECTRON LASERS AT mm WAVELENGTHS[①]

Abstract

A multi-electron-beam Cherenkov free-electron laser can use a multi-dielectric waveguide to generate high-power coherent radiation at mm wavelength. Presented is a mode analysis of the device, allowing a detailed description of the dispersion relations and field distributions of the waveguide for the important modes and the start current for each mode. The microwave loss of the waveguide and the mode competition problem are also discussed. The device is shown capable of producing radiation at millimeter wavelengths with power an order higher than that uses a dielectric-lined circular waveguide.

1. Introduction

High-power microwaves have emerged in recent years for their possible applications ranging from directed energy, radar and high-gradient linear accelerators to plasma heatings. Cherenkov free-electron lasers are attractive because of their potential to use mildly relativistic electron beams ($\gamma < 3$) to generate high-power radiations. In the past 20 years, Cherenkov free-electron lasers have demonstrated their operation at mm wavelengths, with electron beam voltages between 100~500 kV.

A Cherenkov free-electron laser at mm wavelength usually include a dielectric-lined circular waveguide[1]. At high frequency, the field of the slow-wave mode TM_{0n} decays very fast away from the linear surface. For

---

① Coauthers: Qingyuan Wang, Kui Zhao. Reprinted from the Nucl. Instru. and Meth. in Physics Research A. 337 (1993) 224~229.

satisfactory operation of the device, the wavelength should be approximately the diameter of the waveguide, otherwise the beam-wave coupling will be seriously degraded and the high gradient of the axial component of the electric field will arouse large energy spread of the beam and reduce the efficiency of the device. At mm wavelength, the beam channel should be very small thus the beam input power considerably reduced, at certain current density. Also, because of the small cross-sectional area of the waveguide, the electric field, at high power operation, will cause breakdown along the surface of the dielectric linear[2,3].

In order to increase the driven beam current and the power capacity of the waveguide[4], a multi-electron-beam Cherenkov free-electron laser was proposed and a mode analysis presented. In 1991[5], the first operation of an oscillator at mm wavelength was reported. As a preliminary result, a total current of 280 A at 500 kV, in four sheetbeams, was introduced into a multi-dielectric waveguide resonator and 1.7 MW coherent Cherenkov radiation at 33.4 GHz was generated. The motivation of the present work is to study the mode characteristics of the relatively complicated waveguide and discuss the mode competition problem.

2. Dispersion Relations and Field Distribution

The slow-wave supporting structure under consideration is shown in Fig. 1. It is consists of N dielectric plates, spaced by $2a$, of width d and dielectric constant of ε. The inner ones have a thickness of 2Δ and the outer two, which are laid onto metal slabs, have a thickness of Δ. When $d \gg a+\Delta$, the boundary effects can be ignored and for all wave fields we have $\partial/\partial y = 0$. In the present work, we only deal with TM_{0n} modes, which can be efficiently excited by the electron beam propagating along the waveguide axis.

The wave field can be assumed

$$\begin{cases} E = [E_{0x}(x)x + E_{0z}(x)z]e^{i(kz-\omega t)} \\ H = H_{0y}(x)y e^{i(kz-\omega t)} \end{cases} \quad (1)$$

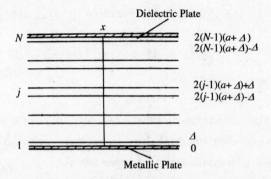

Fig. 1 Slow-wave supporting structure under consideration

In the dielectric plates, $E_{0z}(x)$ and E_{0x} are determined by

$$\begin{cases} \dfrac{\partial^2 E_{0z}}{\partial x^2} + \delta^2 E_{0z} = 0 \\ E_{0x} = \dfrac{ik}{\delta^2} E'_{0z}(x) \\ H_{0y} = \dfrac{i\omega\varepsilon\varepsilon_0}{\delta^2} E'_{0z}(x) \end{cases} \quad (2)$$

and in the beam channels, by

$$\begin{cases} \dfrac{\partial^2 E_{0z}}{\chi x^2} - \eta^2 E_{0z} = 0 \\ E_{0x} = -\dfrac{ik}{\eta^2} E'_{0z}(x) \\ H_{0y} = -\dfrac{i\omega\varepsilon_0}{\eta^2} E'_{0z}(x) \end{cases} \quad (3)$$

where

$$\delta^2 = -k^2 + \frac{\omega^2}{c^2}\varepsilon, \; \eta^2 = k^2 - \frac{\omega^2}{c^2} \quad (4)$$

are the squares of the wave number in the dielectrics and in the channels.

From eqs. (2) and (3), the solutions of axial components of the electric field are

$$E_{0z} = \begin{cases} A_i\cos\delta x + B_i\sin\delta x \\ \quad x \in [2i(a+\Delta)-\Delta, 2i(a+\Delta)+\Delta] \\ C_i e^{\eta x} + D_i e^{-\eta x} \\ \quad x \in [2i(a+\Delta)+\Delta, 2(i+1)(a+\Delta)-\Delta] \\ i = 0,1,\cdots,N \\ x \in [0, 2N(a+\Delta)] \end{cases} \quad (5)$$

Applying the boundary conditions of continuity for the axial electric field and transverse component of the displacement vector at the dielectric-metal and dielectric-vacuum interfaces, we have

$$\begin{cases} C_i = \dfrac{e^{-\eta x_i}}{2}\Big[\Big(\cos\delta x_i + \dfrac{\eta\omega}{\delta}\sin\delta x_i\Big)A_i + \Big(\sin\delta x_i - \dfrac{\eta\omega}{\delta}\cos\delta x_i\Big)B_i\Big] \\ D_i = \dfrac{e^{\eta x_i}}{2}\Big[\Big(\cos\delta x_i - \dfrac{\eta\omega}{\delta}\sin\delta x_i\Big)A_i + \Big(\sin\delta x_i + \dfrac{\eta\omega}{\delta}\cos\delta x_i\Big)B_i\Big] \\ A_{i+1} = e^{\eta(x_i+2a)}\Big[\cos\delta(x_i+2a) + \dfrac{\delta}{\eta\omega}\sin\delta(x_i+2a)\Big]C_i \\ \qquad\quad + e^{-\eta(x_i+2a)}\Big[\cos\delta(x_i+2a) - \dfrac{\delta}{\eta\omega}\sin\delta(x_i+2a)\Big]D_i \\ B_{i+1} = e^{\eta(x_i+2a)}\Big[\sin\delta(x_i+2a) - \dfrac{\delta}{\eta\omega}\cos\delta(x_i+2a)\Big]C_i \\ \qquad\quad + e^{-\eta(x_i+2a)}\Big[\sin\delta(x_i+2a) + \dfrac{\delta}{\eta\omega}\cos\delta(x_i+2a)\Big]D_i \\ i = 0,1,\cdots,N \\ A_0 = 0 \\ B_N = -A_N\tan[2N\delta(a+\Delta)] \end{cases} \quad (6)$$

where $x_i = 2i(a+\Delta)+\Delta$

In the case of a dielectric-lined circular waveguide, the most important mode is TM_{01}, for δ in the range of $0\sim\pi/2$. For other values, which are related with the harmonics, the field decays much faster so the beam-wave coupling is usually ignored. In the present work, we will only deal with that modes for δ between 0 and $\pi/2$. For a specified wave number of δ, eq. (6) can have at most N solutions. We name the modes TM_{01}^1, $TM_{01}^2, \cdots TM_{01}^N$, respectively, in the sequence of their corresponding solutions η.

Fig. 2 gives the synchronous frequency for each mode versus the electron beam voltage, where the waveguide geometry parameters are chosen according to the previous experiment: $N=5$, $\Delta=0.6$ mm, $2a=2.8$ mm. The dielectric material is ceramic (95% Al_2O_3) with a relative dielectric constant of about 10 at 8 mm wave band. Fig. 3 shows the field distributions for each mode in synchronism with beam voltage 500 kV, the frequencies are 35.3, 33.4, 28.1, 11.2 and 0.027 GHz, respectively. Decreasing the beam voltage to its Cherenkov threshold, for velocity $\beta=1/\sqrt{\varepsilon}$, the frequency for all modes approach each other and are tuned asymptotically upwards to infinite. Increasing the beam voltage, the frequencies decrease and become different. From Fig. 3, the TM_{01}^1 mode is the most preferable mode, with fields the same in every channel and relatively uniform in each channel. This is very important for a device to work with high gain and high efficiency. While the other modes provide fields relatively weak in the inner three channels and with a high gradient in each channel.

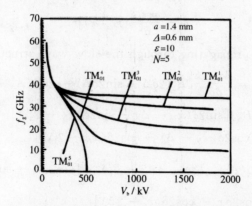

Fig. 2 Synchronous frequency for each mode versus beam voltage

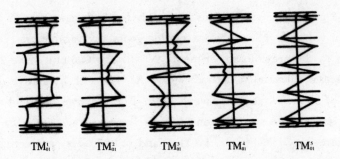

Fig. 3 The field distribution for each mode in synchronism with beam voltage 500 kV, the frequencies are 35.3, 33.4, 28.1, 11.2 and 0.027 GHz, respectively

3. The Resonator Loss, Start Currents for Each Mode and Mode Competition Problem of the Device

In the beam channels, the power density is

$$\bar{\omega} = \frac{\omega\varepsilon_0 k}{\eta^4}|E'_{0z}(x)|^2$$

and in the dielectrics, it is

$$\bar{\omega} = \frac{\omega\varepsilon\varepsilon_0 k}{\delta^4}|E'_{0z}(x)|^2$$

The total power propagating through the slow-wave structure is

$$\begin{aligned}P_p =& d\int \bar{\omega}(x)dx = \frac{d\omega\varepsilon\varepsilon_0 k}{\delta\delta^3}\bigg(B_0^2(2\delta\Delta + \sin 2\delta\Delta) + 2\delta\Delta(A_N^2 + B_N^2) \\
&- (A_N^2 - B_N^2)[\sin 2\delta(x_N - \Delta) - \sin 2\delta(x_N - 2\Delta)] \\
&+ 2A_N B_N[\cos 2\delta(x_N - \Delta) - \cos 2\delta(x_N - 2\Delta)] \\
&+ \sum_{i=1}^{N-1}\{4\delta\Delta(A_i^2 + B_i^2) - (A_i^2 - B_i^2) \times [\sin 2\delta x_i - \sin 2\delta(x_i - 2\Delta)] \\
&+ 2A_i B_i[\cos 2\delta x_i - \cos 2\delta(x_i - 2\Delta)]\} \\
&+ \frac{2\delta^2}{\eta^3 \varepsilon}\sum_{i=1}^{N-1}[C_i^2 e^{2\eta x_i}(e^{4\eta a} - 1) - D_i^2 e^{-2\eta x_i}(e^{-4\eta a} - 1) - 2C_i D_i \eta a]\bigg) \quad (7)\end{aligned}$$

The surface current density amplitude at the upper metal slab is

$$\alpha_u = H_{0y}|_{x=x_N-\Delta} = -\frac{i\omega\varepsilon\varepsilon_0}{\delta}\frac{A_N}{\sin\delta(x_N-\Delta)}$$

and at the bottom one, it is

$$\alpha_b = H_{0y}|_{x=0} = \frac{i\omega\varepsilon\varepsilon_0}{\delta}B_0$$

The total microwave loss in the metal slabs is

$$P_w = \frac{dL(\alpha_u^2+\alpha_b^2)}{2\sigma\delta_p} = \frac{dL}{2\sigma\delta_p}\left(\frac{\omega\varepsilon\varepsilon_0}{\delta}\right)^2\left[B_0^2+\frac{A_N^2}{\sin^2\delta(x_N-\Delta)}\right] \quad (8)$$

where $\delta_p = \sqrt{2/\omega\mu_0\sigma}$ is the penetration depth in the metal. The microwave loss in the dielectrics is

$$P_d = \frac{d\omega\varepsilon''\varepsilon_0 L}{2}\int dx(E_{0x}^2+E_{0z}^2)$$

where L is the length of the interaction region and ε'' the imaginary component of the complex dielectric constant. The integration is performed over all the dielectric cross sections.

$$\begin{aligned}P_d = \frac{dL\omega\varepsilon''\varepsilon_0}{8\delta}&\Bigg(B_0\Big[2\delta\Delta\Big(1+\frac{k^2}{\delta^2}\Big)-\Big(1-\frac{k^2}{\delta^2}\Big)\sin2\delta\Delta\Big]\\
&+A_N^2\Big\{2\delta\Delta\Big(1+\frac{k^2}{\delta^2}\Big)[1+\cot^2\delta(x_N-\Delta)]\\
&+\Big(1-\frac{k^2}{\delta^2}\Big)[\sin2\delta(x_N-\Delta)-\sin2\delta(x_N-2\Delta)][1-\cot^2\delta(x_N-\Delta)]\\
&+2\Big(1-\frac{k^2}{\delta^2}\Big)[\cos2\delta(x_N-\Delta)-\cos2\delta(x_N-2\Delta)]\cot\delta(x_N-\Delta)\Big\}\\
&+\sum_{i=1}^{N-1}\Big\{4\delta\Delta(A_i^2+B_i^2)\Big(1+\frac{k^2}{\delta^2}\Big)+(A_i^2-B_i^2)\Big(1-\frac{k^2}{\delta^2}\Big)\\
&\times[\sin2\delta x_i-\sin2\delta(x_i-2\Delta)]-2A_iB_i\\
&\times\Big(1-\frac{k^2}{\delta^2}\Big)[\cos2\delta x_i-\cos2\delta(x_i-2\Delta)]\Big\}\Bigg)\end{aligned} \quad (9)$$

The microwave attenuation coefficients of the metal slabs and dielectrics are defined respectively as

$$A_w = P_w/P_p, \quad A_d = P_d/P_p$$

Using the single-particle theory [6], the power gained by the wave from the beam can be calculated out

$$P = -\frac{e^2}{8mc^2}nL^3\omega\int\frac{d\theta F(\theta)}{\gamma^3(\theta)\beta^2(\theta)}\frac{\partial}{\partial\theta}\left(\frac{\sin^2\theta}{\theta^2}\right)\int dA|E_{oz}|^2$$

where $\theta=(kv-\omega)L/2v$ is the relative phase change of the wave in the beam coordinate, $F(\theta)$ the electron beam distribution function in θ, which satisfies $\int F(\theta)\mathrm{d}\theta=1$. Suppose the electron beam is mono-energetic, i.e., $F(\theta)=\delta(\theta-\theta_0)$, and perform the integrations with respect to θ and over the beam cross section, the interchanged power is

$$P = 1.69 \times 10^{-2} \frac{e\omega L^3 I}{m(\gamma\beta c)^3 \eta N w} \sum_{i=0}^{N-1} \{c^2 e^{2\eta x_i}[e^{2\eta(a+w)} - e^{2\eta(a-w)}] - D_i^2 e^{-2\eta x_i}[e^{-2\eta(a+w)} - e^{-2\eta(a-w)}] + \delta C_i D_i \eta w\} \quad (10)$$

where I is the total current of the beams. The singlepass gain of the device is defined as

$$G = P/P_p$$

The start current can be calculated from

$$I_{st} = \frac{I\left(A_w + A_d + \frac{1}{R} - 1\right)}{G} \quad (11)$$

where R is the power reflection coefficient of the second cavity mirror. The first mirror has been assumed perfectly reflecting.

Fig. 4a depicts the microwave loss in the metal slabs. From eq. (8) it is proportional to the 3/2nd power of the operation frequency. At high voltage region, the operation frequencies decrease from TM_{01}^1 mode to TM_{01}^2 mode, and the microwave loss for the TM_{01}^1 mode is the largest. Also, for each mode the synchronous frequency increases with decreasing of the beam voltage, so the wave loss becomes larger.

The microwave loss in the dielectrics has a similar characteristic, as shown in Fig. 4b, but much smaller in quantities compared with that in the metal slabs at an assumed dielectric loss tangent of 10^{-4}. At beam voltage 500 kV, the attenuation coefficient of the metal slabs is 0.19 and that of the dielectrics 0.0039. This would be a favorable property for the repetitive and high-power operation of the device. In this case, the metal slabs can be cooled by water system or other method from the outside. But for the inner dielectric plates, cooling is relatively difficult.

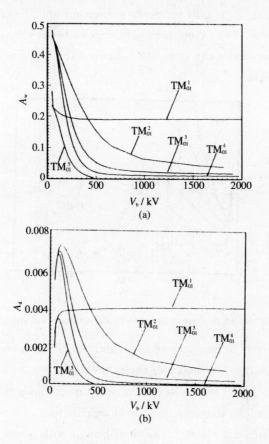

Fig. 4 (a) Attenuation coefficient of the metal slabs versus beam voltage.
(b) Attenuation coefficient of the dielectric plates versus beam voltage

In Fig. 5, the start current of each mode is depicted versus beam voltage. For each mode, there is a minimum start current. At lower voltage, the fields in the channels decay very fast so the beam-wave coupling will be degraded. At higher beam voltage, on the other hand, although the field distribution can be improved, the increase of the electron energy makes the electron relativistic mass larger and the electron beam bunching will be difficult. For each mode, the lowest start current occurs at a beam voltage of about 150 kV, at the specific choice of system parameters. At this beam voltage, however, the field distribution in the channels for the operation mode TM_{01}^1 is still seriously evanescent and the efficiency of the device will be low. For a compromised choice of beam voltage between 200 and 600 kV, the device can generate radiations for frequency ranging from 35.0 to 38.2 GHz.

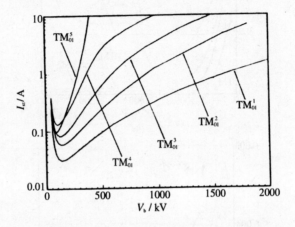

Fig. 5 Start current for each mode versus beam voltage

If the device is to be operated as an amplifier, the operation mode will be TM_{01}^1. The largest influence of the other modes on the operation mode will come from TM_{01}^2, which has a frequency most close to that of the operation mode and a relatively good field distribution. Fortunately, TM_{01}^2 mode has a symmetric electric field distribution with respect to the center plane of the waveguide structure while that of TM_{01}^1 an asymmetric distribution. Single mode operation can be assured by properly adjusting the phases of the input signals. The signal frequency difference between TM_{01}^1 and TM_{01}^2 will also help choose the operation mode.

If the device is to be operated as an oscillator, the source signals may have wide spectrums. There may be several modes that can be excited by the beam-wave interaction and this may lead to the mode competition problem. According to Fig. 5, for a wide range of beam voltage above 150 kV, TM_{01}^1 mode has the lowest start current and the curves separate apart with the increase of the beam voltage. At 500 kV, the start currents are respectively 96, 240, 620 and 2700 mA, for TM_{01}^1, TM_{01}^2, TM_{01}^3, TM_{01}^4

modes and for TM_{01}^5 mode, start current is much higher than them all. After several passes of the wave in the resonator, TM_{01}^1 will dominate the interaction process and the other modes will be effectively supressed. It can be anticipated that TM_{01}^1 will be the operation mode and the mode competition problem will not be severe.

4. The Effective Beam Channel Area and the Power Capacity

In the case of a dielectric-lined circular waveguide, the field for the operation mode TM_{01} can be expressed as $E_{0z} \alpha I_0(\eta r)$, where $\eta = \omega/\gamma\beta c$ and $\gamma = 1 + Vb \text{ (MV)}/0.511$. Define the field uniformity factor K as the field at the waveguide axis over that at the dielectric surface

$$K = E_{0z}|_{r=0}/E_{0z}|_{r=r_d} = 1/I_0(\eta r_d)$$

For a successful device, K ranges from 0.6 to 0.9. If $Vb = 500$ kV, $f = 37.5$ GHz, and $K = 0.8$, the thickness of the dielectric liner will be 0.6 mm and the radius of the beam channel 2 mm. In contrast, the total effective cross-sectional area of the channels and of the waveguide system are considerably enlarged in a multi-dielectric waveguide. For a preliminary estimate, the power capacity of the waveguide is proportional to its cross-sectional area and the driven beam current of the device proportional to the total area of the beam channels. Assuming $d = 20$ mm, $N = 5$, $2\Delta = 1.2$ mm, $2a = 2.8$ mm and a beam-dielectric gap width of 0.5 mm, the power capacity can be increased by a factor of

$$\frac{2(N-1)d(a+\Delta)}{\pi(r_d+\Delta)^2} = \frac{2 \times (5-1) \times 20 \text{ mm}(1.4 \text{ mm} + 0.6 \text{ mm})}{\pi(2 \text{ mm} + 0.6 \text{ mm})^2} \approx 15$$

The beam current can be increased by a factor of

$$\frac{2(N-1)d(a-g)}{\pi(r_d-g)^2} = \frac{2 \times (5-1) \times 20 \text{ mm}(1.4 \text{ mm} - 0.5 \text{ mm})}{\pi(2 \text{ mm} - 0.5 \text{ mm})^2} \approx 20$$

5. Summary

A mode analysis of a multi-electron-beam Cherenkov free-electron laser has been finished. Maxwell's equations and a single-particle theory were combined to give the mode dispersion relations and the field distributions. The microwave loss of the waveguide and of the dielectrics were included to decide the start current of an oscillator and furthermore, to study the mode competition problem. The system parameters of a previously finished experiment were used in the numerical calculation and it is concluded that 1) in the case of N dielectric plates, there exist N TM_{01} modes corresponding with the fundamental mode in a circular waveguide with a cylindrical dielectric liner, in which however, the TM_{01}^1 will be the best operation mode and the competition problem will not be serious, 2) microwave loss in the dielectrics is about two orders lower than in the waveguide and it is beneficial to the device working at a high average power, 3) the power capacity and the driven beam current can be increased by more than an order, which will largely increase the power output of the device.

References

[1] J. E. Walsh, T. C. Marshall and S. P. Shlesinger, Phys. Fluid 20(1977) 709.
[2] E. P. Garate, H. Kosai, W. Peter, A. Fisher, W. Main, J. Weatherall and R. Cherry, in: Intense Microwave and Particle Beams, ed. H. E. Brandt, SPIE Proc. vol. 1226 (SPIE, The International Society for Optical Engineering, Bellingham, WA, 1990).
[3] W. Peter, E. Garate, W. Main and A. Fisher, Phys. Rev. Lett. 65(1990) 2989.
[4] Q. Wang, S. Yu and S. Liu, Int. J. IR & MM Wave 10(1989) 889.
[5] Q. Wang, S. Yu, P. Xun, S. Liu, K. Hu, Y. Chen and P. Wang, Appl. Phys. Lett. 59(1991) 2378.
[6] J. E. Walsh and J. B. Murphy, IEEE J. Quantum Electron. 18(1982) 1259.

强流束传输中束晕与混沌的 Poincare 图象与 Lyapunov 指数分析[①][②]

摘　要

强流离子直线加速器因其可作为次临界装置的驱动器,近来在放射性洁净核能系统等方面的应用中出现了一些新的动向,引起人们极大的关注.同时,这些新的应用对加速器低束损也提出了苛刻的要求.强流束在低能段的传输中,弥散在束核外围的束晕是粒子损失的主要来源之一,对束晕问题的研究是当前加速器物理研究中的热点.本文采用束核-试验粒子模型,通过在哈密顿动力学系统中同时求解 KV 包络方程和 Hill 方程,单向自洽地研究了空间电荷控制的束流的传输问题.借助 Poincare 映射技术和 Lyapunov 指数,对粒子在相空间的运动进行了数值分析.结果表明,失配的包络振荡是引起粒子哈密顿参量共振的前提,而共振条件则只决定于有效空间电荷参数.参量共振的发展导致束核附近的粒子运动出现混沌.随着球形混沌的演化,使更多位于 KV 包络附近的粒子成为束晕粒子.结果同时显示,束晕具有再生性,而且在二维情况下是有界的.

关键词　强流　混沌　束晕　Poincare 映射　Lyapunov 指数

1. 引　言

近年来,强流、高占空因子的离子直线加速器正引起人们越来越大的兴趣,出现了一些新的应用方向,如用于生产洁净裂变核能、嬗变核废料、转换武器钚以及生产氚和医用放射性同位素等.显然,在这样的加速器中,希望加速尽可能大的束流.然而,强流束在传输过程中会产生一个围绕在束核周围的"晕圈(halo)",并伴随有发射度的急剧增长,导致束流的损失.自从美国洛斯阿拉莫斯国家实验室的 Jameson[1]和马里兰大

---

① 合作者:刘濮鲲,方家训.摘自《第六届全国加速器物理学术交流会论文集》,1997,p.279～285.
② 中国博士后科学基金资助项目.

学的 Kehne[2]等人先后在实验中观察到束晕的存在以来,对束晕问题的研究已成为强流粒子束物理和加速器束流动力学中的一个热门课题.

目前在束晕的理论研究中,主要采用了束核-试验粒子模型[3~6]、半解析的参量共振分析法[7]、Fokker-Planck-Poisson 模型[8~9]以及 PIC 粒子模拟的方法[10~12]. 其中,束核-试验粒子模型由于物理图象简单、清晰,因而在束晕的研究中为众多的学者所采用. 束核-试验粒子模型的基本思想是用考虑空间电荷效应的包络方程描述束核的集体运动,而选取试验粒子来描述束晕的形成. 这里束核的运动会影响试验粒子,而试验粒子不对束核产生反作用. 显然,由于束核外围粒子所占比例不大,这样做是合理的.

本文拟采用束核-试验粒子模型,通过在哈密顿动力学系统中同时求解描述包络振荡的 KV 方程和描述 betatron 振荡的包含了聚焦和空间电荷力的 Hill 方程,单向自洽地研究空间电荷控制的束流的传输问题. 借助 Poincare 截面图和 Lyapunov 指数,对粒子在相空间的混沌运动及束晕的形成机制进行讨论,同时也对束晕的一些特点进行分析.

2. 束核运动的 KV 方程与粒子运动的 Hill 方程

在旁轴对称的传输通道内,粒子的包络哈密顿函数为

$$H_e = \frac{1}{2}P_b^2 + V_{KV}(R_b) \tag{1}$$

式中 R_b 为束的平衡半径,(P_b, R_b) 为包络相空间的配对坐标,$V_{KV}(R_b)$ 为 KV 位函数:

$$V_{KV} = \frac{1}{2}k_f(s)R_b^2 - K_b \ln R_b + \frac{\varepsilon^2}{2R_b} \tag{2}$$

此处 s 为纵向坐标,我们用它作时间变量;$k_f(s)$ 为聚焦强度参数;K_b 为空间电荷导流系数;ε 为束流发射度.

定义如下的归一化参数和变量:

$$\frac{s}{L} \to s, \quad \frac{R_b}{\sqrt{\varepsilon L}} \to R, \quad P \to \sqrt{\frac{L}{\varepsilon}}P_b, \quad L^2 k_f \to k, \quad \frac{LK_b}{\varepsilon} \to K$$

(L 为聚焦场的空间周期长度). 本文仅研究均匀聚焦通道,这时 L 为任

意值；聚焦强度参数 k 为常数，即 $k(s)=v^2$，这里 v 相应于横向 betatron 振荡的相移[13]. 于是可将包络哈密顿函数(1)改写为：

$$H_e = \frac{1}{2}P^2 + \frac{1}{2}kR^2 - K\ln R + \frac{1}{2R^2} \tag{3}$$

同样可以写出在旁轴对称的传输通道内，粒子横向振荡的哈密顿函数：

$$H_p = \frac{1}{2}(p_x^2 + p_y^2) + V_p(x,y) \tag{4}$$

式中 (x,y,p_x,p_y) 为拉摩进动体系中[13]的横向相空间坐标，而 $V_p(x,y)$ 为 KV 模型中包含了束流自场的横向聚焦位函数：

$$V_p = \frac{1}{2}k_f r^2 - \frac{K_b}{2R_b^2}r^2\Theta(R_b - r) - \frac{K_b}{2}\left(1 + 2\ln\frac{r}{R_b}\right)\Theta(r - R_b) \tag{5}$$

此处 $r=\sqrt{x^2+y^2}$，Θ 为用来描述在束核内部或外部的空间电荷位的阶跃函数. 将(4)用旋转对称坐标 $(r,\varphi,p_r,p_\varphi)$ 的形式写成

$$H_p = \frac{1}{2}\left(p_r^2 + \frac{p_\varphi^2}{r^2}\right) + V_p(r) \tag{6}$$

由于上式不显含角变量 φ，故角动量 p_φ 为常数. 为简单起见，我们只考虑 $p_\varphi=0$ 的情况.

我们用 y_b 表示 x 或 y，p_b 表示 p_x 或 p_y，s 为时间变量，并同样采用相空间中的归一化坐标 $y=\dfrac{y_b}{\sqrt{\varepsilon L}}$，$p=\sqrt{\dfrac{L}{\varepsilon}}p_b$，则哈密顿函数(6)变成

$$H_p = \frac{1}{2}p^2 + \frac{1}{2}ky^2 - \frac{K}{2R^2}y^2\Theta(R - |y|) - \frac{K}{2}\left(1 + 2\ln\frac{|y|}{R}\right)\Theta(|y| - R) \tag{7}$$

此处 R 遵守 KV 哈密顿流.

可以证明，对均匀聚焦场，哈密顿函数(3)是可积的；如果包络半径 R 为常数，哈密顿函数(7)是非时变的，能量为一运动常数. 然而，如果包络半径 R 是时变的，那么这个非线性系统就是不可积的.

下面我们作一个定标变换. 对时间变量，令 $\bar{s}\to vs$；对包络相空间，$(\bar{R},\bar{P})\to(\sqrt{v}R,P/\sqrt{v})$；对粒子相空间变量，令 $(\bar{y},\bar{p})\to(\sqrt{v}y,p/\sqrt{v})$. 于是得到新的包络与粒子哈密顿函数：

$$\overline{H}_e = \frac{1}{2}\overline{P}^2 + \frac{1}{2}\overline{R}^2 - 2\chi\ln\overline{R} + \frac{1}{2\overline{R}^2} \tag{8}$$

$$\overline{H}_p = \frac{1}{2}\overline{p}^2 + \frac{1}{2}\overline{y}^2 - \frac{\chi}{\overline{R}^2}\overline{y}^2\Theta(\overline{R} - |\overline{y}|) - \chi\left(1 + 2\ln\frac{|\overline{y}|}{\overline{R}}\right)$$
$$\Theta(|\overline{y}| - \overline{R}) \tag{9}$$

式中 $\chi = K/2v$ 为有效空间电荷参数. 由此可见, 经过变换以后的新哈密顿函数只与一个参数 χ 有关. 因此, 均匀聚焦通道内的 KV 空间电荷模型是一个单参数的哈密顿系统.

利用哈密顿正则运动方程

$$\begin{cases} \dot{q}_i = \dfrac{\partial H}{\partial p_i}, \\ \dot{p}_i = \dfrac{\partial H}{\partial q_i}, \end{cases} \tag{10}$$

我们便可由(3)式得到包络运动的 KV 方程:

$$\ddot{R} + v^2 R - \frac{K}{R} - \frac{1}{R^3} = 0, \tag{11}$$

由(7)式得到粒子运动的 Hill 方程:

$$\ddot{y} + v^2 y - \frac{K}{R^2} y \Theta(R - |y|) - \frac{K}{y}\Theta(|y| - R) = 0. \tag{12}$$

如果令 $\dot{R} = 0, \ddot{R} = 0$, 则由(11)式得

$$v^2 R - \frac{K}{R} - \frac{1}{R^3} = 0. \tag{13}$$

由上式便可求得束流包络相空间中的不动点 $(R_m, 0)$, 其中

$$R_m = [(\sqrt{\chi^2 + 1} + \chi)/v]^{1/2} \tag{14}$$

也是匹配束流的归一化平衡包络半径. 同时, 我们定义失配因子

$$M = (R_m - R_{\min})/R_m, \tag{15}$$

式中 $R_{\min}$ 为失配束流的最小包络半径. 对匹配束流, $M = 0$; 而当初始包络不匹配(即 $R_0 \neq R_m$)时, 束流处于非静态, 束流包络作周期性振荡. 由于在包络位函数中的小的非简谐性, 包络振荡相对于 R_m 可能是不对称的. 正是由于束核密度的这种振荡, 破坏了粒子哈密顿函数(9)的可积性, 并最终导致了粒子相运动的混沌与束晕的形成.

3. 粒子相运动的 Poincare 截面分析

我们用数值计算的方法同时求解式(10)和(11)这两个单向耦合的二阶微分方程组. 选取一些具有不同初始条件, 包括核内、核附近以及远离束核的代表性试验粒子, 跟踪它们在外部聚焦场和束核空间电核场作用下的运动. 在每一个束核振荡周期中, 我们在最小包络半径位置作出粒子运动在相空间的 Poincare 截面图, 如图 1~2 所示.

图 1(a)给出了 $M=0.2$, $\chi=1.5$ 时相轨道的 Poincare 截面. 从图中可以清楚地看到存在由 1/2、1/3 以及 1/4 参量共振所产生的共振岛. 图 1(b)将失配因子 M 增大到 0.3, 这时共振强度增大, 这一点从 1/4 共振

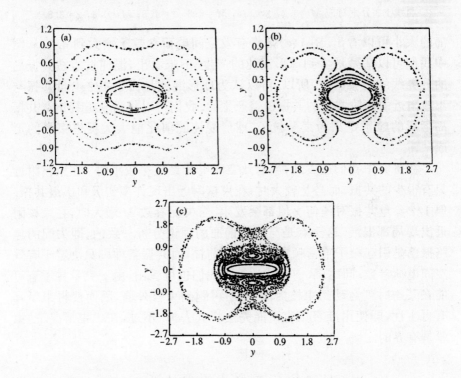

图 1 粒子相轨迹的 Poincare 截面. 此处 $R_m=1.002$,
(a) $M=0.2$, $\chi=1.5$; (b) $M=0.3$, $\chi=1.5$; (c) $M=0.2$, $\chi=4.2$

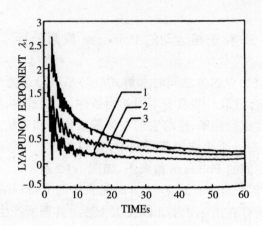

图2 最大Lyapunov指数λ_1随归一化时间变量s的变化,图中$R_m=1.002$,曲线1-3分别对应$M=0.2$,$\chi=1.5$;$M=0.3$,$\chi=1.5$;$M=0.3$,$\chi=3.5$.

岛的大小可以看出. 图1(c)则将有效空间电荷参数χ增大到4.2,从图中我们可以清楚地看到球形混沌的产生. 不难看出,混沌发生在共振岛的分界线的交会区域. 所以,可以认为球形混沌是由于1/2参量共振与束核附近的高阶共振的共振覆盖产生的. 我们可以将沿轨道绕1/2共振岛运行的粒子定义为束晕粒子. 球形混沌的演化能大大促进束晕的形成.

在研究中我们还发现,当有效空间电荷参数χ较小时,在束核附近只有很少的共振,而当χ较大时,在束核附近出现许多初级和次级共振. 但1/2参量共振对任何χ值都会发生. 当失配参数M增大时,在束核附近出现局部混沌. 这一点是与我们前面的理论分析一致的,即失配的包络振荡是引起粒子哈密顿参量共振的前提,而共振条件则只决定于有效空间电荷参数. 同时,从前面的计算中我们也可看出,由于束晕粒子有可能在某一时刻运动到束核内部,随后再跑到束核外面,因而使得束晕具有再生性,即使用准直锥也不能完全将其去掉. 不过,在二维情况下,束晕是有界的.

4. Lyapunov 指数的计算

前面利用 Poincare 截面法对粒子在相空间的混沌运动及束晕的形

成进行了初步的分析,这里我们将进一步计算系统的 Lyapunov 指数.
我们知道,Lyapunov 指数是判定一个动力学系统是否出现混沌的最有
效的诊断手段之一[14]. 任何含有一个以上正的 Lyapunov 指数的系统都
被定义为混沌的,而指数的大小则反映了系统动力学行为变得不可预期
的时间尺度.

Lyapunov 指数谱的定义如下:

$$\lambda_i = \lim_{m \to \infty} \frac{1}{ms} \sum_{k=1}^{m} \log_2 \frac{|d_i(s_k)|}{|d_{i-1}(s_{k-1})|} \quad (i = 1, 2, \cdots, n), \tag{16}$$

式中 $n=4$ 为非线性方程的个数,s 为归一化时间变量的迭代步长,m 是
迭代次数,$|d_i(s_k)|$ 为在时间 s_k 两相邻轨道沿 i 方向的特征距离. 我们取
4 个正交的单位矢量:

$$\begin{pmatrix} 1 \\ 0 \\ 0 \\ 0 \end{pmatrix}, \begin{pmatrix} 0 \\ 1 \\ 0 \\ 0 \end{pmatrix}, \begin{pmatrix} 0 \\ 0 \\ 1 \\ 0 \end{pmatrix}, \begin{pmatrix} 0 \\ 0 \\ 0 \\ 1 \end{pmatrix},$$

它们沿由积分方程(11)和(12)所获得的基准轨道变化. 而其变分可通过
将非线形方程(11)和(12)线形化来得到:

$$\frac{d}{ds} \begin{pmatrix} \delta R \\ \delta \dot{R} \\ \delta y \\ \delta \dot{y} \end{pmatrix} = \begin{pmatrix} \delta \dot{R} \\ -v^2 \delta R - \frac{K}{R^2} \delta R - \frac{3}{R^4} \delta R \\ \delta \dot{y} \\ -v^2 \delta y + \frac{K}{R^2} \left(\delta y - \frac{2y}{R} \delta R \right) \Theta(R - |y|) - \frac{K}{y^2} \delta y \Theta(|y| - R) \end{pmatrix} \tag{17}$$

将上式积分一步到时间 s_1,这四个单位矢量的长度则从开始时的
$|d_i(s_0)|$ 变化到时间 s_1 时的 $|d_i(s_1)|$. 对这四个矢量在时间 s_1 重新进行归
格化正交,然后重复以上过程. 当迭代次数 $m \to \infty$ 时,按(16)式,其极限
值就是 Lyapunov 指数.

图 2 中给出了当 $R_m = 1.002$ 时,最大 Lyapunov 指数 λ_1 随归一化时
间变量 s 的变化,图中曲线 1~3 分别对应 $M=0.2$, $\chi=1.5$; $M=0.3$,
$\chi=1.5$; $M=0.3$, $\chi=3.5$. 从图中可以看出,最大 Lyapunov 指数的极限
值分别约为 0.07, 0.17 和 0.24,随失配因子 M 和有效空间电荷参数 χ
的增大而增大. 当失配因子 $M=0$ 时,最大 Lyapunov 指数的极限值趋于

零.这再次证明了失配的包络振荡是导致粒子相运动出现混沌的前提,而粒子相运动混沌程度的大小则只取决于有效空间电荷参数.

5. 结　语

本文在哈密顿动力学系统中采用单向自洽的束核-试验粒子模型,通过求解 KV 包络方程和 Hill 方程,对强流离子直线加速器中束流在均匀聚焦系统中的传输问题进行了研究.结果表明,失配的包络振荡是引起粒子哈密顿参量共振的前提,而共振条件则只取决于有效空间电荷参数.其中 1/2 参量共振在任何有效空间电荷参数下均会发生.参量共振的发展导致 KV 包络附近的粒子运动呈现球形混沌状态.混沌的演化,使更多位于束核附近的粒子成为束晕粒子.

这里我们只对均匀聚焦通道内空间电荷控制的束流中的混沌与束晕问题进行了初步的研究.需要指出的是,对束晕问题的研究目前在加速器物理中正方兴未艾,尚有许多问题有待解决.我们需要更多的实验研究,同时在理论上也要对诸如非 KV 分布束流、纵向对横向的耦合、均温效应发生的温差阈值、均温破坏 KAM 面后束晕是否还有限、束晕的产生与加速器参数的关系等问题作进一步的研究,以对强流离子加速器中为减免束晕应当遵循的规则提出建议,并更好地预见束流的损失.

参 考 文 献

[1] Jameson R A. Beam halo from collective core/single-particle interaction, Los Alamos National Laboratory Report, LA-UR-93-1209, Mar. 1993

[2] Kehne D, Reiser M, Rudd H. Experimental studies of emittance growth due to initial mismatch of a space charge dominated beam in a solenoidal focusing channel. Conf. Rec. IEEE Particle Accelerator Conf., 1991, San Francisco, Vol. 1, p. 248～250

[3] O'Connell J S, Wangler T P, Mills R S, et al. Beam halo formation from space charge dominated beam in uniform focusing channels. Proc. 1993 Particle Accelerator Conf., Washington D. C., May 1993, p. 3657～3659.

[4] Gluckstern R L. Analytic model for halo formation in high current ion linacs. Phys. Rev. Lett., 1994, 73(9): 1247～1250.

[5] Lagniel J M. Chaotic behaviour and halo formation from 2D space-charge dominated beams. Nucl. Instrum. Methods Phys. Res., 1994, A345(3): 405~410
[6] Qian Q, Davidson R C, Chen C. Halo formation induced by density nonuniform in intense ion beams. Pyhs. Rev., 1995, E51(6): R5216
[7] Lee S Y, Riabko A. Envelope Hamiltonian of an intense charged-particle beam in periodic solenoidal fields. Phy. Rev., 1995, E51(2): 1609~1612
[8] Bohn C L. Transverse phase-space dynamics of mismatched charged-particle beams. Phys. Rev. Lett., 1993, 70(7): 932~935
[9] Bohn C L, Delayen J R. Fokker-Planck approach to the dynamics of mismatched charged-paeticle beams. Phys. Rev., 1994, E50: 1516
[10] Cucchetti A, Reiser M, Wangler T. Simulation studies of emittance growth in RMS mismatched beams. Conf. Rec. IEEE Particle Accelerator Conf., 1991, San Francisco, Vol. 1, p. 251~253
[11] Lagniel J M, Piguemal A C. On the Dynamics of space-charge dominated beam. Proc. of 1994 Int Linac Conf., Aug. 1994, Tsukuba, Japan, p. 529
[12] Ryne R D. Advanced computers and simulation. Proc. 1993 Particle Accelerator Conf., Washington DC, Vol. 5, p. 3229~3233
[13] Chen C, Davidson R C. Nonlinear properties of the Kapchinskij-Vladimirskij equilibrium and envelope equation for an intense charge-particle beam in a periodic focusing field. Phys. Rev., 1994, E49(6): 5679~5687
[14] 郝柏林, 分岔、混沌、奇怪吸引子、湍流及其它——关于确定论系统中的内在随机性, 物理学进展, 1983, 3(3): 329~415

第四部分 射频离子直线加速器研究

螺旋线波导直线加速器简介[①]

摘　要

利用螺旋线波导中的电场加速荷电粒子的概念早在1948年就提出来了[1].但此后由于Alvarez直线加速器的成功运行,螺旋线波导直线加速器的发展[2~4]没有引起人们的注意.60年代后期重离子和超导加速技术的发展要求寻找一种尺寸小、结构简单的低频、低速直线加速结构,这就重新引起了人们对螺旋线加速器的广泛兴趣[5~8].为了探索重离子的加速技术,我们曾对这种加速器进行了一些调研,并对螺旋线波导腔作了一些初步的测试.下面将在第一节中介绍螺旋线加速器的基本原理,第二节中分析这种加速器的性能特点,第三节中介绍二种典型常温螺旋线加速器的设计方案,第四节中简要介绍超导螺旋线加速器的发展情况并在最后一节中简要介绍我们对于螺旋线波导腔的初步测试结果和一些看法.

一、基　本　原　理

螺旋线波导的加速结构如图1所示.在高频电源的激励下,电磁场以相当于光速的速度沿导线表面传播.此时沿着轴线便有一个速度小于光速的所谓"慢化"了的轴向电场向前进行.慢波场的相速度

$$v_p \approx c \cdot \sin\psi \approx c \cdot \frac{p}{2\pi a} \tag{1}$$

其中c系光速,p螺距,a螺旋线半径,ψ螺旋角.适当地选择ψ可使慢波电场与比光速慢几十倍的低能离子同步行进.这样的慢波可以用来连续地加速离子.(实际的螺旋线波导加速器往往采用驻波加速结构.此时除了同向行进波外还存在着与粒子反向运动的慢波,后者对粒子加速的贡献很小,常常可以忽略.)

螺旋线波导中的电磁场是TM和TE两种模式的混合场.上述可以

[①] 摘自《自然科学通讯》,北京大学汉中分校,No.3,1973.8.

用来加速离子的是场的 TM 成分,它由导线上周期性地分布着的电荷所建立,此外线上的高频电流还感生 TE 场.后者很弱,对离子运动的影响很小.

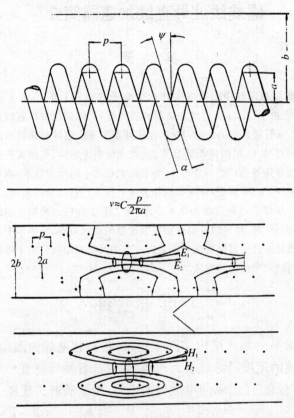

图 1　螺旋线结构及其电磁场分布的示意图

原则上,螺旋线波导中慢波的传播速度、电磁场的空间分布等都可通过解麦克斯韦方程求得.在实际计算中则常常采用所谓"导面模型"[9]或"薄带模型"[10]来简化边界条件(图 2).前者把螺旋线当作一个无限薄的柱面,面上沿着线圈导体的方向(图上 ψ 的方向)完全导电,其他方向都不导电.这样的模型相当于一个用许多股互相绝缘的细丝紧密绕制而成的螺旋线.当导波波长 $\lambda_g \gg p$ 时,这种模型能相当准确地给出慢波的相速度及轴线周围的场分布.'薄带模型'把螺旋线当作无限薄的导电扁带来处理.这样的模型比"导面模型"更接近于物理实际.由此得到的解包括了无穷多个由边界条件规定的空间谐波,能够反映出螺旋结构的周期性,较准确的表述场的传播特性和空间分布.不过在加速器的工作频率范围内,两种模型所给出的结果差异不大.因为只有在很高的频率或相当大的 ψ 下,结构的周期性才对传播特性有明显影响.由于这个缘故,在许多场合中人们往往宁愿使用计算较简便的导面模型.

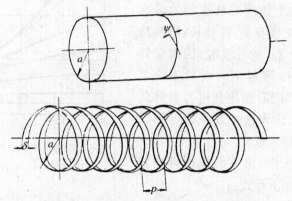

图 2 "导面模型"和"薄带模型"示意图

由导面模型得出的波的轴向相速度 v_0 随高频频率的变化情况可由色散方程写出:

$$\beta\gamma\cot\psi = \left[\frac{I_0(g_0a)I_1(g_0b)}{I_1(g_0a)I_0(g_0b)} \cdot \frac{K_0(g_0a)I_0(g_0b) - K_0(g_0b)I_0(g_0a)}{K_1(g_0a)I_1(g_0b) - I_1(g_0a)K_1(g_0b)}\right]^{\frac{1}{2}} \quad (2)$$

其中

$$g_0^2 = \left(\frac{w}{v_0}\right)^2 - \left(\frac{w}{c}\right)^2 = \left(\frac{2\pi}{\beta\gamma\lambda}\right)^2, \quad \frac{v_0}{c} = \beta, \quad \gamma = (1-\beta^2)^{-\frac{1}{2}}$$

I,K 是修正的贝塞尔函数.

由于导面模型的轴对称性,(2)式所给出的只是所谓'基模'(薄带模型的 '0' 模)的情况,而不包括方位角不对称的高次模式.图 3 上的曲线画出了不同 b/a 下,$[v_0/c\sin\psi]$ 随 ω 的变化情况.由图可见,慢波的相速先随 ω 的升高而下降,但是很快就趋于 $c\sin\psi$.此后曲线很平坦,色散较小.

实验上测得的 v_0 比式(2)给出的要小[11].产生偏差的因素有二,首先是导面模型本身不准,由此产生的误差约为 10^{-2};其次是所谓末端效应的影响. 因为螺旋线不是一个严格的周期性结构, 线上找不到一个结构的对称平面,任何终端都将引起场的畸变, 主要是 TE 成分的畸变, 结果导致 v_0 的偏离. 这个效应对加速腔的影响与螺旋线的长度有关, 线越长, 影响越小. 对于长为 $2\lambda_g$ 的驻波腔, 因末端效应引起的频移约为 3%; 而对于一个长为 λ_g 的超导腔, 相速竟可因而降低 40%. 也有人认为这一效应恰可用以降低注入器的能量[12].

在柱坐标中, 由导面模型得到的场的基模有如下形式:

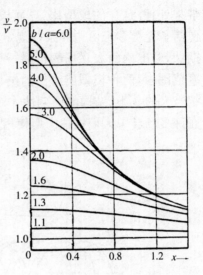

图 3 螺旋线波导的色散曲线
($x = g_0 a$, $v' = c\sin\psi$)

$$r < a$$

$$E_z = E_0 I_0(g_0 r), \qquad E_r = iE_0 \gamma I_1(g_0 r)$$

$$E_\phi = -E_0 \frac{I_0(g_0 a)}{\cot\psi} \frac{I_1(g_0 r)}{I_1(g_0 a)}, \qquad B_z = -iE_0 \frac{g_0}{\omega\cot\psi} \frac{I_0(g_0 a)}{I_1(g_0 a)} I_0(g_0 r)$$

$$B_r = E_0 \frac{\gamma g_0}{\omega\cot\psi} \frac{I_0(g_0 a)}{I_1(g_0 a)} I_1(g_0 r), \qquad B_\phi = iE_0 \frac{\omega\varepsilon\mu}{g_0} I_1(g_0 r)$$

$$a < r < b$$

$$E_z = E_0 I_0(g_0 a) \left[\frac{I_0(g_0 r)K_0(g_0 b) - I_0(g_0 b)K_0(g_0 r)}{K_0(g_0 b)I_0(g_0 a) - I_0(g_0 b)K_0(g_0 a)} \right]$$

$$E_r = iE_0 \gamma I_0(g_0 a) \left[\frac{I_1(g_0 r)K_0(g_0 b) + I_0(g_0 b)K_1(g_0 r)}{K_0(g_0 b)I_0(g_0 a) - K_0(g_0 a)I_0(g_0 b)} \right]$$

$$E_\phi = -E_0 \frac{I_0(g_0 a)}{\cot\psi} \left[\frac{I_1(g_0 r)K_1(g_0 b) - I_1(g_0 b)K_1(g_0 r)}{K_1(g_0 b)I_1(g_0 a) - K_1(g_0 a)I_1(g_0 b)} \right]$$

$$B_z = -iE_0 \frac{g_0}{\omega\cot\psi} I_0(g_0 a) \left[\frac{I_0(g_0 r)K_1(g_0 b) + I_1(g_0 b)K_1(g_0 r)}{K_1(g_0 b)I_1(g_0 a) - K_1(g_0 a)I_1(g_0 b)} \right]$$

$$B_r = E_0 \frac{\gamma g_0}{\omega\cot\psi} I_0(g_0 a) \left[\frac{I_1(g_0 r)K_1(g_0 b) - K_1(g_0 r)I_1(g_0 b)}{K_1(g_0 b)I_1(g_0 a) - K_1(g_0 a)I_1(g_0 b)} \right]$$

$$B_\phi = iE_0 \frac{\omega\varepsilon\mu}{g_0} I_0(g_0 a) \left[\frac{I_1(g_0 r)K_0(g_0 b) + K_1(g_0 r)I_0(g_0 b)}{K_0(g_0 b)I_0(g_0 a) - K_0(g_0 a)I_0(g_0 b)} \right]$$

式中 E_0 是轴上行波电场的振幅.

上式各项都省略了因子 $e^{i\omega(t-z/v_0)}$.

场的主要成分随 r 的分布如图 4 所示,E_z、B_r 在 $r=a$ 处达最大值；E_r 在导面的外表面处,B_z 在导面的内表面处分别达到最大值.

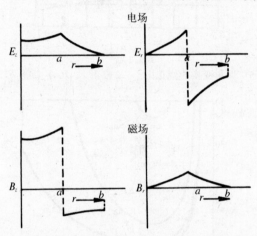

图 4 螺旋线波导中电磁场沿半径的分布

上述导面模型的结果在 Z 轴附近是较准的,但是在导线邻近($r\sim a$) 就不可靠.因为无限薄的导面完全不同于有限尺寸的导线.实际上在曲率半径相当小的导线周围,局部的场强必然增大.增加的量和 d/p 有关.d/p 大时 E_z 和相应的 B_r 首先增大,同时由于通量守恒,B_z 和 E_r 也随之增大；d/p 小时主要是 E_r 和相应的 B_z 增加.这些情况可用保角变换方法算出(近似假定场处于似稳态).并用以修正导面模型的结果.修正因子和 d/p 的关系如图 5 所示[13].

导线表面的磁场决定着螺旋线腔的高频损耗.我们用经保角变换修正的表面磁感应强度 B_{Sr} 和 B_{Sz} 算得的铜质螺旋线体的高频功率损耗可以写成

$$P_a = 1.81 \times 10^{-4} \times E_0^2 \times [a \cdot \sqrt{f} \cdot \xi] \quad (W/m) \tag{3}$$

(其中 E_0、a、f 分别以[MV/m], [cm], [MHz]为单位). 式中 ξ 是 d/p,

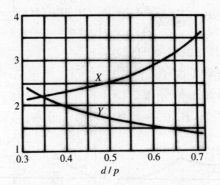

图 5 保角变换的修正因子;X 修正 E_z、B_r,Y 修正 E_r、B_z

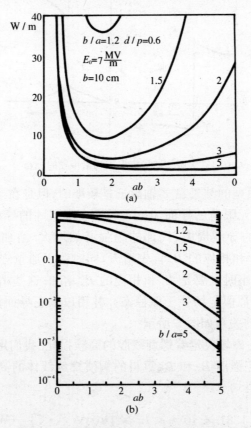

图 6 (a)螺旋线功率损耗随 b/a,$g_0 b$ 变化的情况($R_A \sim 10^{-8}\ \Omega$)
(b) 外壳损所占的比例

b/a 和 g_0a 的函数. 如 $b/a > 2$, 当 $g_0a \sim 0.7$, $d/p \sim 0.5$ 时 $\xi_{\min} \sim 7$. 法兰克福大学由实验求得的最佳在值 $g_0a = 0.69$, $d/p = 0.414$ 附近[14].

螺旋线腔外壳的功耗 P_b 通常约为 P_a 的 $1\% \sim 10\%$, 或更小些, 腔耗 P 和驻波分路阻抗 Z_{SW} 随腔体参量、能量的变化情况如图 6、7. 行波分路阻抗 $Z_{TW} = 2Z_{SW}$.

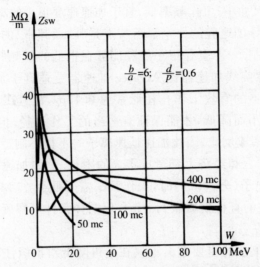

图 7 $b/a=6$, $b=12.5$ cm, $d/p=0.6$ 时 Z_{SW} 随加速能量变化情况

导线表面的最大电场强度 E_{smax} 比加速场强 E_0 高出约 $5 \sim 10$ 倍. 事实上 E_{smax} 必需小于击穿场强 (或超导体的电破坏场强), 因此比值 $\dfrac{E_{smax}}{E_0}$ 规定了加速腔的最大能量增益梯度. $\dfrac{E_{smax}}{E_0}$ 也是 g_0a, d/p, b/a 的函数. 当工作频率为 100 MHz, $b/a = 6$ ($b = 12.5$ cm) 时; 表面电场 $\dfrac{E_{sr}}{E_0}$、$\dfrac{E_{sz}}{E_0}$ 的数例如下:

W	750 keV	20 MeV	40 MeV
g_0a	1.1	0.21	0.16
d/p	0.63	0.4	0.3
E_{sz}/E_0	7.8	2.24	2.39
E_{sr}/E_0	1.56	4.08	5.66

Dome[14] 估计 $g_0a = 0.6$, $d/p = 0.42$ 时, E_{smax}/E_0 有极小值, 此时允许的

最大加速场强将达到极大值.

二、性 能 与 特 点

常温螺旋线波导加速器的结构与工作性能有如下特点：

(1) 横向尺寸小，工作频率低，宜于加速速度很小的离子.

螺旋线波导加速腔的一个显著特点是其"通频带"的下限不受腔体横向尺寸的限制. 一个直径很小的腔能够在很低的射频频率下工作. 由于这个缘故，螺旋线加速器比 Alvarez 加速器更适宜于加速低能离子. 因为直线加速器的有效工作孔径通常为 $0.15\beta\lambda$，故其工作频率必须随离子速度的减小而降低，才能保证有适当的工作孔径. 但 Alvarez 加速器的直径与频率成反比，因此加速低速离子时不得不制造尺寸很大的加速腔，这在技术上和经济上都造成很大的困难. 例如加速能量为每核子 0.15 MeV 的离子 ($\beta=0.05$) 时，工作频率需降低至 25 MHz 左右. 此时 Alvarez 加速腔的直径就得做到 8 m! 但同样条件下螺旋线加速腔的直径只有 25 cm 左右!

加速低速离子的重要意义不仅在于可降低对前级注入器的能量要求，更重要的是可利用相空间体积的"绝热收缩"效应提高相空间的束流密度. Montague[15]曾建议用螺旋线加速器作为高能加速器直线注入器的前级，以降低注入器的起始工作能量，提高束流品质，改进高能加速器的束流俘获效率.

(2) Q 值较小，调谐容易，对公差要求较低.

常温螺旋线加速腔的 Q 值约为 2000，比 Alvarez 腔小约 40 倍. 这是因为前者的电场有效地集中在轴线附近的小体积中，总的储能比体积庞大的 Alvarez 腔小得多的缘故.

Q 值小的腔其谐振曲线较为宽阔，故易于调谐、对机械公差的要求较低、工作时也不需要精密的温度控制.

(3) 群速度较高，填充时间短，宜于加速强流离子束.

在"0模"状态下工作的 Alvarez 腔的群速度为零，输入的射频功率全靠主模与邻近模的混波传递. 因此瞬态的相位振幅发生许多"波折". 高频场的填充时间也很长，往往达 150 μs. 由于这些原因 Alvarez 腔的瞬态性能不好,不适宜于加速强流离子束. 螺旋线波导腔则不同,它的群速度较高. 在相当大的范围内群速度接近于相速度. 高频场的填充时间也短得多,仅约 10 μs. 因此它的瞬变性能较好,与杆耦合飘浮管结构相当,适宜于加速强流离子束.

此外由于螺旋线腔的填充时间短,瞬态过程所消耗的能量比 Alvarez 腔小得多. 因此在脉冲状态下工作时,尽管前者的峰值功率与后者相当或稍大,但其平均功率则常常比后者小得多.

(4) 工作能量范围较窄,一般 $0.15 \frac{\text{MeV}}{\text{A}} \leqslant W \leqslant 30 \frac{\text{MeV}}{\text{A}}$.

螺旋线加速器工作能量的下限受到击穿场强和功率损耗的限制. 降低工作能量时,螺距 p 随着减小. 但因机械强度和冷却方面的要求,导线的直径不能无限减细,通常 d 不小于 $3\sim 4$ mm. 结果 E_{sz}/E_0 迅速升到击穿值. 加大螺旋线半径 a 可在一定程度上减低 E_{sz},但因而引起的导线长度和表面磁场的增强却导致高频损耗 P_a 的急剧上升. 因此一般 $W_{\min} \geqslant 0.15$ MeV/A.

螺旋线加速腔的能量上限受到高频功耗的限制,当 $W > 30$ MeV/A 后,分路阻抗降到 ~ 18 MΩ/m(参阅图 10),显著的低于 Alvarez 结构. 故通常以此为上限.

迄今提出了几种扩大螺旋线加速器能量范围的可能途径. 一是采用所谓双螺旋线结构[13],即在外筒和螺旋线之间增加一个反向绕制的螺旋线. 此时原来的色散曲线就分解成快慢两种模式(图 8). 当两个螺旋线中的电流同相时,慢波速度大于单螺旋线的 v_0,称"快模";二者反相时波速比单螺旋线小称"慢模". 快模状态下二螺旋线的最大磁场相消,沿轴的电场则相加故损耗减小. Sierk 估计,利用快模可使螺旋线加速器的能量上限增高至 150 MeV 左右. 反之慢模则可用以降低注入能量. 此时磁场相加、电场相消故损耗增加. 但因导体表面的最大电场强度比单螺旋线低,故能量的下限可降至 100 keV 左右而不至击穿. 缺点是双螺旋线在结构上复杂得多,为此在工艺上必需付出代价.

采用超导技术同样可以增加螺旋线腔的加速能量. 因为超导螺旋线腔体的高频损耗将比常温腔减低 $10^5 \sim 10^6$ 倍,故分路阻抗的变化不再是限制能量范围的关键因素.

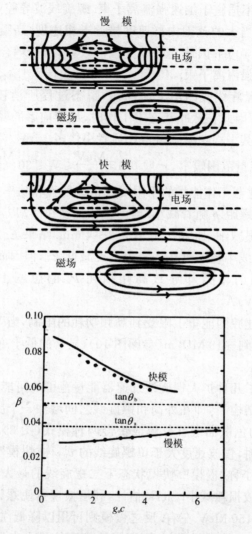

图 8　双螺旋线的色散曲线与力线分布

三、两个典型螺旋线加速器的设计方案

(1)"Helac"[5]：这是法兰克福大学在模型试验的基础上设计的一个可变能量、全粒子(由 P 至 U 全部元素的离子)螺旋波导直线加速器方

案. 其最大能量为 7 MeV/A, 加速器的参数如表 1.

表 1

	源→	Helix 27.12 MHz	→剥离器→	Helix 108.48 MHz	
能量/MeV·A^{-1}	0.13	1.4		4.5	7.0
荷质比 $\frac{Q}{A}$	11/238	11/238		25/238	25/238
加速电压/MV	2.8	27.5		29.5	23.8
$\overline{Z}_{SW}$/MΩ·m^{-1}		25.6		33.5	26.5
功耗 P/MW/脉冲		1.3		1.35	1.7
场强/MV·m^{-1}		1.06		1.22	1.63

这一加速器的特点是: ① 注入能量很低仅 0.13 MeV/A. ② 加速器的外径尺寸小, $\phi = 25$ cm. ③ 尽管尺寸小注入端的频率仍很低 $f = 27.12$ MHz. 这样即使对 0.13 MeV/A 的低能离子 ($\beta \sim 0.017$) 横向接收面积仍达 52 cm·mrad; 且四极透镜的最高梯度适中 (1.5～3.5 kGs/cm). ④ 考虑到分路阻抗随能量降低的情况在 1.4 MeV/A 后改用 108.48 MHz 的频率工作, 这样由 1.4～4.5 MeV/A 的有效分路阻抗达 33.5 MΩ/m 整个加速器的平均分路阻抗接近 30 MΩ/m 与"杆耦合"Alvarez 加速器完全可以媲美. ⑤ 加速粒子束的能量精度为 3.7×10^{-3}, 束流脉冲宽度 τ_b 为 0.39 ns, 加散相器 (匀能器) 后 $\Delta\omega/\omega \sim 3 \times 10^{-4}$; $\tau_b \sim 3$ ns. ⑥ 加速腔的 $Q \sim 1700$, 脉冲重复率 600 Hz (脉冲上升时间 ~ 12 μs).

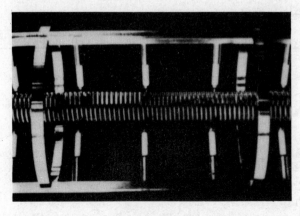

图 9 法兰克福大学的螺旋线驻波加速腔

⑦ 加速腔采用驻波结构,每节长约 1 m. 螺旋线上波节处支有 Al_2O_3 支架,保证结构的机械强度. 整个加速器共有 75 节腔,每节都有自己的功率源和独立的相位振幅控制装置,这样做虽在控制上较为复杂,但可方便地在 4~7 MeV/A 范围内连续调节能量. 例如关掉最后几节就可使能量逐步降低;还可调节最后一节的能量和振幅,使之处于不同程度的加速或减速状态,以对能量进行细调. 也可把这些节当作聚束器或散相器以调节束流的相位结构和能量精度. 总之,这种结构具有很大的灵活性. ⑧ 加速器的公差要求比较低. 螺旋线长度的累计误差每米小于 1 mm,电场振幅容差±2%,相位容差±2°. ⑨ 各节加速腔的结构以及各节高频功率源、控制系统等都采用规范设计. 这样不仅可以成批加工,降低造价,加快建造速度,而且即使建成后,若要更动工作范围,也只需简单的更换螺线芯柱,方便易行.

美 Blann 估计上述螺线加速器,每节包括高频设备和控制系统在内,总造价仅需 5 万美元[6],而且建造速度快,二年内便可建成. 由于加速器的尺寸小、重量轻,基建费用也可大大节省.

(2) 20~30 MeV 螺旋线加速器

这个能量范围的加速器适宜于作高能加速器的注入器,或作直线注入器的前级. 这方面 CERN 做了不少工作[15~17],但因无有关资料,这里只介绍法兰克福 1967 年[18]作的一个方案. 方案的参数如表(2). 这个方案的主要特点是外壳尺寸特小,仅 12 cm,上升时间也很快,仅 6 μs. 为了适应在较高能量范围中工作的要求,工作频率升至 200~250 MHz. 不过平均分路阻抗仍只有~18 MΩ/m,比 GERN 同样能量的 Alvarez 注入器低 33%左右. 因此单位长度的能量增益选择得稍低(约 0.65~0.8 MeV/m). 此时每节高

表 2

注入能量	2 MeV	2 MeV
终能量	22.3 MeV	34 MeV
外壳半径	6 cm	6 cm
螺旋线半径	2.0~1.55 cm	2.0~2.5 cm
导管直径	6 mm	4 mm
工作频率	200 MHz	250 MHz
能量增益梯度	0.65 MeV/m($\varphi_s=30°$)	0.8 MeV/m($\varphi_s=30°$)
填充时间($=3\tau$)		8 μs
束流脉冲宽度 τ_b	连续工作	25 μs
每节高频功率	30 kW	50 kW
总高频功率	960 kW	2 MW(脉冲)

频功率为 30～50 kW,不算太高,可以在连续态或高负载因子状态下运转而得到高的平均束强. 值得注意的是在脉冲状态下运行时,它的填充时间比 Alvarez 结构短 10～20 倍,故每个脉冲中束流的负载因子～$(\tau_b/\tau_b+3\tau)$ 比一般 Alvarez 结构高出约 3～6 倍! 由于这个缘故它的平均腔耗要比后者低得多.

四、超导螺旋线波导加速器

螺旋线加速器适宜于发展成超导加速器. 因为它的结构简单,方便于超导腔体的加工制造. 它的体积很小,易于冷却至低温. 它的工作频率低,利于降低超导体的高频功耗. 由于这些原因螺旋线加速器成了目前质子超导低能直线加速器的重要生长点之一.

与常温螺旋线加速器相比,超导螺旋线加速器的突出优点是:功率消耗大大节省,高频功率(包括束流功率在内)可节省约 27 倍,总的电源功率(包括制冷机,低温设备功率在内)节省约 7 倍(参阅表 2);工作性能稳定,束流的能散度预期 $\gtrsim 10^{-4}$;工作能量的上限可扩展到 30 MeV 以上;加速器的长度可比常温缩短约 3 倍. 超导螺旋线加速器与常温螺旋线加速器一样宜于加速速度小的离子并有可能达到高的束强. 联邦德国卡茨鲁的 kuntz 等经过计算和有关试验后认为超导螺旋线加速器的束强可达到 1 ma 以上[19～20].

表 3 联邦德国所做常温螺旋线加速器与超导螺旋线加速器的比较

($\Delta T = 1$ MeV.)

	常 温	超 导
加速场强	1.2 MV/m	3 MV/m
电长度	0.83 m/MV	0.33 m/MV
频率	108 MC/sec	90 MC/sec
高频系统	1.85×10^5 联邦德国马克	5.0×10^4 联邦德国马克
加速结构	7.5×10^4 联邦德国马克	3.5×10^4 联邦德国马克
He—冷却		1.35×10^5 联邦德国马克
总固定投资	2.60×10^5 联邦德国马克	2.20×10^5 联邦德国马克
高频功率	80 kW	3 kW
总设备容量	165 kW	30 kW
运转费用	27 联邦德国马克/小时	4 联邦德国马克/小时

 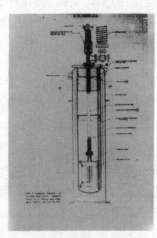

图 10　(a) 卡茨鲁的铌制超导螺旋线加速腔　　(b) 美橡树岭的涂铅螺旋线腔

国外迄今进行了两种超导腔的试验,一种是在铜螺旋线上涂铅的腔[21],一种是纯铌的腔[22]. 前者做起来简便,腔体损耗的改进因子也接近于 10^5,但能达到的最大加速场强比后者低. 铌腔一般用杂质小于 5×10^{-4} 的纯铌作. 为了得到高的破坏场强预先要对螺旋线进行一系列精心的清洁处理,包括化学处理、电解抛光、高温高真空排气等等. 在 1.8 K 的低温下,联邦德国卡茨鲁的铌腔的表面电阻改进因子约达 $10^5 \sim 10^6$. $\beta=0.06$ 时行波成分的最大振幅达到 3 MV/m[19](最大表面场强为 30 MV/m). 从前面提到的 E_S/E_0 随能量变化的关系来看,对较大的 β, $E_{0\max}$ 还有可能提高 50% 或更高一些.

铌制超导螺旋线的机械强度很低,容易发生电与机械的耦合振动,妨碍场的建立. 这个问题曾一度引起对超导螺旋线腔发展前途的争论. 后

图 11　卡茨鲁的超导螺旋线加速节

来美国橡树岭实验室用自激振荡器的方法解决了这个问题,联邦德国卡茨鲁中心则用自动锁相系统克服了这一困难.

尽管质子超导加速器的工作开展得较晚,如联邦德国于1969年才动手进行试验研究,但进展速度很快.卡茨鲁原子核研究中心和法兰克福大学已计划建造一台60 MeV超导螺旋线加速器作为超导介子工厂(1 GeV, 1 mA)的注入级.1972年初他们试验了注入级的第一节腔.这是由五个互相强耦合着的短螺旋线驻波结构(直径 $2a=7\sim9$ cm)构成的加速腔,参阅图11.腔体外壳直径40 cm,长54 cm.腔体的品质因素 $Q\approx3\times10^7$.试验结果,由高压倍加器注入的 $\beta=0.04$ 的质子,获得了 $400\text{ keV}\pm10\%$ 的能量增益,符合理论的预计[23].加速过程中由腔移走的热功率共4 W.下一步他们将试验二节超导腔的联合运转,并将提高束流强度.

目前国外除卡茨鲁以外还有5~6个单位从事着超导螺旋线加速器的研究.据报导美国阿贡国立试验室[24~25]也于今年初制成二节超导重离子螺旋线加速腔.从工作开展的势头来看,超导螺旋线加速器的前途是乐观的.当然在目前还要花很大的努力来解决超导体的老化("时效"),辐照损伤以及提高超导结构的机械强度等等问题.

五、初步的实践和体会

为了探索适合于重离子的加速结构,我们在调查研究基础上制备了三个不同参数的螺旋线波导试验腔(图12).对腔体的品质因数,相速以

图12 北京大学螺旋线波导试验腔

及分路阻抗等进行了初步的测试.试验的目的在于检验现有理论并掌握腔体物理性能的实际变化规律.试验腔的螺旋线由直径为 4 mm 的镀银黄铜丝制成,半径分别为 25.4,28.6,31.1 mm,螺距 7.010~7.017 mm,长度分别为 5、6、7 个半波长.外壳内径 24 cm,试验腔都在驻波态下工作,波节处支有聚乙烯柱.高频讯号由一自制 98~104 MHz(频率稳定度 2×10^{-6}/h)1 W 发生器供给,腔内电场分布则由一直径为 12 mm 的有机玻璃微扰小球测得[26].典型的场分布曲线如图 13.

测量典型结果如下:

表 4

$\bar{a}$	$\bar{b}$	$\bar{d}$	$\bar{p}$	f	Q	$\bar{\beta}$	Z_{SW}
28.6 mm	240 mm	4 mm	7 mm	100.2 MHz	1.5×10^3	$0.0453\pm1\times10^{-4}$	16 MΩ/m
					1540*	0.0461*	17.3 MΩ*/m

(字母上面的"—"号表平均值;* 号表由导面模型给出的理论值.)

关于测试的详情及有关结果将另文发表。

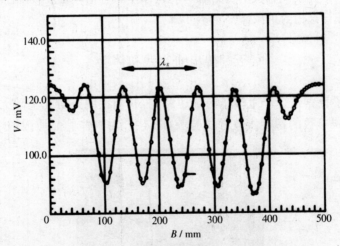

图 13 1 号腔轴线上场分布的微扰曲线

经过这一段的初步实践我们感到① 对于长螺旋线腔,导面模型可以相当准确的预言基模的相速;理论值与实践结果的偏差约 1%~2%.看来这种模型可以方便地作为加速腔设计的初步基础;② 我们曾在 3 个月的时间内,对同一腔反复进行 12 次测量,各次测量间的环境温度、探针和天线的位置等都有差异(先后环境温度的差异最大达 6°C). 在这样的条件下,调谐频率的变化约为 0.26%,而相速的涨落则在测量统计误差 3×10^{-3} 之内. 这一情况表明由于螺旋线腔的公差要求低、色散小,腔体的工作性能比较稳定. 对温度等的控制比 Alvarez 腔的要求低得多. ③ 螺旋线波导腔的结构较简单,尺寸也小. 制备时除螺旋线的绕制工艺需要重点掌握外,其他加工比较简单,涉及的材料、工种比较少,不需大型设备. 同时螺旋线腔高频电源的频率正好是电视设备(或短波机)的波段,国内已有条件生产. 由于这些缘故这种加速腔有易于制备、上马较快的优点.

以上的调研和实践都是初步的,计划在今后深入地研究不同条件下腔体性能的变化规律;尤其是注意研究末端效应等现有理论未能准确预言的问题并在此基础上进行粒子加速试验. 另一方面,我们注意到超导螺旋线波导加速器是常温螺旋线波导加速器发展的必然趋势. 尽管前者在表面工艺,机械振动等问题上仍有相当大的困难,但前者终将胜过后者,因此将适时地开展超导螺旋线波导加速器的研究.

参 考 文 献

[1] W. Walkinshaw & Wyllie, Math. Memo 57(1948).
[2] K. Johnsen Chr. Michelsens Inst. Breter 14(1951).
[3] W. Muller et al., N.I.M. 4(1959) 202.
[4] D.R. Ghick et al., Proc. IEE Part. C Vol. 108 13(1961) 425.
[5] H. Klein et al., Proc. Heidelberg Gonf. on Heavy Ions 1969 p.540.
[6] M. Blann N.I.M. 97(1971) 1.
[7] H. Klein et al., N.I.M. 97(1971) 41.
[8] Physics Today, (1972) 7, p.19.
[9] J.R. Pcirce Travelling Wave Tubes 1950.
[10] S. Sensiper, Proc. I.R.E. 43(1955) 149.
[11] J. Crouzet, P.A. Vol. 1, 2(1970).
[12] A.J. Sierk, IEEE Trans. NS—19, 2(1971) 309.
[13] A.J. Sierk, P.A. Vol. 2, 2(1971) 149.
[14] G. Dome, Linear Accelerators p.725~727.
[15] B.W. Montague, CEAL—2000, p.174 (1967) CERN ISR—300/w/Jw/pd (1969).
[16] B.W. Montague, CERN, ISR—300/w/67~65 (1967).

[17] B. W. Montague, CERN, ISR—300 Lin/67~34 (1967).
[18] H. Danzer et al., Z. Naturforsch 21a, (1966) 1761.
[19] H. Klein, M. Kuntze, IEEE Trans. NS—19, 2(1972) 304.
[20] M. Kuntze et al., IEEE Trans. NS—183 (1971) 137.
[21] C. M. Jones P. A. Vol. 3, 2(1971) 103.
[22] A. Cirton et al., IEEE Trans. NS—19, (1972) 278.
[23] CERN Courier, No. 4, Vol. 12(1972) 134.
[24] A. H. Jaffey, IEEE Trans NS—19, 2(1972) 238.
[25] A. H. Jaffey et al., Bull. Am. Phy. Soc. Feb. (1973).
[26] 李坤等,螺旋线波导腔的测试,北京大学技物系内部报告.

螺旋波导加速腔的不载束高功率试验[①]

摘　要

本文简述了一个在 19 kW 功率下净电压增益 400 kV 以上的螺旋波导加速腔,从实验上研究了该加速腔的静态频移和动态不稳定性等工作特性.这个腔计划用作 400 kV 高压倍加器的后加速增能器.

近年来国际上广泛地用常温或超导的螺旋波导加速腔,"盘香"式和"分离环"式螺旋线谐振腔等作直流高压型加速器的后加速增能器,扩大加速粒子的能量和种类,取得了良好的效果.这类结构尺寸小、运用灵活、容易加工.在我国发展这类增能器的技术,将有利于充分发挥现有高压倍加器和静电加速器等设备的效用.

鉴于国内现有高功率源的频率较低(30 MHz 以下)和注入器的能量较低,我们选用螺旋线型加速腔作 400 kV 高压倍加器的后加速增能器.为取得必要的经验,我们制备了一个初始能量约 0.2 MeV/A 的螺旋波导加速腔.加速腔的结构如图 1 所示.主要参数如下:外罩筒内径 240 mm、长 650 mm,腔内装有两根耦合的半波长螺旋线,螺旋线总长 339 mm、螺距 5.65 mm、直径 80 mm.实验测得加速腔的谐振频率 f_0 为 28.4685 MHz,品质因素 Q_0 为 965,净分路阻抗 η_0 为 22 MΩ/m,电长度 L 为 383 mm.为了能加速低能粒子,设计的相速较低,螺旋线的几何参数只得偏离最佳值,因而 Q_0 和 η_0 均较低.用小球微扰法测得加速腔轴向电场分布见图 2,由于场的边缘效应,两边场分布较低,实际的场分布相当于三个相速分别为 0.0214、0.0309、0.0203 的半波长驻波场.

[①] 合作者:梁仲鑫,李纬国,林揆训,韩崇霈,钱伟述.摘自《高能物理与核物理》,Vol. 7, 4 (1983) 515~518.

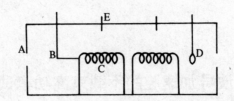

A. 外罩筒 B. 馈送头 C. 螺旋线 D. 调谐环 E. 探针

图 1　加速腔结构示意图

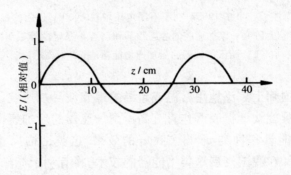

图 2　场分布曲线

由净分路阻抗及场分布推算得 19 kW 下净电压增益 $V_0 = \int |\varepsilon_s(z)| \, dz = 400$ kV,最大能量增益 $\Delta W_{max} = 312$ keV.

试验装置见方块图 3. 运行时,系统的真空度不低于 2×10^{-5} mmHg.

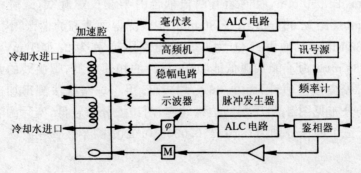

图 3　实验装置方块图

在脉冲状态下进行了不载束的高功率试验,实验表明:负载周期1/6

时,加速腔可在轴上平均加速场强 0.82 MV/m 的水平下正常运转,此时轴上驻波场强峰值达 1.92 MV/m,累积考验时间 700 小时以上,其中连续运转时间最长为 18 小时,轴上平均加速场强最高曾加到 0.92 MV/m,此时进腔功率为 25 kW,轴上驻波电场峰值达 2.27 MV/m.

在不载束的高功率条件下,我们研究了螺旋波导加速腔的一些工作特性,主要结果如下:

一 静态频移和"病态"Q 曲线

螺旋线受到电磁场压力作用而产生压缩形变,可引起其谐振频率 f_0 向低频方向移动,称之为静态频移 Δf_s. 它满足关系式[1] $\Delta f_s/f_0 = -KQP$,式中 P 为进腔功率,K 为静态电磁、机械耦合常数. 在不同负载周期 D_f 下,实验测得 $P\sim f$ 曲线见图 4. 结果如下:

D_f	$\dfrac{1}{4}$	$\dfrac{1}{6}$	$\dfrac{1}{8}$
$\Delta f_s/P$(kHz/kW)	−5.55	−3.73	−2.78
$\Delta f_s/\overline{P}$(kHz/kW)	−22.2	−22.4	−22.2

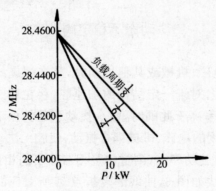

图 4 静态频移曲线

实验表明:静态频移正比于进腔的平均功率 $\overline{P}$,$\Delta f_s/\overline{P} = -22.3$ kHz/kW.

静态频移引起 Q 曲线向低频方向歪斜——"病态"Q 曲线. 我们在负载周期为 1/8、进腔功率 15.7 kW 条件下,测量了"病态"Q 曲线见图 5.

曲线表明:"病态"Q曲线的实验值和理论值基本相符.

随着螺旋线温度的升高,加速腔的谐振频率要向低频方向移动.实验测得 $\Delta f_s/\Delta T = -1.2\,\mathrm{kHz/°C}$.

电磁压力和温升引起的频移,由一个调谐能力为 220 kHz 的调谐环来补偿.

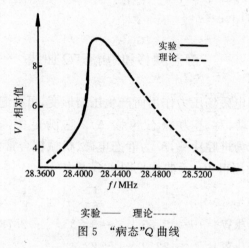

实验—— 理论------
图 5 "病态"Q曲线

二 动态不稳定阈功率

当外界存在电磁、机械或其他扰动时,易在螺旋线谐振频率的高侧引起动态不稳定性,对于一定结构的加速腔,存在着一个动态不稳定阈功率 $P_u^{d[2]}$.进腔功率高于此阈时,微小扰动引起的振动将自动地连续增强,直至发生大幅度的电磁、机械耦合振动,使腔无法正常工作.

实验时改变进腔功率 P,测量由鉴相器输出的相误差电压 $|\Delta V|$,可画出 $P \sim |\Delta V|$ 曲线如图6,曲线的转折点 A 所对应的功率定义为动态不稳定阈功率.测量的结果如下:

D_f	$\frac{1}{2}$	$\frac{1}{3}$	$\frac{1}{4}$	$\frac{1}{6}$
P/kW	7.0	10.2	13.3	20.2
$\overline{P}$/kW	3.5	3.4	3.4	3.4

由此可见:动态不稳定阈功率 P_u^d 系由平均功率所决定的.对于本试

验的加速腔,动态不稳定阈功率 $P_u^d \equiv 3.4\ \text{kW}$.

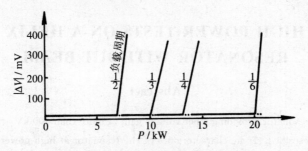

图 6 相误差电压与进腔功率关系曲线

三 "有害"调制频率

利用功率调制的方法可使脉冲功率高出平均阈功率好几倍,所以功率调制是保证腔稳定运行的有效措施.但是调制本身是一种周期性的扰动,当调制频率为螺旋线的机械振动基频及其倍频时,同样将使螺旋线的机械振动的振幅越来越大,造成腔的场强发生严重的不稳定,我们称这种调制频率为"有害"调制频率.工作时必须避开它!

实验时固定进腔功率,改变调制频率,在 200 Hz 以内,我们测到的"有害"调制频率有 19.6,40.0,59.5,79.5 和 99.5 Hz.

实验还观察到:当进腔功率为 10～20 W 时,有二次电子共振现象,此时真空度下降,测不到加速腔的 Q 值.当进腔功率为 18 kW 时,在加速腔外罩筒表面可测到很强的 X 射线,局部表面竟达 800 毫伦/h. 用 3 mm 厚的铅板阻挡后,离腔 1 m 以外的地方可达安全剂量.

高功率运行中最大的困难是来自电磁、机械耦合振动,为此,应提高螺旋线的机械强度,改善电场稳幅控制.

在试验中,高频相控组的同志曾参加了部分实验工作,并对试验提过很多宝贵意见,特此感谢.

参 考 文 献

[1] 陈佳洱,螺旋线波导馈电问题,内部资料,1976,3.
[2] A. Schempp, W. Rohrbach and H. Klein, *Nuclear Instruments and Methods*, Jan-

uary 1, No. 1, Vol. **140**(1977).

HIGH POWER TESTS ON A HELIX RESONATOR WITHOUT BEAM

Abstract

A helix resonator providing a net voltage gain of more than 400 kV at power levels of 19 kW is lescribed. Some characteristics of the resonator at high power levels, such as static frequency shift and dynamic instability are studied experimentally. This helix resonator will be used as a booster for a 400 kV Cockcroft-Walton generator.

螺旋波导加速腔载束全功率试验[①]

摘 要

本文扼要叙述了螺旋波导加速腔载束全功率试验装置. 着重描述了加速腔体结构、静态试验和不载束全功率试验测得的参数和特性. 较详细地叙述了载束全功率试验测量能谱的方法及试验结果, 并与理论计算做了比较. 在试验过程中遇到的稳定性问题也做了初步的试验研究和分析.

当入射粒子能量为 250 keV 时, 在 19.3 kW, 负载因子为 1/6 的高频功率激励下, 测得增能为 $\Delta W = 322$ keV, 比理论计算结果 312 keV 偏高 4%. 测得腔的净分路阻抗为 21.6 MΩ/m, 这比静态测量结果 21.4 MΩ/m 只偏高 1%, 比理论计算结果 21.95 MΩ/m 只偏低 1.5%. 说明理论与试验结果符合得很好.

试验测得渡越时间因子 T 的相对变化. 结果表明入射粒子能量大于 200 keV 后, T 的变化缓慢. 说明短螺旋波导加速腔对注入粒子能量的变化适应能力还是较强的.

一、引 言

继螺旋波导加速腔的聚束试验[1]、重粒子模拟加速试验[2]和不载束全功率试验后[3,4], 我们又进行了载束全功率试验. 载束全功率试验的目的在于为建造中小型重离子加速器及为倍加器、静电加速器等小型加速器后接加速增能器提供根据和经验. 试验完成后, 试验腔本身即成为一小增能器, 可以把北京师范大学低能所现有的 360 keV 的质子能量提高到 676 keV.

[①] 合作者: 梁仲鑫、李纬国、林撰训、何佩伦、韩崇霈、钱伟述. 摘自《全国直线加速器会议报告论文集》, BDJN (1982).

全功率载束试验分两个方面进行. 一是在高压倍加器束流能量不变（我们分别在 250 keV 和 360 keV 下试验）情况下，改变加速腔的馈入功率，测进腔功率与输出粒子能量关系曲线. 二是固定馈入腔的功率（我们取 15.2 kW），改变高压倍加器束流能量，测注入腔粒子能量与从腔引出粒子能量关系曲线. 即渡越时间因子 T 随注入能量 W_0 变化曲线.

以下本文分别就试验腔的结构和特性，载束试验装置，试验方法及结果等方面进行讨论.

二、试 验 装 置

载束试验装置由高压倍加器，束流输运系统，束流能谱测量系统，高频功率馈送及控制系统等所组成.

1. 高压倍加器，应用北京师范大学低能所高压倍加器，最高电压 400 kV，高压稳定度约为 $2\sim4\times10^{-3}$. 在靶处束流强度可达 100 μA[5].

2. 束流输运系统见图 1，由倍加器引出的质子束经 90°偏转磁铁后进入水平方向输运线，线上设有一对导向磁铁，四个石英萤光片和两个法拉第筒供对中调束和监测束流之用. 线上还设有三对静电四极透镜，可在腔中心附近和分析磁铁前 $2R=500$ mm 处提供两个光腰. 静电四极透镜的几何孔径有 35 mm 和 40 mm 两种，几何极长 80 mm，最大场梯度为 6.5 kV/cm². 整个输运线输运效率优于 80%.

3. 能谱测量系统由分析磁铁、束流相图仪和 X-Y 记录仪构成. 分析磁铁由电流稳定度为 1×10^{-4}，输出电流从零到 10 安培连续可调的稳流电源激励. 分析磁铁为双聚焦型，曲率半径 R 为 250 mm，偏转角为 90°；有效入射角 α 和出射角 β 均为 26.5°；极间隙 25 mm. 当物缝为 1 mm 时，能量分辨率 $W_0/\Delta W_{\frac{1}{2}}=500$，$W_0$ 是中心能量，$\Delta W_{\frac{1}{2}}$ 为强度减半时的能散宽度. 励磁电流 I 与磁场强度 B 间有较好的线性关系. （见图 7）

4. 高频功率馈送及控制系统，系统框图见图 3. XFG-7 讯号源产生 28 MHz 高频讯号，XC-13 脉冲发生器产生方波讯号分别输入到调制放大器中，由此得到 1/6 调制的高频讯号输入到北京广播器材厂生产的 XFD-D5 型高频线性放大器中，输出的大功率调制讯号通过 75 Ω 馈线馈入腔中，腔呈临界耦合工作状态.

自动调谐电路由电平自动调整电路 ALC[6]、鉴相电路、电调移相器及调谐驱动电路[7]等部分组成. 从串联于功率馈送线上的 40db 反射计取出入射波讯号与由腔提取的讯号均分别经 ALC 电路后输入到鉴相器中比较相位,鉴相器输出误差讯号反映了失谐量 $\Delta f/f_0$. 电调移相器串联于一个支路上,当腔处于谐振点附近(我们取比谐振频率略高一点频率为工作点,驻波比在 1.1 至 1.2 间),调电调移相器使鉴相器输出讯号为零. 整个电路闭环工作时,腔的失谐量 $<\pm 1.5$ kHz. 调谐环动作的响应时间约在 10.1 s 左右.

试验是在没有腔电场幅度稳定装置下进行的.

三、螺旋波导加速腔

腔的合金铝外罩筒内径为 240 mm,长 650 mm,内装两条耦合的半波长螺旋线,工作于 π 模式. 螺旋线由外径为 6 mm 的不锈钢管压扁后绕成,表面镀有银层. 设计参数为:直径 80 mm;螺距 5.62 mm;半波长螺旋线几何长度 169.5 mm. 用小球微扰法测得的静态特性为:谐振频率 $f_0=28.47$ MHz;品质因素 965;净分路阻抗 21.4 MΩ/m.

螺旋线通以普通自来水冷却,出水口后接精密温度计和流量计. 由测量流经螺旋线的水温升和流量来标定进腔功率. 螺旋线一端加有调谐能力为 220 kHz 的调谐环.

由于末端效应的影响,二个半波长线耦合起来形成了相速分别为 0.0205, 0.0309, 0.0200 的三个半波长驻波场(见图 2). 在不载束全功率试验中测得其全功率特性为:

静态频移 $\Delta f_s/\overline{P}=22.3$ kHz/kW,Δf_s 为由功率增加引起的谐振频率向低频方向移动量. $\overline{P}$ 为进腔平均功率.

动态不稳定性阈功率 $\overline{P}_d=3.4$ kW(平均功率).

静态不稳定性阈功率 $\overline{P}_s>3.4$ kW.

螺旋线振动本征频率 $f_m\approx 19.6$ Hz.

有害调制频率为 19.6;40.0;59.5;79.5;99.5 Hz 等. 即是 f_m 的倍频. 试验中应避开上述频率.

在高功率试验中,功率大于 10 kW 后,X 射线剂量明显增强. 在

15 kW 时,剂量竟达 800 毫拉得/小时,用 3 mm 薄铅板将腔屏蔽后,剂量降至安全值.

我们根据静态测量的场分布相对值(图 2),和功率试验中腔内轴向电场值,用辛卜生方法计算了"净阻抗"$\eta_0 = 21.95\ \text{M}\Omega/\text{m}$,计算了初能量 $W_0 = 250\ \text{keV}$,初相位 $\varphi_0 = 20°$;高频功率 19.3 kW 下增能 $\Delta W = 0.31206\ \text{MeV}$. 我们也用龙格库塔方法解相运动微分方程,并用优选法求最佳注入相位下增能,结果为 $W_0 = 250\ \text{keV}, \varphi_0 = 21°, P = 19.3\ \text{kW}, \Delta W = 0.31494\ \text{MeV}$ 两种计算方法符合很好.[8]

四、能谱测量及结果

首先与倍加器配合调束流满足试验要求:束流中心与系统机械中心合轴;流强达到 100 μA 左右. 后让束流通过分析磁铁,由束流相图仪的在垂直方向放置的探针接受束流. 此束流产生的电压讯号经放大输入到 X-Y 记录仪的 Y 输入上. 而稳流源输出的电压讯号接到 X-Y 记录仪的 X 输入上,则改变激励电流强度就可在 X-Y 记录仪上画出能谱了. 能谱中的中间峰(见图 4)为腔内不存在电场的 5/6 调制周期中飘浮过来的束流的能谱曲线,即为倍加器输出束流的能谱. 增能峰或减能峰为经过腔电场加速或减速的粒子的能谱. 由于倍加器输出质子能量是已知的,则我们可以从能谱曲线上标定粒子的能量增加或减小 ΔW.

为减少测量误差,试验中注意了以下几点.

1. 在束流接受探针接受中间峰束流条件下,进一步调整三个四极透镜,使聚焦更好,束流变强.

2. 在束流接受探针接受增能峰(或减能峰,因减能峰最弱,故对减能峰调整好后,一般就可以一次画出三个峰的能谱曲线了.)条件下,再调第一个四极透镜,使增能峰变强,宽度变窄,即能散减小. 说明此时束流光腰已投入加速腔中心附近. 由于通过腔束流横向束宽变小,即减小了横向散焦,增强了束流强度,又减小了由于束流偏离中心轴向电场而引起的能散,故使测得峰变高而窄. 然后再调整腔后面第二和第三个四极透镜,使增能峰束流增强. 说明此时第二个光腰已投入到分析磁铁前 $2R = 500\ \text{mm}$ 处的物点,则使放置于双聚焦型分析磁铁的像点处的接受探针接受束流变强。在物点处设置一两爪光阑,使水平方向通过的束宽限制在 1 mm 以内,以保证磁分析器能量分辨率好于 500。束流接受探针后装有石英萤光片,可以从观察窗口直接观察聚焦的好坏及能散大小。对束流调整很方便。

3. 测量前分析磁铁的磁场是先经过反复磁化后再测量的,从而保证了较好的线性和重复性。

4. 功率测量,我们以螺旋线冷却水的温升与流量来标定进腔功率。这较真实的反映了进腔功率。但由于水引出口处与合金铝腔壁接触,故散失一部分热量,故测得水温略低于实际水温,即使测的进腔功率略低于实际进腔功率。

试验测量结果及理论计算结果见图 5 和图 6。图 5 为倍加器输出能量不变(分别为 $W_0=250$ keV 和 $W_0=360$ keV)条件下测进腔功率与增能关系曲线。图 6 为进腔功率 $P=15.2$ kW 条件下,改变注入能量 W_0,测注入腔粒子能量与腔引出粒子能量关系曲线。实线为理论曲线,虚线为测量曲线,圆圈为实验点。

主要结果如下:

注入能量 W_0	高频功率 P	增能(测量)	能量(计算)
250 keV	18.5 kW	319 keV	305 keV
360 keV	19.3 kW	316 keV	308 keV

理论计算的渡越时间因子为 0.78。这是因设计的腔的相速 β_c 与粒子的速度 β_p 相差较大造成的。试验中测量了渡越时间因子 T 的相对变化。当 W_0 在 250~300 keV 间时,T 几乎不变。W_0 小于 250 keV 时 T 开始下降,并且下降较快。当 $W_0=170$ keV,T 下降 13%。当 W_0 大于 300 keV 时 T 也下降,但下降较慢,$W_0=360$ keV 时,T 只下降 5%。说明短螺旋线腔对入射粒子能量变化适应能力较强。

根据能谱测量计算腔分路阻抗为 21.6 MΩ/m,比理论值 21.95 MΩ/m 低 1.5%,比静态测量值 21.4 MΩ/m 高 1%。说明理论与实验值符合的很好。

五、高功率下运行的稳定性

动态和静态不稳定性问题[3]这里不再重述。只叙述在试验过程中与稳定性有关的几个问题.

1. 真空度的影响,全功率试验过程中真空一直保持在 $1\sim 3\times 10^{-5}$ mmHg 范围,试验可正常进行. 但先后由于腔与馈管联结处"O"圈压偏,水接头密封铅垫漏水等原因造成的与温度有关的真空下降现象. 即不加功率时真空正常,但加到一定功率水平时真空开始下降,如不及时发现并降下功率,真空可降到 10^{-3} 量级,此时腔内出现明亮的放电闪光,功率被全反射,曾使直径为 1.2 mm 的六根铜线并成的馈线烧熔.

2. 接馈管的腔口输出阻抗的影响. 以前是以腔工作在临界耦合工作状态为准来调馈点的,此时 $Q_l=Q_0/2$. Q_l 叫有载品质因素. 现在我们采用 8405 矢量伏特计用 $(1+\rho)$ 方法测腔口输出阻抗为 75 Ω 来定馈点的. ρ 为反射系数. 此方法精度更高. 试验表明腔口输出阻抗为 75 Ω 时可以稳定运行. 偏离较大时稳定性变坏.

3. 馈管长度的影响[9]. 试验表明馈管长度为半波长的整数倍数、稳定性很差. 这可能是高频讯号在馈管中共振引起的. 试验发现长度为半波长整数倍再加上 0.7 半波长时,稳定性好.

入射功率在 5 kW 左右,改变频率 f,测下面三条曲线,即 $V\text{-}f$ 曲线、$P\text{-}f$ 曲线和 $s\text{-}f$ 曲线,V 为腔提取讯号幅度,P 为进腔功率,s 为驻波比. 试验发现,在谐振频率附近,$V\text{-}f$ 曲线取极大,$P\text{-}f$ 曲线取极大,$s\text{-}f$ 曲线取极小;且三条曲线对称性也较好. 如图 9,则工作稳定,反之,如出现图 8 的曲线,则稳定性极差. 功率根本加不上去. 这只是找到了实验规律,理论解释尚在研究之中.

尹利生、韩秀清同志在实验中给予大力协助,本试验是在北京师范大学低能所完成的. 在试验过程中得到了该所的积极配合和大力协助. 在此深表谢意.

参 考 资 料

[1] 螺旋波导腔聚束试验. 北京大学加速器组"7661"会议报告,1976.5
[2] 螺旋波导重离子的模拟加速试验. 北京大学加速器组"7661"会议报告,1976.5
[3] 螺旋波导加速腔不载束全功率试验. 北京大学加速器组"7661"会议报告,1976.5
[4] 螺旋波导加速腔不载束高功率试验. 梁仲鑫等,中国粒子加速器学会第一次代表大会学术报告 1980.10
[5] 倍加器高压稳压总结. 北京师范大学物理系
[6] 螺旋波导加速腔控制系统用鉴相器. 李坤、钱祖保,中国粒子加速器学会第一次代

表大会学术报告 1980.10
[7] 螺旋波导腔谐振频率自动调整系统. 庞继藻,中国粒子加速器学会第一次代表大会学术报告 1980.10
[8] 螺旋波导增能器的计算器模拟. 陈佳洱,北京大学加速器组内部报告 1981
[9] 馈线长度对高功率腔稳定性的影响. 林揆训等,北京大学加速器组内部报告 1981

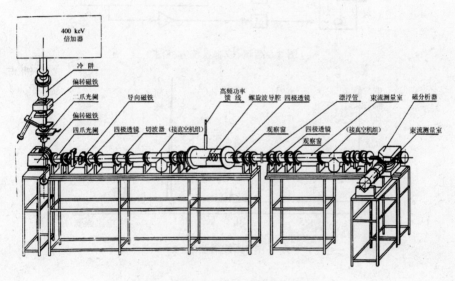

图 1　螺旋波导载束试验线示意图

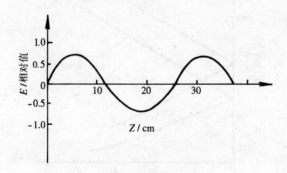

图 2　场分布曲线

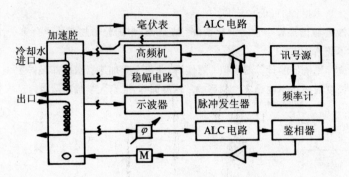

图 3　高频、控制系统方框图

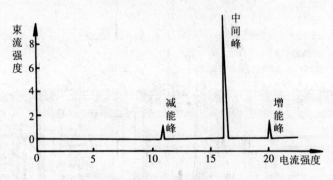

图 4　能谱曲线

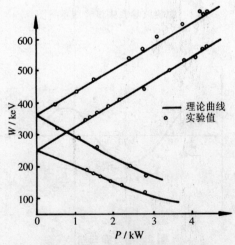

图 5　输入功率与增能关系曲线

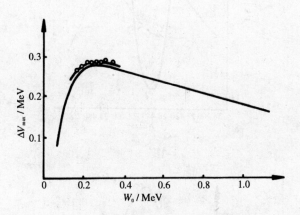

图 6　注入能量与增能关系曲线

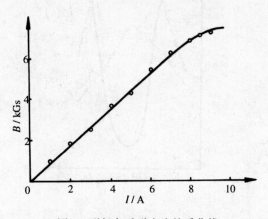

图 7　磁场与励磁电流关系曲线

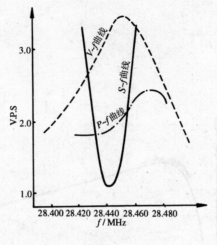

图 8

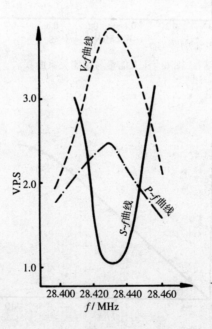

图 9

AN INTEGRAL SPLITRING RESONATOR LOADED WITH DRIFT TUBES & RF QUADRUPOLES [1]

Summary

In order to improve the mechanical stability, the coupled splitring (also spirals) resonators, loaded either with drift tubes or RF Quadrupoles, are integrated together through conducting bars. Investigations on 1/2 and full scale models (50 cm in tank diameter) show considerable improvement on the overall rigidity of the structure while keeping the RF efficiency high. The operating frequency can be greatly reduced by the integration to 24 and 14 MHz for loading with drift tubes and RFQ respectively. The integration also flattens the accelerating voltage distribution and enhances the mode separation and thus facilitates the assembling and commissioning of the accelerating structure. An equivalent circuit of the integral splitring, which agrees well with the experiments, has been developed.

Introduction

The conventional splitring and spiral resonators with drift tubes have the merits of moderate size, simple structure and good efficiency and have been successfully used as the accelerating structures of post linear acceler-

---

[1] Coauthers: J. X. Fang. Reprinted from the IEEE. Trans., NS-32, 5(1985) 2891~2893.

ators in many laboratories[1~4], while RFQ structures with spiral stems have also been developed in a number of laboratories in recent years. [5]

They are very suitable to operate at low frequency for accelerating heavy ions. To achieve stable operation, both the splitring (or spiral) resonator and the spiral RFQ need good enough mechanical rigidity so that the mechanical vibration caused by pondermotive force and mechanical noises can be negligible. However for the case of low frequency the rigidity could be weakened due to the long arms of the structure. In order to keep high stability even though at low frequency, an integral splitring resonator has been developed. This type of resonator is actually a modified coupled structure of conventional splitring or spiral resonators and it can be loaded either with drift tubes or RFQ. A series of 1/2 and full scale models have been constructed and tested. Meanwhile a lumped equivalent circuit was developed and compared with the experiments.

Structure and Properties

The integral splitring resonator is actually a coupled resonator. Usually a multi-cell coupled resonator is good for increasing the RF efficiency of the structure, though the number of the cells must be choosen as a compromise between the RF efficiency and the reduced flexibility of energy variation. However the integral splitring structure not only leads to sufficient RF efficiency, but also higher stability in operation as well as other performances. Actually this structure differs from the conventional coupled splitring mainly in that the cells are directly connected together into an integral structure by several conducting bars. The bottom bar is used to combine all the conventional splitrings together at the lower ends of each arm, while two top bars, which play most important role in improving rigidity and RF properties, are connected to the splitring loops at their top end; one bar for right-wound arms and another bar for left-wound arms. Between the top and bottom bars several additional pairs of combining bars can be added if it is necessary to further increase the rigidity of the structure. Thus the integral structure is formed like a trussed frame. In this structure, all the arms are made perpendicular to the combining bars and the drift tubes or RFQ electrodes can be mounted at their ends near the top bars as in the case of conventional spirals. As a result of integrating, the rigidity of structure is considerably improved. In addition, the bottom bar also offers a better conduction to RF current for all cells so that one can even leaves out the original legs from each ring loop if the cooling water goes through the bottom bar. The full scale model of in-

(a) Resonator with drift tubes

(b) Frame with RFQ

Fig. 1 Integral splitring resonator

tegral splitring loaded with drift tubes and RFQ are shown in Fig. 1(a) and (b) respectively. It can be seen that the structure can be conveniently assembled outside tank as a whole and easily mounted into the tank through a number of legs. Apart from the mechanical property, the integration also improves the RF behaviour of the structure. Usually there are N modes in a conventional coupled resonator containing N electromagnetically coupled cells, and the accelerating voltage distribution between drift tubes is not flat (Fig. 2). But in the integral N-cell splitring resonator with top bars, only π and 0 mode exist and the accelerating voltage distribution is flattened (Fig. 2). Meanwhile the integration keeps RF efficiency reasonably high. These will be described in detail later on.

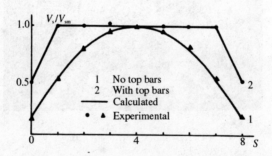

Fig. 2 Accelerating voltage distribution in gaps

Lumped Equivalent Circuit

The RF behaviour of the integral splitring resonator can be briefly explained by using a simple lumped circuit. The lumped equivalent circuit for the case without the top bars is shown in Fig. 3(a). For simplicity, the cells are considered uniform and each couples only with its closest neighbours. Each of the N cells is corresponding to an arm with a drift tube. The end cells of both sides are changed a little by the end effect:

$$L' = L(1 - M^2/L^2), \quad C' = C(1 + D/C)$$

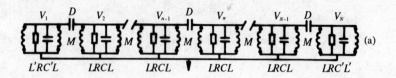

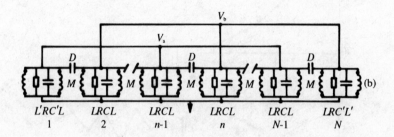

Fig. 3 Equivalent circuits

(a) Without top bars,   (b) With top bars

When Q value is high enough, i.e. $Q \gg \omega_s/\omega$, where $Q = \omega_s(C + 2D)R$,

$$\omega_s^2 = 2L/((C + 2D)(L^2 - M^2))$$

the mode frequencies and voltage distribution along the chaincircuit can be solved from the Kirchhoff's equations as follows[6]:

$$\omega_q^2 = \omega_s^2(1 - K_i\cos\phi_q)/(1 - K_c\cos\phi_q)$$

$$V_{qn} = A_q\sin n\phi_q$$

$$K_i = M/L, \qquad K_c = 2D/(C + 2D)$$

$$\phi_q = q\pi/(N + 1), \qquad A_q = \text{const}$$

$$q, n = 1, 2, \cdots, N$$

The voltage in the S-th gap is:

$$U_{qs} = B_q\cos((2S + 1)\phi_q/2)$$

$$S = 0, 1, \cdots, N. \quad B_q = \text{const}$$

The results above indicate that there are N modes for a resonator which contains N coupled cells. As an example, the calculated mode frequencies and voltage distribution of an integral splitring resonator of 8 cells without top combining bars are illustrated in Fig. 4. It agrees well with the experimental results.

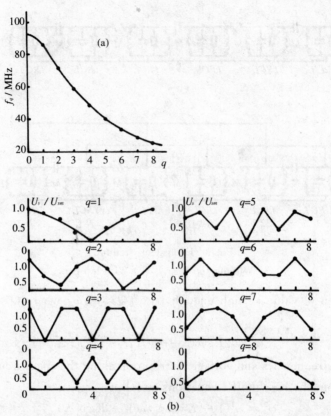

Fig. 4 (a) Mode frequencies   (b) Mode voltage in gaps

$N=8$, no top bars

——— Calculated   • Experimental

As for the function of the top combining bars, if concentrating on accelerating mode, it can be simply considered as adding the conducting wires to the corresponding cells as shown in Fig. 3(b). Thus only π and 0 modes can exist, the mode frequencies and voltages are 6:

for π mode, $\omega_\pi^2 = \omega_s^2(1+FK_i)/(1+FK_c)$

$$V_a = -GV_b$$

for 0 mode, $\omega_0^2 = \omega_s^2(1-FK_i)/(1-FK_c)$

$$V_a = GV_b$$

where $F=1-1/N$, $G=1$,   if N is even

$F=G=(1-1/N)/(1-1/N^2)^{1/2}$, if N is odd

The chaincircuit shown in Fig. 3(b) can also simply be used to RFQ loading, if the D is considered as the piecewise coupled capacitances between the two pairs of RFQ electrodes.

There might be more factors that should be taken into account in practice. For instance, the finite cross section of the top bars may reduce the resonant frequencies. In addition, the assumption of closest neighbouring coupling is not met quite well in some cases. Hence the RF behaviour of the integral splitring resonator can not be wholly derived from a simple lumped chaincircuit like Fig. 3, nevertheless the principal behaviour, especially for the accelerating mode, have been illustrated well by these chaincircuits. So the equivalent circuit should be helpful to the study and design of the integral structures if the parameters have been determined previously.[7,8]

Experimental Studies

Extensive studies were carried out experimentally to show various effects of the combining bars on the RF properties of the cavity. As the first phase, a series of small model resonator (24 cm in tank diameter) of both integral and conventional coupled splitring were constructed, tested and compared carefully with each other so as to determine if the integral structures are suitable to be used as an accelerator structure and whether its RF properties can be competitive with the conventional ones. After a great deal of systematic measurements, the answer is affirmative and satisfactory. Then the second phase of the experiment was followed to get accurate data for the actual integral splitring resonators. Finally, a series of full scale model resonator (50 cm in tank diameter) were constructed and tested.

The tube diameter for the full scale loop arms is 2.2 cm, and the dimension for the drift tubes are: length 5.85 cm, outer diameter 3.85 cm, inner diameter 1.8 cm, inner gap 2.0 cm and end gap 1.0 cm. All the structures are either made of copper or aluminium. In the experiments, great care were given to ensure accurate assembling and good RF conduction so as to achieve a nice stable structure at lower frequency with high efficiency. In addition to the structure loaded with drift tubes, trapezoidally modulated RFQ electrodes loading was also tested. Some typical results of the experiments are listed in Table 1~3 and illustrated in Fig. 2, 4, 5. The field distribution in all cases were measured by well known method of bead perturbation.

Table 1 Experimental Results with Drift Tubes

No.	1	2	3	4	5*
D of tank/cm	50.0	50.0	50.0	50.0	50.0
L of tank/cm	64.0	64.0	64.0	64.0	64.0
N of arms	8	8	8	8	8
L of arms/cm	44.2	44.2	182.3	182.3	44.2
f_π/MHz	102.66	98.14	24.95	24.28	101.15
Q	5800	5248	2626	2529	5813
R_p/MΩ	53.0	50.2	43.9	40.3	49.3
Z/M$\Omega \cdot$m^{-1}	82.8	78.4	68.6	63.0	77.0
Top bars**	No	Yes	No	Yes	No

D—Diameter, L—Length, N—Number
*Conventional coupled splitring
**With cross section of 1.8×1.0 cm^2

Table 2 Experimental Results with RFQ

No.	1	2	3	4	5
D of tank/cm	24	50	50	50	50
L of RFQ/cm	28.8	66.4	66.4	66.4	66.4
N of arms	6	4	4	4	4
L of arms/cm	25.8	44.2	179.0	179.0	179.0
D of aper.*/cm	1.0	1.0	1.0	1.25	1.5
f_π/MHz	87.59	33.15	14.67	15.54	16.27
Q	1406	1571	1202	1233	1261
ρ/k$\Omega \cdot$m	43.2	110.4	190.0	235.6	243.8

*aper.—aperture

AN INTEGRAL SPLITRING RESONATOR LOADED WITH DRIFT TUBES & RF QUADRUPOLES

Table 3 Influence of combining bars on f_π and Z*

No.	bars**	single tube		double tubes	
		f_π/MHz	$Z/\text{M}\Omega \cdot \text{m}^{-1}$	f_π/MHz	$Z/\text{M}\Omega \cdot \text{m}^{-1}$
1	a	190.73	85.6	204.79	73.3
2	ab	189.07	74.4	204.25	65.5
3	abc	188.91	73.7	203.10	58.1
4	abc	191.79	76.3		

*On 1/2 scale model resonators (with 6 drift tubes)
**Cross section of each bar:
  No. 1,2,3—5×2 mm²
  No. 4—Dia. 1.5 mm wire

The parallel resistance R_P, shunt impedance Z and specific R_P-value ρ in the list are defined as follows.

$$R_P = (\int |E|dZ)^2/P = Zl, \quad \rho = lV^2/P$$

P—dicipated RF power in the resonator
V—the voltage between two pairs of RFQ electrodes
l—the length of the cavity

For the case of integral splitring loaded with drift tubes, R_P or Z are nearly as high as that of the conventional coupled splitring cavity, provided that its combining bars are not too thick (Table 1, No. 2 and 5). The ratio of R_P to the number of drift tubes, N, are nearly kept constant as N varies, while the frequency f_π reduces with increasing N (Fig. 5). As for the integral structure with RFQ loading, its ρ value are good enough at low frequencies (Table 2, No. 3~5). The experiments also showed that the RF efficiencies and frequencies decrease slowly with the increase of either the number or the cross section of the combining bars (Table 3, No. 1~4). In addition, the frequencies f_π can also be reduced by increasing the equivalent parameters including inductance L, capacitance C, mutual inductance M or coupled capacitance D, as the equivalent circuits show. As for the gap voltage distribution along the axis, it has been considerably flattened by top bars as can be seen from Fig. 2. The axial and radial field distribution for the RFQ loading structure are also very satisfactory.

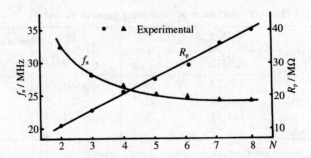

Fig. 5 R_P and f_π vers. number of drift tubes
(with top bars except $N=2$)

The RF behaviour for an integral structure consisted of two parallelly wound tubes, with a higher rigidity, does not show much change from the case of single tube (Table 3, No. 1~30).

Conclusion

The integral splitring resonator loaded with drift tubes or RFQ electrodes improves considerably the overall rigidity of the structure comparing to the conventional coupled splitring or spiral resonators. This will be in favour of a stable operation with good RF efficiencies especially for low frequency structures, which are most suitable to accelerate low energy heavy ions either by drift tubes or RFQ. In addition, the integral splitring structure has the merits of flattened voltage distribution, enhanced seperation and the convenience of assembling and comissioning. The power test for the further study of this integral structure is in progress.

Acknowledgement

The authors wish to thank Prof. F. C. Yu for his kind advices. The effective cooperation of Mrs. J. L. Yuan is also gratefully acknowledged.

References

[1] J. M. Brennan, C. E. Chen, et al., Bulletin of American Physical Society, Vol. **28**.

2(1983) 91.

[2] J. M. Brennan, C. E. Chen, et al., IEEE Trans. **NS-30**, 2798 (1983).

[3] L. M. Bollinger, IEEE Trans. **NS-30**, 2065 (1983).

[4] M. Grieser, et al., IEEE Trans. **NS-30**, 2095 (1983).

[5] H. Klein, IEEE Trans. **NS-30**, 3313 (1983).

[6] J. X. Fang, "An Equivalent Circuit Model of the Integral Splitring Resonator", Int. memo. Peking University, 1984.

[7] E. Mueller, J. X. Fang, Int. Rep. 81～15, Institut fuer Angewandte Physik der Universitaet Frankfurt a. M., 1981.

[8] E. Mueller, H. Klein, Nucl. Inst. Meth. **224**, 17(1984).

A RFQ INJECTOR FOR EN TANDEM LINAC HEAVY ION ACCELERATOR[1][2]

Summary

The conceptional design of a RFQ to be used as an injector of heavy ions for the Tandem-Linac system is discussed. Singly charged ions from the negative ion source are to be accelerated to more than 300 keV by the RFQ at a frequency of 26 MHz. The pulse width of the output beam is designated as <2 ns and the energy dispersion ⩽2% with a beam utilization efficiency of >95%. The bunching process was carefully studied with a model of a sequence of RF gaps simulating the cells of the RFQ structure so as to optimize the design parameters. A novel structure with variable longitudinal drift space in-between cells was then developed. The result given by the analytic expressions agree well with those computed by the Monte Carlo simulation. The computation also shows that the total length of the RFQ can be shortened by additional double drift harmonic buncher set ahead of it. The profile of the RFQ electrodes was investigated in the light of the above discussion.

Introduction

The RFQ combines the function of radial focusing, bunching and acceleration into one structure. Hence it has many applications in various

---

[1] Coauthers: Pan Ou-jia, Fang Jia-Xun. Reprinted from the *Proc. 3-rd China Japan Joint Symposium on Accelerators for Nuclear Science and their Applications*, Riken, Saitama, Japan, 1987, p. 116~119.
[2] Work supported by NSFC.

fields! Compared with the conventional D.C. injectors with bunchers, a RFQ injector has the merit of higher injection energy, better beam quality, higher beam utilization efficiency and smaller dimensions. The Peking University RFQ Group initiated studies on the feasibility of a RFQ injector soon after the transfer of the Oxford EN Tandem to the Group in 1986.

For a heavy ion injector of a Tandem-Linac system, more emphases are put on injection energy, beam quality and transmission efficiency rather than beam intensity. The concrete requirements include: to accelerate ions directly from the source to an energy sufficiently high so that the tube bulge lens at the entrance of the Tandem can be neglected. The Tandem injection energy ranges from about $250 \sim 500$ keV. In general, 25 keV. per nucleon is thought to be reasonable; to bunch the D.C. beam into bursts of $1 \sim 2$ ns. With an energy dispersion $\leqslant 2\%$ and beam utilization efficiency $>95\%$; to enable the emittance of the output beam matching with the Tandem acceptence, (12 mm · mrad · MeV), at the entrance of the LE accelerating tube; to operate at the same frequency as the Linac booster, i.e. 26 MHz. See also Table 1.

Table 1

Frequency	26 MHz	Pulse Length	<2 ns
Input Energy	$20 \sim 30$ keV	Energy Spread	$\leqslant \pm 2\%$
Output Energy	$250 \sim 500$ keV	Radial Emittance	<12 mm · mrad · $\sqrt{\text{MeV}}$
Transmission	$>95\%$		

In order to optimize the parameters so as to meet the requirements appropriately, the bunching process in the RFQ was carefully studied using a multi-gap model. It was found that longitudinal drift space can be used as a proper parameter to modify the energy dispersion and facilitate adiabatic damping of the phase excursion. Further computation shows that the structure length can be shortened by setting additional double drift harmonic buncher in front of the RFQ. The dynamic parameters are determined in the light of the model studies and three examples are given. So far, the 4-Vane structure appears to be the most popular version of the RFQ accelerators. However, for a heavy ion injector operated at a frequency as low as 26 MHz, the size would be too big to be accepted. Encouraged by the success of the 4-rod RFQ at Frankfurt and else where, an integral split-ring supported 4-rod structure working at 14~88 MHz has been developed in Peking University since 1984.[3,4] Based on these experiences, the resonator as well as the rod-elecrode were designed. Accordingly, the energy gain, transmission efficiency and output beam emittance in both longitudinal and radial spaces are computed.

Multi-Gap Model Studies

To bunch a D. C. beam into short pulses in a reasonably short distance with energy dispersion least possible is really a demanding yet hard task. Typical phase and energy spread of a RFQ with 90% of transmission efficiency is about $\mp 20°$ and ∓ 11 keV respectively.[5] Which is obviously twice as big as what is demanded. This shows that a careful study on the bunching and accelerating process is very desirable. We looked into the problem by observing at first a beam passing through a sequence of RF gaps holding a voltage of $V \sin\omega t$ and separated with each other by a distance of $d=(N\beta\lambda)/2$ (N integrals) as shown in Fig. 1.

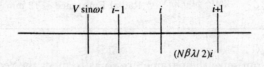

Fig. 1

The longitudinal motion of an ion in the system can be expressed in terms of phase and energy excursion ($\Delta\Phi$, ΔW) increments per gap as:

$$d\Delta\Phi/dn = - N\pi\Delta W/(2W_s) \qquad (1)$$

$$d\Delta W/dn = QV_a \cdot (\sin\Phi - \sin\Phi_s) \qquad (2)$$

where $\Phi=\omega t$ and the foot note 's' denotes the synchronus particle. Then the Hamiltonian of the motion:

$$H = \frac{N\pi}{4W_s}\Delta W^2 - QV_a \cdot (\cos\Phi + \Phi\sin\Phi_s) \qquad (3)$$

Hence the effect of the drift length on the rate of phase advance per gap, oscillation frequency and energy dispersion can been seen from the following:

$$\left[\frac{d\Delta\Phi}{dn}\right]^2 = \left[\frac{d\Delta\Phi}{dn}\right]_0^2 + N\pi\varepsilon_m[(\cos\Phi - \cos\Phi_0) + (\Phi - \Phi_0)\sin\Phi_s] \qquad (4)$$

where $\varepsilon_m = QV_a/W_s$ and the foot note '0' denotes the initial state; and the period of phase oscillation:

$$n_T = \frac{4}{\sqrt{N\pi\varepsilon_m}} \int_{\Phi_0}^{\Phi_s} \frac{d\Phi}{\sqrt{(\cos\Phi - \cos\Phi_0) + (\Phi - \Phi_0)\sin\Phi_s}} \qquad (5)$$

the maximal energy dispersion:

$$\left(\frac{\Delta W}{W_s}\right)_{max} = \pm\sqrt{\frac{8\varepsilon_m}{\pi}} \cdot \sqrt{\cos\Phi_s + \left(\Phi_s - \frac{\pi}{2}\right)\sin\Phi_s} \qquad (6)$$

For $\Phi_s = 0$, the frequency $\Omega = \dfrac{\Omega_s}{(2/\pi)F[\sin(\Phi_0/2),\pi/2]}$ where Ω_s is the frequency of small amplitudes, and $F\left(\sin\dfrac{\Phi_0}{2},\dfrac{\pi}{2}\right)$ the complete elliptical function. The frequency Ω varies by a factor of five with Φ_0 as shown in figure 2, and causes some complexity in the process of bunching.

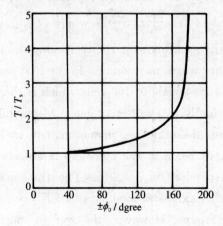

Fig. 2 Period of phase oscillation versus Φ_0

Both $d\Delta\Phi/(dn)$ and Ω appear to be proportional to $\sqrt{N}$, which implies that reasonably long drift distance might result in less number of gaps needed for the same bunching. It might eventually lead to the possibility of extending the range of inpution species if so few gaps are needed that individually controlling Rf phase is then feasible. Longer drift space may also enable higher voltage to be used in the first section of the RFQ which in turn would ease the manufacture and the tolerance on the rod electrodes. In addition, long drifts can help with limiting the energy dispersion. For adiabatic change of N, V and Φ_s the damping of the phase and energy excursion can be found from the action integral associated with the Hamiltonian. For small amplitudes the resulting invariants are

$$I_\Phi = (\Delta\Phi) \cdot \left(\frac{V_a \cdot W_s \cdot \cos\Phi_s}{N}\right)^{1/4} \qquad (7)$$

$$I_W = (\Delta W) \cdot \left(\frac{N}{V_a \cdot W_s \cdot \cos\Phi_s}\right)^{1/4} \qquad (8)$$

The actual damping process of the beam is much more complicated than what is indicated above. The resultant phase and energy excursion may depend considersbly on the way how the parameters change. To illustrate the actual process, phase plots in $\left(\Delta\Phi, \frac{\Delta W}{W_s}\right)$ space were computed numerically gap by gap. In order to help parameter optimization a figure of merit was defined as $M = \frac{360° - \Delta\Phi_{max}}{\Delta W_{max} \cdot D}$ where $D = \sum_i \left(\frac{N\beta_s\lambda}{2}\right)_i$ is the total length drift by the sychronous particle. Among the results computed, three typical cases are worth mentioning. In the first case the drift length after the first four gaps equals to 100 cells which is the longest in the series. It reduces gradually to 1 cell in 10 gaps. Meanwhile the gap voltage and the synchronous phase inceases monotonously to 15 kV and 70° respectively. The ouput beam at the 20th gap is already bunched to $\pm 9°$ with an energy dispersion of $\pm 1.9\%$. The transmission efficiency is 99%. It is shown that extensive modulation of N leads to a minimal need on the number of RF gaps. However, the over all length of the structure is proportional to N, and so it is too long to be accepted. (Fig. 3) To reduce the structure length, N is limited to be <25 in the second case. The phase excursion of the particle with smaller amplitudes ($\pm<90°$) is limited by considerably increasing the V_a immediately after the first focus. To facilitate the damping, V_a/N is periodically set lower than the average whenever the maximal energy excursion tends to be enhanced by the gap voltage, and set higher whenever it tends to be offset by V_a. In addition, the acceleration starts promptly with $\Phi_s=15°$ when the beam bunched to $\pm45°$ and Φ_s increases linearly from then on till reaching 70°. The total number of cells are 212 and the phase and energy spread at the output are ±0.94 ns and $\pm2\%$ respectively with 99% transmission. (Fig. 4)

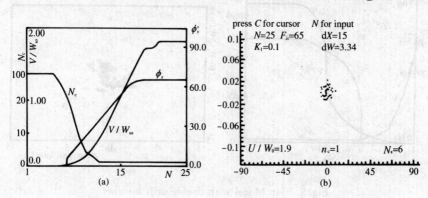

Fig. 3 Gap Model 1
(a) parameters versus gaps   (b) Phase plot of the output beam

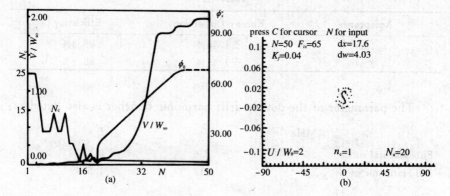

Fig. 4 Gap Model 2
(a) Parameters   (b) Phase plot

A double drift harmonic buncher is set in front of the RFQ in the 3rd case. As the speed of the phase advance $d\Delta\Phi/dn$ of the boundary paricles ($\Phi_0 = \pm 179°$) is very much enhanced by the 2nd harmonic voltage, so the length of the shapper reduces drastically. The total number of cells equals to 140. In order to examine the effect of energy dispersion of the injected beam, the bunching process was computed by Monte-Carlo simulation using 1000 particles distirbed randomly in 358° with energy dispersion of $\mp 0.3\%$. The transmission efficiency and the feature of the output beam are listed in Table 2 and Figure 5 respectively.

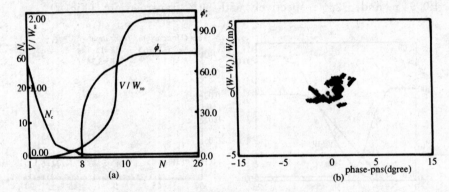

Fig. 5 Gap Model with double drift buncher
(a) Parameters (b) Monte-Carlo simulation

Table 2

Acceptance	Energy Dispersion	Efficiency
1 ns	$\pm 1.45\%$	98.3%
2 ns	$\pm 1.45\%$	99.2%

The parameter of the double drift harmonic buncher is also listed as:

	MHz	V	L
Fundamental mode	26	236	0.273
2nd Harmonic mode	52	112	1.539

RFQ Parameters

The electric cmponents in the RFQ can be expressed as:

$$E_r = -\left[\frac{FV}{a^2} \cdot r\cos 2\Psi + \frac{kAV}{2} I_1(kr)\cos kz\right]\sin\Phi$$

$$E_\Psi = \frac{FV}{a^2} \cdot r\sin 2\Psi \cdot \sin\Phi \quad (9)$$

$$E_z = \frac{kAV}{2} \cdot I_0(kr) \cdot \sin kz \cdot \sin\Phi$$

$$\Phi = (\omega t + \Phi_0)$$

Where accelerating coefficient $A=(m^2-1)/[m^2 I_0(ka)+I_0(mka)]$ Focusing coefficient $F=1-AI_0(ka)$, wave number $k=\frac{2\pi}{\beta_s\lambda}$, a is the radius of minimum aperture and m the profile modulation, V the potential difference between adjacent rod tips.

For the longitudinal motion in each cell,

$$\frac{dW}{dz} = \frac{QkAV}{2} \cdot I_0(kr) \cdot \sin kz \cdot [\sin\Phi - \sin\Phi_s] \quad (10)$$

$$\frac{d\Delta\Phi}{dz} = k\left(\frac{\beta_c}{\beta} - 1\right)$$

The results of the gap model can be applied to the design just by setting $A=(4Va)/(\pi \cdot V)$ and N equals to the number of cells between adjacent segments, where $Ez=0$.

The smoothed transverse equation of motion is

$$\frac{d^2r}{dz^2} + \frac{1}{2W} \cdot \frac{dW}{dz} \cdot \frac{dr}{dz} + \frac{1}{8\pi^2\beta^2\lambda^2}(B^2 + 8\pi^2\Delta_{rf}) \cdot r = 0 \quad (11)$$

Where RF defocusing strength $\Delta_{rf} = \frac{Q\pi^2 V A\cos\Phi}{2m_0 C^2 \beta^2}$ and radial focusing strength $B=\frac{Q\lambda^2 V}{m_0 C^2 r_0^2}$ the charateristic radius $r_0=a/\sqrt{F}$. Both b and r keep constant over the whole structure. Taking the peak surface field $E_{smax}=$ 11. 4 MV (1. 58 times of the Kiplatrick's limit) into account the parameters are determined with a vane voltage $V=80$ kV.

Typical design parameters are listed in Table 3. The lonitudinal and transverse emittance are shown in Fig. 6.

Table 3

Ions (q/A) min	1/14
Operating Frequency/MHz	26
Input Energy/keV	20~30
Output Energy/keV	687
Total Number of Cells	265
Cell Number of Radial Matching	5
Structure Length/cm	337
Charateristic Radius r_0/cm	1.09
Minimum bore radius a/cm	0.695
Maximam Modulation M_{max}	2.02
Focusing Strength	6.87
Maximum Defocusing	−0.0721
Synchronous Phase	64°
Inter Vane Voltage/kV	80
Maximum Field E_{sm}/kV·cm^{-1}	114
Transmission	0.963
Phase Spread	+9.7°
Energy Spread	+9.9°
Output Beam Size/cm×cm	0.6×0.8

The Frankfurt group has successfully developed 4 rod RFQ with trapezoidal segments. The profile of the rod-electrodes in the present design is determined essentially in a similar way as theirs, except in the interval of longitudinal drift space, where $a=r_0$. In order to investigate the field distribution with respect to the rod profile, a number of measurements were made on a 4 rod RFQ model excited by integrated split-ring resonators as shown in Fig. 7. The model is operated at 27.26 MHz with a specific R_ρ value $\rho=135.6$ kΩ·m and $Q=1190$. Typical field distribution for electrodes with trapezoidal segments separated from each by a number of cells is shown in Fig. 8. It appears that each segment causes 3 peaks; with the main peak there associate two small peaks corresponding the fringe field. The fringe peak has about the same width as the main one yet its height is about 45% of that of the latter. This means that the effect of each segment is enhanced by a factor 1.9 due to the fringe field. However, the figure of merit M of the RFQ can still be high as long as the effect of the fringe field is taken in to account in the design of electrodes.

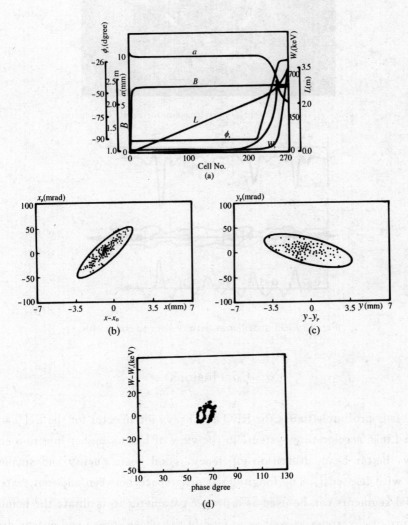

Fig. 6 (a) RFQ Parameters (b) Transverse emittance x-x'
(c) Transverse emittance y-y' (d) Longitudinal phase plot

Fig. 7　A view of integral split-ring supported RFQ structure

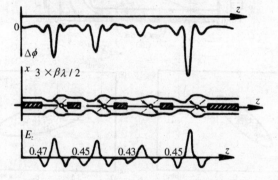

Fig. 8　Field distributon with respect to rod profile

Conclusions

It is profitable to use the RFQ as a heavy ion injector for the EN Tandem-Linac accelerating system, in the view of having higher injection energy, better beam utilization efficiency, good beam quality and smaller size with less cost. The longitudinal drift space between adjacent trapezoidal segments can be used as a proper parameter to facilitate the manufacture of electrodes as well as to modify the phase spread and energy dispersion.

Acknowledgements

Thanks to Prof. H. Klein and Dr. S. H. Wang for valuable informations and suggestions.

References

[1] H. Klein IEEE Trans. Nucl. Sci. vol. NS-30 (1983) 3313.
[2] A. Shempp et al 1987 Particle Accelerator Conf. U.S.A., March 16~19.
[3] Fang Jia-xun, Chen Chia-erh IEEE Trans. Nucl. NS-325 (1985) 2891.
[4] J. X. Fang, C. E. Chen This Symposium.
[5] S. H. Wang DESY 84-092, ISSN 0418-9833 Hamburg, September 1984.

重离子整体分离环高频四极场(RFQ)加速结构的研究[①]

关键词　高频四极场(RFQ)、整体分离环、π 模

高频四极场(RFQ)加速器是 1970 年卡帕钦斯基等提出的一种新型直线加速结构[1]. 它将加速、聚束、横向聚焦与匹配等几种作用集中于一个加速腔之中, 能直接加速从离子源引出的低能离子, 束流利用效率可达 90% 以上, 极限流强高达数百毫安. 它是一种强流、高效、束流品质好、体积小巧、使用方便因而具有广泛应用前景的新型直线加速器. 1980 年美国洛斯阿拉莫斯实验室(LANL)试验第一台加速质子的四翼型(4 Vane) RFQ 样机获得成功. 自此以来, RFQ 加速器的研制便成了国际上低能加速器发展的一个热点[2]. 现在四翼型的轻离子 RFQ 已广泛的取代体积庞大的高压倍加器, 用作高能加速器的注入器及强流中子源等, 甚至还被送上太空进行空间武器试验[3].

近年来, 在重离子物理研究与技术应用的推动下, 加速重离子的 RFQ 已成为发展的新热点. 其中 MeV 级强流重离子束在新型微电子器件制作、材料改性、惯性压缩热核聚变研究等各项应用, 更是国际上竞相研制重离子 RFQ 的巨大动力. 但它们要求 RFQ 具有工作频率低、能量可变和束流负载因子高等特性, 而现有的四翼型 RFQ 难以满足要求. 这就推动了四杆型、同轴分离型等新结构的研究[2]. 1984 年北京大学首次提出了采用整体分离环激励的 RFQ 新结构[4]. 在国家自然科学基金的支持下, 通过系统的理论、试验和工艺研究制成了全尺寸高频结构样机, 还发展了微扇型水冷电极. 经过满功率试验证实这种新结构的工作频率可在 15～100 MHz 内大范围调变, 负载因子可高达 25%. 这不仅证实了

[①]　合作者: 方家驯, 李纬国, 潘欧嘉, 陆元荣, 李德山, 袁敬琳, 王丽珊, 曾葆青. 摘自《自然科学进展——国家重点实验室通讯》, Vol. 4, 3(1994) p. 271～277.

新结构的可行性,还说明具有自己特色的新结构非常适合于重离子加速的要求.

1 粒子动力学的研究

静电四极透镜是低能离子束传输中常用的聚焦元件.如果使它的电极面沿离子运动方向起伏调制,就可以在保持径向聚焦的同时,还产生一个使离子加速或减速的轴向电场.此时,若用适当的高频电场取代静电场,就有可能使处于这个高频四极场中的离子既受到强的径向聚焦,又得到连续的加速.这就是RFQ的基本概念.四翼型RFQ加速腔及其电极形状分别如图1及图2所示.腔中高频四极场的电位分布可近似地写成[1]:

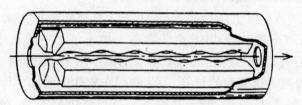

图1 四翼型RFQ加速腔

$$U = -(V/2)[F(r/a)^2\cos(2\theta) + AI_0(kr)\sin(kz)]\cos(\omega t + \varphi), \quad (1)$$

其中,加速因子 $A=(m^2-1)/[m^2I_0(ka)+I_0(mka)]$,聚焦因子 $F=1-AI_0(ka)$,$k=2\pi/\beta\lambda$,V——相邻电极间电位差,a——束流孔半径,I_0——修正的贝塞尔函数.由(1)式得[5]:

$$d^2x/dz^2 = (e/m_0c^2\gamma\beta^2)[(FV/a^2)\cos(\omega t + \varphi)$$
$$+ (k^2AV/4)\sin(kz)\cos(\omega t + \varphi)]x, \quad (2)$$
$$d^2y/dz^2 = (e/m_0c^2\gamma\beta^2)[(-FV/a^2)\cos(\omega t + \varphi)$$
$$+ (k^2AV/4)\sin(kz)\cos(\omega t + \varphi)]y, \quad (3)$$
$$(d/dz)[\beta^3\gamma^3 d(\varphi-\varphi_s)/dz] = [-\pi^2 eAV/(m_0c^2\beta\lambda^2)](\cos\varphi - \cos\varphi_s). \quad (4)$$

当粒子的运动偏离参考粒子较小时,其轨迹可用正弦或余弦函数作光滑近似. 此时纵向和横向运动的聚焦作用可分别用每加速单元上的相移 σ_{0l} 和 σ_{0t} 来表示：

$$\sigma_{0l}^2 = (-e\pi^2 AV\sin\varphi_s)/(m_0 c^2 \beta^2), \tag{5}$$

$$\sigma_{0t}^2 = [(eFV/\pi m_0 c^2)(\lambda/a)^2]^2/8 - \sigma_{0l}^2/2, \tag{6}$$

其中 e, m_0 是离子的电量与质量. σ_{0l} 和 σ_{0t} 是实数时运动才是稳定的.

为了研究 RFQ 的粒子动力学，洛斯阿拉莫斯实验室发展了一个著名计算软件 PARMTEQ，可以方便地求出粒子束在 RFQ 中运动的包络、相图和传输效率. 它已成为国际上 RFQ 动力学研究的一个标准工具. 为了方便、直观地进行动力学计算，我们

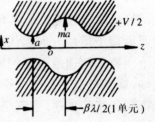

图 2　四翼型 RFQ 电极示意图

将原来要在 VAX 机上运行的 PARMTEQ 移植到微机上来，并增加了人机对话和动态显示等功能. 离子在每一加速单元中的轨迹、相位、能散、相图及丢失情况都能在屏幕上动态地显示出来. 然而为了确定结构参数，还要经过多次繁复的试算. 为此我们在软件中增加了计算 σ_{0l} 和 σ_{0t} 以及极限流强 I_1 与 I_t 的功能，并编写了 OPTIMUM 软件，通过 I_1 和 I_t 计算最佳结构参数，大大地简化了参数优化的过程. 在这些工作的基础

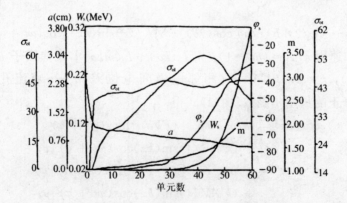

图 3　26 MHz RFQ 粒子动力学设计参数曲线

上,我们对低频重离子 RFQ 的动力学及参数设计进行了系统的研究,得到了比较满意的设计参数组. 图 3 画出了有关粒动力学参数的曲线. 它在工作频率 26 MHz 和极间电压 75 kV 的条件下能将 N^+ 离子从 20 keV 加速到 300 keV, 流强 5 mA, 束流传输效率可达 95％[6]. 此外, 现有的 PARMTEQ 实际上只是在小振幅近似下求解粒子的纵向运动, 为了搞清其局限性我们求出了大振幅下的严格解, 结果将另文发表.

2 整体分离环 RFQ 加速腔及其高频特性

普通分离环(Split Ring)是 70 年代末出现的一种有效的加速结构. 它由一对彼此反向的环臂一端相联、接地, 另一端各负载一个漂浮加速管构成. 我们将 N 对分离环通过一根或数根联杆构成了整体分离环, 发现它不但机械强度增加, 而且表征高频效率的分路电阻也提高到 N 倍. 为了形成低频而稳定的 RFQ 结构, 我们用两对(4 根)RFQ 电极取代整体分离环的漂浮管, 得到整体分离环 RFQ 结构[7]. 它的示意图见图 4. 从高频电路上看, 它的每个环臂都近似是一个 1/4 波长谐振线, 相邻环臂通过互感和电容耦合. 它们负载的两组 RFQ 电极也形成了一对谐振线. 整个 RFQ 结构的集中参数等效电路见图 5. 从这个等效电路推导、算得的模式频率、电压分布、比分路电阻、品质因素等各项重要高频特性参数, 均与整体分离环 RFQ 的实测结果符合较好[8]. 其中我们感兴趣的 π 模, 两组电极的电位反相, 极间电压分布均匀, 满足 RFQ 高频四极场(1)

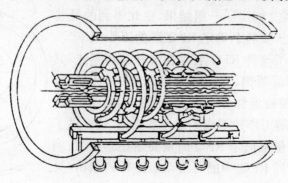

图 4 整体分离环 RFQ 结构示意图

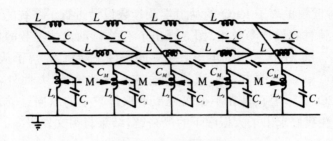

图 5 整体分离环 RFQ 等效电路图

式的要求. 算得 π 模的圆频率 ω_π、比分路电阻 ρ 及品质因数 Q 为[9]

$$\omega_\pi^2 = [K_C(1+FK_L)/[L_SC_M(1-K_L^2)(1+FK_C+2C_QK_C/NC_M)], \tag{7}$$

$$\rho = V^2/PS_Q = (16L_SSS_Q\sin^2\psi/NR)/[(1+KS)(C_S+4Q\sin^2\psi/N)], \tag{8}$$

$$Q = (2\omega_\pi SL_S/R)/(1+KS), \tag{9}$$

其中, $K_L=M/L_S$, $K_C=2C_M/(C_S+2C_M)$, $F=1-1/N$, S_Q, C_Q——电极长及其等效电容, N, S, L_S, C_S, R——卷臂数及其长度与等效电感、电容、电阻值. 以上各式可作为整体分离环 RFQ 高频结构分析与设计的基本依据. 为了同时从实验上对整体分离环 RFQ 的高频特性进行研究, 我们在 1∶1 模型上作了一系列测量. 表 1 列出一些典型结果[10]. 它表明工作频率主要由卷臂长度决定, 卷臂长时, 电感 L_S 大, 测量和(7)式表明得到的频率低. 测量还发现当仅仅改变卷长度 S 时, 频率 f_π 和 S 的乘积随 S 为线性关系. 表 2 列出了一组结果, 由此得到的经验公式可用于从实验上确定 RFQ 腔的工作频率[11]. 实验还证实了这种 RFQ 腔存在多种高频模式. 其中 π 模是最低频模, 它与邻近模式频率的间隔较大, 因而高频稳定性好[8]. 可以证明, RFQ 输出的离子能量可随工作频率成平方关系变化. 这提供了实现可变能量重离子 RFQ 的重要途径. 为此我们提出了在这种 RFQ 中的可滑动短路杆的结构(见图 6), 通过它改变卷臂的有效长度 S_E 以变化频率, 从而改变离子能量. 表 3 列出了在 1∶1 变频

图 6 改变频率机构图

RFQ 模型腔上测量的变频结果. 它表明这样的 RFQ 可在 20～40 MHz 乃至更大的范围内大幅度地改变频率以及离子的能量. 实验和理论研究表明整体分离环 RFQ 的比分路的能量值较高, 即有较高的高频效率. 而且它还可以通过减小极间电容和结构的表面与接触电阻, 进一步提高比分路电阻. 从表 3 还可看出 ρ 值近似与 f_π 成反比, 这表明在频率变化时它仍保持较高的电压或能量增益效率.

表 1 RFQ 特性测量结果

No	N	S/cm	a/mm	f_π/MHz	Q	ρ/k$\Omega\cdot$m
1	8	44.2	5	41.30	1639	91
2	6	44.2	5	37.88	1612	95
3	4	44.2	5	33.15	1517	110
4	6	116.0	7.5	24.01	1086	169
5	6	116.0	6.5	23.03	1028	180
6	6	116.0	5	21.54	1016	197
7	4	179.0	7.5	16.27	1261	244
8	4	179.0	6.5	15.54	1233	236
9	4	179.0	5	14.67	1202	190

表 2 RFQ 频率随卷臂长变化的测量值与经验公式[a]

S/cm	84.5	107.8	135.1	179.9
f_π/MHz	26.27	21.54	18.14	14.85
$f_\pi S$/MHz·cm	2222	2322	2451	2672
$f_\pi S^*$/MHz·cm	2218	2327	2455	2666

a) 由经验公式算得: $f_\pi S = 1820 + 4.7S$ (MHz·cm)

表 3 可变频率 RFQ 腔的特性(腔直径 50 cm)

No	N	S_E/cm	a/mm	f_π/MHz	Q	ρ/k$\Omega\cdot$m
1	6	29.5	9	40.82	1201	99
2	6	39.5	9	36.51	1217	101
3	6	49.5	9	33.38	1113	102
4	6	64.5	9	28.62	1022	113
5	6	74.5	9	26.15	1046	135
6	6	84.5	9	24.79	1078	154

3 微扇电极

RFQ电极是加速结构的关键部件,现在常用的有四翼电极和四杆(4 Rod)电极,它们的形状分别见图2和图7,前者用在四翼型RFQ上,后者则用在四杆型RFQ上.四翼电极给出接近理想高频四极场的电位分布(见(1)式),电极面为三维曲面,加工复杂.四杆电极则呈轴对称形,加工方便,但它的强度差,且杆细不易水冷,因此我们根据整体分离环RFQ的特点和要求,发展了一种"微扇"新型电极(图8).这种电极的横断端面为等半径R的圆弧形,因而加工比较方便,且强度好易于水冷.微扇电极形成的电位与理想四极场电位(1)式比较,出现了高次多极场及空间谐波.我们对它的粒子动力学特性进行了大量计算,证明这3种电极的粒子动力学特性非常相近.图9给出了一组对三种电极表征聚焦特性的相移常数σ_{01}与$\sigma_{0\alpha}$的计算结果[6].考虑到微扇电极具有易加工、水冷和强度好的优点,并且它在整体分离环RFQ结构中还易于更换,可满足加速不同种类离子和能量范围等要求.我们认为微扇电极是适用于整体分离环RFQ的优良电极.

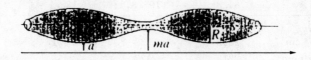

图7 四杆型电极形状

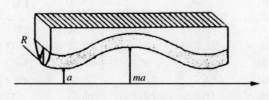

图8 微扇电极形状

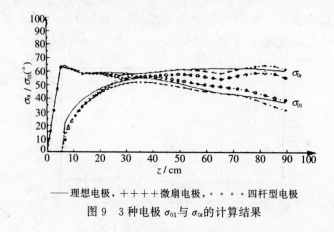

——理想电极,++++微扇电极,。。。。四杆型电极

图9 3种电极 σ_{01} 与 σ_α 的计算结果

4 全尺寸高频结构样机及高功率试验

为了研究整体分离环 RFQ 加速结构的可行性及满功率下它的各项性能,我们在1:1模型试验及结构工艺研究的基础上,建成了一台全尺寸的高频结构样机.这台样机的设计有两个特点,一是实现了电极和各主要部件的有效水冷以达到高负载因子(15%～25%),以取得高平均束流强度;二是实现了可更换的电极结构,发挥了整体分离环 RFQ 可采用多种电极以加速多种离子与大范围改变能量范围的特点.

全尺寸样机的结构示意图见图4.它的腔筒直径为50 cm,长90 cm.电极用螺钉固定在基衬上可更换.整个加速结构也可以从腔内卸下取出,因而便于在腔外装配、调整以及整个加速结构的更换.冷却水由腔筒外进入卷臂、电极支撑环、电极基衬再流出腔外,保证了电极及各部分的充分冷却.周密的水路设计和完善的 O 圈密封系统确保了水路和真空的良好密封.

高功率试验中高频功率源为 XFD-D5 型脉冲调制运行的高频机.为取得与 RFQ 腔的匹配,我们在磁耦合环馈送高频功率的装置中采取了电容补偿.极间电压是通过测量电极发射电子产生的轫致辐射谱的最高能量来确定的,这种新测量方法可靠、准确.辐射的 X 射线由 712005 型高纯平面锗探头测得,讯号最后经 1024 多道分析器分析.图10画出了

43.2 kW 高频功率下测得的辐射能谱图. 用 Am^{241} 和 Eu^{152} 的 γ 射线标定能量求得这时极间电压为 82.3 kV[6].

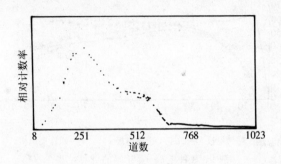

图 10　韧致辐射谱

表 4 列出了 1/6 负载因子下测得的极间电压、谐振频率、出口冷却水温度及比分路电阻值随高频峰值功率的变化. 从该表看出,在 40 kW 功率下极间电压即已超过了设计值 75 kV. 表中数据还显示该样机的频移和冷却水温升都不大,这表明它的稳定性好,水冷系统效率高. 高功率试验中系统的真空达到 1.0×10^{-3} Pa,符合系统工作要求,说明真空密封良好. 样机还在 40 kW 功率下进行了 8 h 的连续稳定运行. 然后将负载因子提高到 1/4,在 35 kW 功率下运行仍然正常[12]. 上述高功率试验的结果表明,高频样机的设计、建造是成功的,这种 RFQ 加速结构是完全可行的.

表 4　高功率试验数据

功率/kW	19.66	24.62	29.70	39.57	44.36
电压/kV	62.3	66.9	71.6	78.5	81.7
频率/MHz	25.739	25.737	25.733	25.732	25.728
水温/°C[a]	16.5	18.0	19.0	22.0	23.0
ρ/kΩ·m	168	155	147	132	128

a 初始水温 11.0°C.

重离子 RFQ 在当前国际上正处在发展阶段. 四翼型 RFQ 在加速重离子中已充分暴露了它的弱点. 以日本岛津公司建造的一台迄今频率最低的 70 MHz 重离子四翼型 RFQ 为例,它的腔直径已达 90 cm,频率还不能变化,若为了提高束流强度将频率降到 26 MHz,则这种 RFQ 的腔直径需增大到约 2.5 m！另一类同轴分离型(SCR)重离子 RFQ 结构仍在改进，日本东京大学的 25.5 MHz 同轴分离型 RFQ 腔直径为 90 cm，负载因子为 4%，频率还不能改变．四杆型 RFQ 在向低频发展采用与整体分离环类似的卷臂结构后，一台 27 MHz 的加速腔正在德国法兰克福大学建造、试验[13,14]，它的正方形腔截面的边长为 60 cm．与这几种重离子 RFQ 相比，整体分离环 RFQ 显示了自己的特色：工作频率范围宽，频率、能量可变，体积不大，负载因子高，稳定性好，及可采用能更换的微扇电极结构．这种水平先进、很有发展前途的新型加速器，填补了我国的空白，为日后重离子 RFQ 加速技术的实用化和应用打下了良好的基础．

致谢 作者们对国际合作研究单位德国法兰克福大学应用物理研究所的 H. Klein 教授和 H. Deitinghoff 博士及 A. Schempp 博士的有益讨论表示感谢．

参 考 文 献

[1] Kapchinskij, I. M., Teplyakov, V. A., *Prib. Tech. Eksp.*, 1970, (4): 19.
[2] Klein, H., *IEEE Trans. Nucl. Sci.*, 1983, **NS-30**: 3313.
[3] Headline News, *Aviation Week & Space Technology*, 1989, **131**: 31.
[4] Xie Jia-lin, *Proc. of 1984 Linear Accelerator Conf.*, Gesellschaft füer Schwerionenforschung mbH, Seeheim, Germany, 1984, 14.
[5] Crandall, K. R, et al., *IEEE Trans. Nucl. Sci.*, 1979, **NS-26**: 3469.
[6] Chen Chia-erh, Fang Jia-xun, Li Weiguo, et al., *Proc. of 1992 European Particle Accelerator Conf.*, Editions Frontiers, Berlin, 1992, 1328.
[7] Fang Jia-xun, Chen Chia-erh, *IEEE Trans. Nucl. Sci.*, 1985, **NS-32**: 2981.
[8] Fang Jia-xun, Schempp, A., *Proc. of 1992 European Particle Accelerator Conf.*, Editions Frontiers, Berlin, 1992, 1331.
[9] 陈佳洱、方家驯、潘欧嘉等，中国粒子加速器学会第四次学术年会论文集，中国粒子加速器学会，西安，1988，161．
[10] Fang Jia-xun, Chen Chia-erh, Schempp, A., *IEEE Trans. Nucl. Sci.*, 1989, 89CH2669-0: 1725.
[11] Chen Chia-erh, Fang Jia-xun, Schempp, A., *Proc. of 1990 European Particle Accelerator Conf.*, Editions Frontiers, Nice, France, 1990, 1225.
[12] 陈佳洱、方家驯、李纬国等，中国粒子加速器学会第五次学术年会论文集，中国粒子加速器学会，北京，1992．
[13] Schempp, A. et al., *Nuclear Instruments and Methods*, 1989, **A278**: 169.
[14] Deitinghoff, H. et al., *Particle Accelerators*, 1992, **37~38**: 47.

RFQ加速器同时加速同 q/M 正负离子的设想[①][②]

摘　要

提出了用RFQ加速器同时加速同荷质比正负离子的设想,阐述了原理可行性与将进行的实验验证.

关键词　RFQ加速器　同时加速　正负离子

1　引　言

自RFQ型加速器问世以来,以其体积小、束流传输率高、可加速脉冲束流强等优点,日益获得了广泛的应用.为满足干净核能源计划和大剂量氧离子埋层注入对高达数百毫安加速流强的需要,鉴于RFQ加速器目前只利用了馈入高频功率的正或负半周的现状,提出了RFQ加速器同时加速同荷质比正负离子(尤其是同一元素的正负离子)的设想,原则上可使RFQ加速器的极限载束能力约提高一倍,因而可作为上述问题的一种解决办法.在高剂量氧离子埋层注入应用研究中,正负氧离子同时注入,可避免硅片上由于电荷积累造成的局部击穿,从而可提高埋层硅片的质量和设备的可靠性.

2　RFQ加速器同时加速同 q/M 正负离子原理与可行性

2.1　原理

RFQ加速器同时加速同一荷质比的正负离子原理,主要基于荷质比相同的正或负离子经过电位差值相等的电场时,所获速度相等,即

[①] 合作者:于金祥,李纬国,任晓堂,方家驯,摘自《核物理动态》,Vol. 13, No. 2(1996), p. 34～36.
[②] 本工作是两次国家自然科学基金资助课题的继续.

$$v = (2qV/Am_0)^{1/2} \tag{1}$$

式中,q/A 为荷质比,m_0 为质子的静止质量,V 为电压(单位 MV). 当初速度相等的同一荷质比离子,如:H^+、H^-、O^+、O^- 等,进入 RFQ 加速器后,如正负离子间的互相作用小,可以忽略,则在高频电场作用下,在聚束段分别各自聚拢成移向 Φ_s、$\pi+\Phi_s$ 平衡相位的束团,并在各自的平衡位相下进一步边加速、边聚焦,直至加速到所需能量. 有关同一荷质比的正负离子在 RFQ 加速器中的动力学过程,可用本所编制的模拟计算程序(NPRFQ)描述. 图 1 和图 2 给出了在 RFQ 加速器中,同一荷质比正负离子通过不同加速单元 N 时的聚束、加速过程的相图和密度分布.

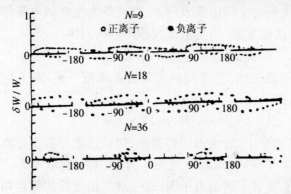

图 1　同荷质比正负离子束通过不同加速单元 N 时的相图

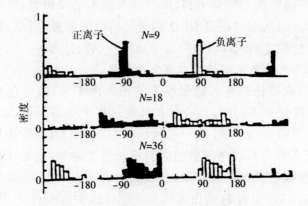

图 2　同荷质比正负离子束通过不同加速单元 N 时的密度分布

2.2 可行性

早在 80 年代初,美国洛斯阿拉莫斯国立实验室(LANL)就已用同一台 RFQ 加速器分别加速 p 和 H$^-$ 离子,并获得良好结果. 通常,可将初速度相等的同荷质比的正负离子同时注入 RFQ 加速器加速过程视为初速度相等的同荷质比正负离子分别注入 RFQ 加速过程的迭加,其差别仅在于正负离子束在注入 RFQ 前后直至分成正负束团过程中的空间电荷效应不同及存在正负离子碰撞造成的束流损失. 这两者都是影响 RFQ 加速器极限流强大小的重要因素.

2.2.1 空间电荷效应

强流低能束的空间电荷效应,主要表现为在空间电荷排斥力的作用下,离子束将逐渐发散. 引起束流发散的电场力为

$$F_r = qE_r = qI/2\pi r\varepsilon_0(2q/Am_0V)^{1/2} \tag{2}$$

与离子束流强成正比. 对于同时注入的正负离子束,离子束流强应为两者的代数和,即

$$I = I_+ + I_- \tag{3}$$

显然,在理想情况下,当注入同荷质比、初始速度与密度分布相同及流强相等的正负离子束,合轴后总流强 I 为零,无空间电荷力. 实际过程中,由于正负束流来自不同的离子源,这一点在相当长的输运路程中是难于实现的,但当 RFQ 加速器注入等速等流强同一荷质比的正负离子束时,可近似于空间电荷的完全中和,即使注入束不完全相等,也可显著降低空间电荷效应,从而可提高 RFQ 加速器的极限流强. 当然,正负离子束团分开后,它们之间的库仑力将影响束团与纵向分布.

2.2.2 正负离子束互碰撞的电荷交换损失

同向等荷质比和注入能量相等的正负离子束,由于存在能量歧离 $\pm \Delta E$ 及密度分布不均匀,在合轴和群聚过程中会有部分正负离子相互碰撞复合,导致束流损失. 这一过程可视为正离子束流处在负离子束流气氛中(反之亦可),以与能散相应的相对速度运动,其过程和性质与带电离子束和残余气体碰撞类似,在较低的束流能量下,可采用同一电荷交换截面 σ. 初始值为 I_0 的正或负束流经过长度 L(cm)后,其流强为

$$I = I_0 \exp^{-(n+N)\sigma L} \tag{4}$$

式中,σ 为电荷交换截面(cm^2)、束流能量低于 50 keV 时,σ 约为

10^{-16} cm², n 为与负或正束流相应的离子流密度, N 为残余气体密度, 分别为

$$n = 4.52 \times 10^6 J(A/V)^{1/2} \qquad (5)$$
$$N = 4.6 \times 10^{18} T_0 P/T \qquad (6)$$

式中, A 为原子质量单位、V 为加速电压(MV), J 为电流密度 (mA/cm²), $T_0 = 273$ K, T 为环境温度(绝对温标)、P 为系统真空度 (Pa). 通常对于 RFQ, 如注入束流密度为 100 mA/cm², 能量为 20~50 keV 的 H、N、O 离子束, 则 n 值在 10^9 范围内. 与束流输运线中, 真空为 1.3×10^{-3} Pa 时的残余气体密度相比小两个量级. 因此, 正负束相互碰撞引起的束流损失和它们与系统内残余气体相碰撞引起的束流损失相比, 可以忽略.

此外, 正负束流同时注入 RFQ 腔后的群聚过程中的平均等效束流, 约为注入的正或负束流的 1/2, 因此在聚束段束流与腔的相互作用, 较正或负束单独注入时为小, 这一点对强流束(数十 mA 以上)的加速是有益的.

综上所述, 用 RFQ 同时加速同荷质比正负离子, 并不存在机理性障碍. 如果适当选配正负离子源并解决好注入系统与 RFQ 的匹配, 这一思想应是可行的.

3 实验装置与目的

实验装置拟由正负双束注入器和一台现有的 300 kV、可加速 C、N、O 的 RFQ 加速器及分析测量等装置组成, 并将在今后的实验中逐步完善. 通过对 O 的正负离子分别和同时注入与加速的实验研究, 将了解和验证如下问题: (1) 在不同位置测量束流的发射度与流强变化, 比较单双束注入时空间电荷效应的变化与影响; (2) 注入段正负离子的复合损失大小与聚束段正负束相互作用引起束流纵向束流品质的变化; (3) RFQ 腔的束流负载效应以及正负束同时加速的迭加原理模型等, 并最终给出有价值的可行性报告, 有关详尽实验结果可望一两年后给出.

如果实验验证无误, 用 RFQ 加速器同时加速同荷质比的正负离子, 对前面提到的一些实际应用是非常有益的.

参 考 文 献

[1] Kapchinskij M, et al. Prib. Tech. Eksp., 1970, 4: 19.
[2] Chen Chiaerh, Fang Jiaxun, Li Weiguo, et al., Proc. of 1994 EPAC, London, England.
[3] 李纬国,于金祥,陆元荣等. 第八届三束学术年会论文集,宜昌,1995,21.

INVESTIGATION ON NEGATIVE AND POSITIVE IONS WITH EQUAL q/m IN A RFQ ACCELERATOR AT THE SAME TIME

Abstract

A tentative plan of accelerating negative and positive ions with equal q/m in a RFQ accelerator at the same time is introduced, its feasibility in principle and draft tests are also given.

Key Words　RFQ accelerator　accelerate together　negative and positive ions

EXPERIMENTAL STUDIES ON THE ACCELERATION OF POSITIVE & NEGATIVE IONS WITH A HEAVY ION ISR RFQ[①②]

Abstract

With a 26 MHz water cooled Integrated Split Ring (ISR) RFQ, N^+ beam was accelerated to more than 300 keV at Peking University. Experimental studies on the acceleration of positive and negative Oxygen ions have been carried out since then and the operating parameters were optimised respectively. Feasibility study of accelerating both positive and negative ion beam simultaneously in the same RFQ is also performed. The latter has the merit of enhancing the total number of ions and compensating the space charge both in the process of beam injection as well as on the target.

1. Introduction

Based on the investigation of conventional split ring resonator, an ISR RFQ with water cooled mini-vane electrodes has been developed at Peking University[1~2] since 1984. The properties of this type of structure have been explored by a series of rf measurements on full scale models together with theoretical analysis. It turns out that the ISR type RFQ suits well for low frequency operations and for heavy ion acceleration. A 26 MHz ISR RFQ for accelerating N^+ ions up to 300 keV was then de-

---

① Coauthers: J. X. Fang, J. F. Guo, W. G. Li, D. S. Li, Y. R. Lu, X. T. Ren, Z. Y. Wang, J. X. Yu. Reprinted from the Proc. 5th European Particle Accelerator Conf., June 1996, Sitges (Barcelona), IOP Publishing, Bristol & Philadelphia p. 2702

② Work Supported by National Natural Science Foundation of China.

signed, and constructed for the purpose of ion implantation and it was tested to full power successfully both with and without beams since last year[3]. In order to enhance the total number of accelerated ions in one RF cycle and to compensate the space charge both in the process of injection as well as on the target, a new test bench capable of accelerating of both positive and negative ions was suggested and constructed[4]. The acceleration of positive and negative oxygen ions was first performed separately and the operating parameters were optimised respectively. The average current maximum reached 17.5 μA (with a duty cycle of 1/6) and the corresponding micro-peak current is of 1 mA order. The feasibility study of accelerating simultaneously both positive and negative ions in one RFQ was then carried out. The result is quite encouraging. It turned out that the positive and the negative half period of an RF cycle can be used to accelerate both sign of ions at the same time and the interactions between negative and positive ion bunches are negligible. The above contents will be presented as follows.

2 The Experimental Layout

The layout of the experiments is schematically shown on Fig. 1. The positive and negative ion sources are located at $\pm 45°$ with respect to the beam axis. The ion beams extracted from both sources are to be focused by the Einzel lenses (EL) next to the source. The ions can be either bent one by the other or funnelled simultaneously on to the beam axis by a combining magnet (CM) and then focused by an matching Einzel lens (MEL) in front of the RFQ. The beam current of positive or negative ions accelerated by the RFQ is to be deflected by a small magnet (DM) and measured by two Faraday cups (FC) located at a range of ± 8 cm off the central axis respectively. Two additional beam monitors (BM) are mounted at the RFQ entrance and exit to measure the input and output beam intensity. The energy spectrum of the beam is to be measured by using the analysing (AM) magnet.

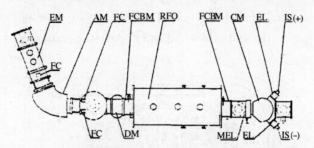

Fig. 1   Schematic layout of the experiment

2.1 Ion Source and Injection

Cold cathode PIG ion sources with permanent magnets have been developed at our lab for producing positive and negative oxygen ion beams. The one used for producing O^+ ions is of side extraction type, while the other one for O^- ions is an end extraction sputtering type PIG ion source. About one mA of negative and positive oxygen ions can be extracted from both sources at 20 kV. The percentage of O^- and O^+ are nearly 80% at a discharge current of 150 mA. The extracted O^- or O^+ beam is focused by a tri-cylinder Einzel lens with a diameter of 45 mm. The O^- and O^+ beams are then funnelled by a $2 \times 45°$ combining magnet with a radius curvature of 20 cm. In order to compensate the undesired deflection caused by the magnetic fringe field of the side extraction PIG source, a pair of electrostatic deflection plates is placed between the einzel lens and the combining magnet of the positive ion beam line, so as to ensure that the O^- and O^+ beams are funnelled right to the beam axis.

2.2 ISR RFQ Cavity

The structure of an ISR RFQ resonator operating at 26 MHz is shown in Fig. 2. It consists of 3 pairs of right wounded and left wounded spiral arms connected to a common ground plate. The drift tubes in the conventional split ring cavities are replaced here by 4 mini-vane electrodes. The structure can be assembled as a whole outside the cavity and the electrodes can be replaced whenever necessary with fast easy. The whole structure is cooled by water flowing through the spiral tubes, supporting rings and quadrupole electrodes, so as to have high duty factor and hence high average beam current. The end section of the electrodes were specially shaped so as to minimize the end effect and to improve the transverse beam quality.

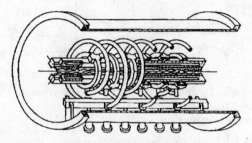

Fig. 2   View of a 26 MHz ISR RFQ

The diameter and the wall thickness of the spiral tubes are 30 mm and 1.5 mm respectively, which turn out to be strong enough to ensure mechanical stability. The parameters of this RFQ are listed in Table 1.

The rf power is fed by a linear power amplifier (XFDD5) with maximum power output of 30 kW (CW) or 50 kW (pulsed) through a water cooled loop. A distributing capacitance of about 30 pf was added to the rf feeder so as to compensate the inductance of the input impedance. The amplitude of the field gradient in the RFQ cavity is stabilized by a feedback loop with a Double Balance Mixer.[5~7]

Table 1   Principal parameters of a 26 MHz ISR RFQ

f	MHz	26
Charge/Mass		~1/14
W_{in}	keV/u	1.4
W_f	keV/u	21.4
Diameter	cm	50
Length	cm	90
V_0	kV	75
ρ	k$\Omega \cdot$m	204
Q		1300

Table 2 Results of the high power test

P_{in}	kW	19.7	24.6	29.7	39.6	44.4
V_0	kV	62.3	66.9	71.6	78.5	81.7
f	MHz	25.7	25.7	25.7	25.7	25.7
T_{water}	°C	16.5	18	19	22	23
ρ	k$\Omega \cdot$ m	168	155	147	132	122

3 Experimental Results

3.1 Acceleration of O^+ and O^- Beams

The DC current of O^+ ions obtained at the entrance of the RFQ after a deflection of 45° ranges from 200～490 μA, which was extracted under 17～20 kV from the PIG source and focused by the Einzel lens. The beam energy after acceleration was determined by the analyzing magnet and was shown in Fig. 3 as a function of RF power. It can be seen that the highest energy gain of 306 keV was obtained at a RF power of 30 kW. The output beam was quite sensitive with the matching lens in front of the RFQ and is shown in Fig. 4 as the beam transmission efficiency versus focusing voltage. The former is defined as the ratio of deflected output current received by the offset Faraday cup to that of received by the cup in front of the RFQ entrance. As the aperture of the latter is of 25 mm, which is considerably larger than that of the entrance diaphragm (15 mm) located at a distance 10 cm downstream. So the efficiency thus measured could possibly be underestimated. The beam efficiency increases with the vane voltage as shown in Fig. 5. With a duty cycle of 1/6, the highest average current measured was 17.5 μA, and the highest transmission is more than 43%, the DC equivalence of which is 105 μA and the corresponding microscopic peak current is estimated to be more than 1 mA.

On the acceleration of O^-, the transmission efficiency seems to be a little higher than that of the positive ions. The DC current obtained at the RFQ entrance was 175 μA. The transmission efficiency reaches a maximum of 48% at a matching lens voltage of 13 kV and tends to increase further with focusing voltage.

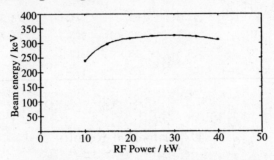

Fig. 3　Beam energy of O^- versus RF power

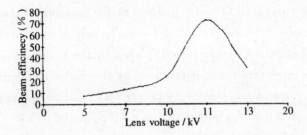

Fig. 4　Beam transmission efficiency of O^+ vs. lens voltage

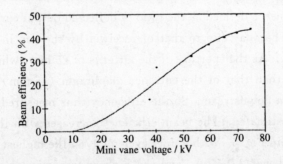

Fig. 5　Beam transmission efficiency of O^+ vs. vane voltage

3.2　Simultaneous Acceleration of O^+ and O^- Beams

For simultaneous acceleration of O^+ and O^- ions, the ion sources and the focusing lenses related are set according to their own characteristics. The extraction voltage for O^+ is normally ~20 kV, while for O^- is limited to 17 kV because of sparking. However, the optimum voltage of the matching lens for O^+ is about 11 kV while it should be $\sim$18 kV for O^- ions. So the voltage setting has to be a kind of trade off, in our case $\sim$13 kV was chosen for the time being. The simultaneous acceleration was performed at various RF power levels with the average current output varied from 0.1 to 21 μA under a duty factor of 1/6. The current ratio of O^+ to O^- varied from 0.1 to 5. The sum of $O^+ + O^-$ current I_{sim} for simultaneous acceleration is compared with that of I_{sep} where O^+ and O^- were accelerated separately under the same condition in all cases. It appears that I_{sim} is about the same as I_{sep} in all the current ranges as can be seen from Fig. 6. This means that both the positive and the negative half period of the RF cycle can be used to accelerate both sign of ions at the same time and the interactions between negative and positive ion bunches are negligible, as it was expected, so far as the micro-peak current concerned is of 1 mA order. It is also worth mentioning that the result implies that dual species of ions of different amount can be implanted at once with deliberate ratio by using the present set-up.

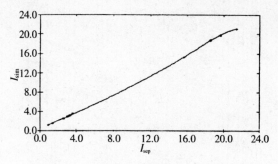

Fig. 6 I_{sim} vs. I_{sep}

Acknowledgements

The authors wish to thank Prof. Dr. H. Klein, Prof. Dr. A. Schempp, Dr. H. Deintinghoff and Dr. R. Thomae of Frankfurt University for valuable discussion.

References

[1] J. X. Fang, C. E. Chen, IEEE NS-32 (1985) 2981.
[2] C. E. Chen, O. Pan, J. X. Fang, 3-rd Japan-China Symposium on Accelerators (1987) 116.
[3] C. E. Chen, J. X. Fang, et al., The Progress of Natural Science.
[4] J. X. Yu, C. E. Chen, et al., Trends in Nuclear Physics Vol. 13, No. 2 (1996) p. 34.
[5] C. E. Chen, J. X. Fang, A. Shempp, EPAC 90(1990) 1225.
[6] C. E. Chen, J. X. Fang, W. Li, O. Pan, H. Klein, H. Deitinghoff, A. Shempp, et al., EPAC 92(1992) 1328.
[7] C. E. Chen, J. X. Fang, W. Li, O. Pan, Y. Lu, D. Li, J. Yuan, 5-th Japan-China Symposium on Accelerators (1993) 116.

第五部分 射频超导直线加速器研究

FIRST OPERATION OF THE STONY BROOK SUPERCONDUCTING LINAC[1][2]

This talk will review the first few months of operating experience with the new superconducting LINAC booster at Stony Brook. Many previous status reports to SNEAP and other accelerator conferences have for more than five years summarized the planning, development, construction and installation of this machine and related improvements to the FN tandem. So it is an exciting moment to finally have an opportunity to discuss actual operation!

LINAC operation to deliver beams for the regular laboratory research program began in May of this year (1983) and now (October 1, 1983) totals 1200 hours. During this period the emphasis has been on achieving routine reliable operation of the whole system. This means much more to us than that the hardware must function well and reliably. Stony Brook is a University laboratory in which the tandem and ion-source was traditionally started-up by technical staff but otherwise operated around-the-clock by the experimenter-users. It has been the intention and hope that informal operating style could be maintained with the much more complex coupled tandem-LINAC system. So a critical part of the first operating effort has been to develop devices, procedures and documentation "user-friendly" enough that routine operation could be handled by the faculty, graduate student and post-doc users. This has been achieved by extensive

---

[1] Coauthers: J. M. Brennan, B. Kurup (Tata Institute, Bombay, India) and J. W. Noé. Reprinted from the Proc. Symposuim of Northeastern Accelerator Personel, Oct. 1983, University of Rochester, U.S.A.

[2] Supported in part by the National Science Foundation.

use of real-time computers in the control system and for set-up calculations, and by the implementation of effective beam diagnostic devices.

Fig. 1 shows an overall view of the LINAC control console. (The tandem controls are a separate system out of sight to the left, which has changed relatively little during the upgrading.) The main features are the two identical control stations at the left and middle and the console LSI-11/23 with RL01 disk drive and keyboard at the right. (Further to the right, out of sight in this picture, is the remote control box for the helium refrigerator.) NIM bins are used for the various bunching system control boxes and the SENTEC NMR for the dipole magnets, and at the top are several small video monitors for fixed displays of cryogenic system and resonator status information and a computerized message/announcement board. Finally, the fast-out-of-lock (FOOL) board contains a compact display of the operating status of all 40 resonators in the system via bi-color LED's which show green for an operating phase-locked resonator, red

Fig. 1 Overall view of the superconducting LINAC control console. At the time of this picture (April 1983), the color monitor for the right-hand control station had been removed and the FOOL (fast-out-of-lock) annunciator panel and NIMbin had not yet been installed.

for an unlocked resonator and flashing red for a resonator which has returned to locked status. The FOOL signal itself is the hardwired "OR" of the immediate status of all of the operating controllers, while the LED's are controlled through the computer network. At the resonator field levels which have been used so far all resonators are in lock 100% of the time. At higher fields, where shaking will be more of a problem, the FOOL signal could be used to gate off data collection for an experiment. Redundant protection is provided by the switching magnet in the target area, which would deflect a beam of the incorrect energy.

Fig. 2 is a closer view of one of the control stations. This contains two computer-assigned "knobs" and a related LED readback for knob functions, a small keyboard to initiate knob and screen actions, and a color graphics terminal with an overlaid touch-sensitive screen. The color display is the AYDIN 5217 in which images are built up from character-size blocks of various shapes and colors. The last element of the system is the Tektronix modular oscilloscope system at the top. This is used with an "analog bus" which connects the scope display to the analog phase and amplitude error signals of a resonator controller when the resonator is selected. Thus when one of the 40 resonators is selected its real-time signals appear on the scopes while the two knobs are assigned to the reference phase and amplitude. Turning a resonator "on" or "off" is accomplished by simply touching a "switch" symbol on the screen which symbolically opens the feedback loop shown on the screen. Attaching, for example, the slow tuner to a knob is done in a similar way. Other pages of the control system provide access to and information on the vacuum systems, the beam transport magnets and the cryogenics. On the cryogenic page filling a module with liquid helium or nitrogen can be controlled by touching a symbol of a module. Similarly, on the magnet page (Fig. 3) touching the appropriate symbol attaches a quad, for example, to the two knobs for tuning. By first hitting the "SAVE" key on the keyboard, it is possible to immediately return to the previous values if the tuning proves unproductive. This is a very popular and useful feature.

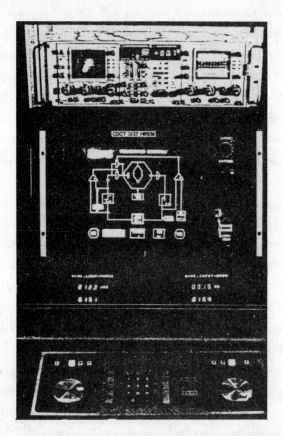

Fig. 2 Close-up view of one of the two identical control stations. The video screen shows a schematic diagram of the feedback circuit of one of the 40 resonators, and by touching the screen various parameters such as the RF coupler or slow tuner can be attached to the programmable knobs shown at the bottom. The small oscilloscopes at the top display various resonator signals.

These functions are carried out through a "star" network of eight satellite LSI-11/2 computers attached to the console 11/23 via conventional 9600 baud serial lines. One satellite is connected to each pair of the 12 LINAC modules, and the remaining two operate the cryogenics and the magnets. The refrigerator LSI-11/2 directly measures the LN2 and LHe levels, converting depth readings into liters of liquid. This computer executes various strategies for equalizing the helium levels in the 13 cryostats. For example, if one crysotat threatens to run dry, proportional control is suspended and all but the one helium valve is closed off. The computer controls the helium and nitrogen valves by adjusting the air pressure in a manifold which is cyclically connected to each of the approx. 30 valves.

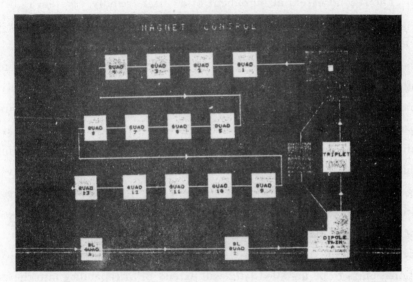

Fig. 3 View of the magnet control "page" of the control system. The boxes change from blue to yellow (active) when they are touched, or to red if the "save" key is depressed.

This type of interactive computer control system with symbolic (rather than keyboard-based) controls is typical of the large accelerator systems like FERMILAB but up to now has not been seen in small university laboratories. For Stony Brook it was of course virtually mandatory because of the very large multiplicity of identical components — 40 resonators, 30 magnets, etc. Our system-hardware and software-represents in all about 3 man-years of effort. This substantial initial investment means however that we can now relatively easily implement some quite attractive and powerful features, as required by continuing experience with the accelerator. For example the capability of storing or printing a complete log of all system parameters was recently implemented. The hardware has proven to be extremely reliable and cynics who argue that such a complex system is bound to be not working much of the time should not be believed. Furthermore, the response of the large group of users is invariably favorable.

The second essential ingredient of a "user-friendly" system is beam diagnostics which are easily used and interpreted. The principle such device at Stony Brook is the modified NEC beam profile monitor [BPM]. The modifications are a change to direct mechanical coupling via a ferrofluid rotating vacuum seal and the addition of optical isolation for the output signal. With these changes the BPM's give dependable quantitative position readouts at beam currents under one nA. The ferrofluid seal is definitely an improvement over the original NEC magnetic coupling, which frequently slipped or stalled altogether. It does have the disadvantage of high cost ($500) and is not bakeable, so we are currently investigating an improved magnetic coupling before we convert the final few of our units. The BPM's provide a continuous picture of the presence, intensity, position and quality of the beam at the 9 points shown in Fig. 4. Additional units to be installed later will extend BPM capability into the target area. The beam "pictures" have made it possible to very quickly teach new users how to check for problems along the beam path.

Figure 4 shows the beam path from the ion source through the tandem and LINAC to the LINAC target room. (A second target room adjacent to the LINAC is still available for tandem-alone experiments.) A critical aspect of the coupled accelerator system is the need to form the inherently DC tandem beam into microbunches to match the 200 psec time acceptance of the 150 MHz RF LINAC. This is accomplished in two stages. A 9/18 MHz double-drift buncher in front of the tandem compresses 50% of the DC beam into nsec bunches every 106 nsec (see Figure 5). The unbunched 50% of the beam is then removed by an RF sweeper in the LINAC injection beam line. Figure 6 shows the sweeper unit before installation; a second set of plates (horizontal deflection) visible in this photograph will be used for a future down-count deflector to produce arbitrarily long pulse spacings. In the second stage of bunching a superconducting resonator situated immediately before the 180 degree turn in front of the LINAC squeezes the 1 nsec tandem bunches by a further factor of about 5. The width and arrival time of the "superbunch" can be monitored by a channel plate time detector just in front of the LINAC. A second time detector is located after the LINAC, and a third is planned for the target room. Pulse widths just after the LINAC are well under 100 psec.

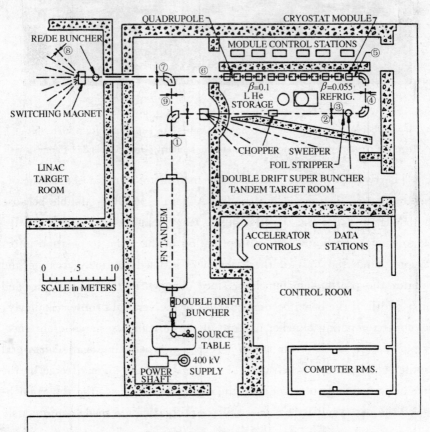

Fig. 4 Overall plan of the Nuclear Structure Laboratory, showing the locations of the 9 beam-profile monitors and various control slits.

Fig. 5 View of the double-drift buncher in front of the tandem as it was being installed. The two buncher tubes are supported from high-voltage RF ceramic feedthroughs mounted on standard 6-inch Dependex flanges.

The two BPM units immediateley in front of, and at the center of, the 180 degree turn preceding the LINAC are especially valuable because the 5 MHz sweeper and the first dipole magnet in the turn, respectively, act to convert time and energy information in the beam to position information. Thus the first of these scanners is used routinely to tune and monitor the pre-tandem buncher, since changes in the arrival time and pulse width of the beam become changes in the vertical profile. Similarly, the superconducting buncher in front of the turn imposes a coherent energy modulation on the beam which is seen as a broadening of the horizontal profile at the center of the turn. The energy modulation is affected by the time of arrival and pulse width of the nsec pulsed beam at the superbuncher, so the net result is that these phase properties are made continuously visible. Small variations in the tandem or along the beam path can cause phase jitter or shifts of the tandem buncher relative to the LINAC master clock. The stability of the beam transport has generally been so good that these fluctuations had only a slight effect on the LINAC transmission, and usually only occasional manual phase corrections are required. We are, however, developing a phase-feedback system for the LINAC injection which will send a feedback signal from slits in the 180 degree turn to the tandem buncher to correct automatically for flight-time variations.

Fig. 6 Testing of the clean-up sweeper and down-count pulsing unit.

Phasing (tuning) of the LINAC is presently done sequen-tially one resonator at a time using a silicon surface-barrier detector to measure the energy of the beam scattered from a thin gold foil. The reference phase of each resonator is first adjusted for zero energy gain. The operating phase is then 90 degrees away less 20 degrees to give a margin of phase stability.

(That is, there is some bunching in addition to the predominant acceleration.) The tedium of this procedure, which takes typically two to three hours, is somewhat relieved by a 90 deg. phase advance key on the control keyboard. Fortunately we have usually been able to return, for the same beam and charge state, to a logged previous setting of the phases by using the channel plate in front of the LINAC as an absolute time-of-arrival reference. The whole phasing procedure can actually also be carried out with the post-LINAC channel plate to determine the phase corresponding to zero energy gain.

A major deficiency of the surface-barrier detector is that it gives poor absolute energy information due to the large pulse-height defect for heavy ions. This is a serious problem because the experimentalists aren't always satisfied with 200 plus/minus 10 MeV!! We are fortunate that the target area contains a large switching magnet capable of bending any LINAC beam to 45 degrees. The field of this magnet is measured with NMR and by bending the beam along a fixed trajectory at 45 degrees an absolute energy measurement can be made with an accuracy of better than 0.5%. At present the 45 degree trajectory is defined by the fiducial of a beam scanner. Slits will soon be implemented immediately downstream of this scanner and these will be used to stabilize the magnet field in such a way that the beam is held centered in the slits. This system will be worked into new software to give a more precise and semi-automatic phasing of the LINAC. The dispersion at 45 degrees is also sufficiently large that we can place an upper limit of 2×10^{-3} on the fractional energy spread of the beam. A somewhat better limit comes from the modest time spread of approx. 1 ns as the beam drifts about 100 feet from the LINAC to the target position.

During this initial few months of operation no attempt was made to approach the performance limits of the resonators. For the typical 8~12 MV equivalent booster voltage so far employed the high-beta resonators operate at 2.0~2.5 MV/m accelerating gradient and the low-betas at 1.0~1.5 MV/m. The latter quite low field does not significantly affect the overall energy performance for Si and S beams, but begins to be a serious limitation for nickel and heavier ions. This operating field was adopted in order to ensure that the low-beta resonators, and hence the whole LINAC, would remain continuously 100% in phase lock. While the high-beta resonators are ultimately limited by dissipation, the low betas are much more vulnerable to vibration and can be sometimes difficult to stabilize even for fields as low as 1.5 MV/m. Although some improvement may result from reducing sources of vibration (which are known to mostly be in the liquid helium system), the best long term solution may be to replace the present low-beta split-loop resonators by the much stiffer quarter-wave resonators of the type which will be used in the Seattle booster. Dissipation performance of the high-beta resonators responds favorably to helium gas and/or RF power conditioning and considerable further improvement is possible if enough time is spent on it. To date such conditioning has been a very low priority, however.

A frequent question is whether there is deterioration of the resonator properties from extended operation. This is now known to be certainly not the case for our lead-plated copper resonators, at least for 1000 hours. If anything the improvements due to conditioning are progressive and cumulative. We have however had a significant difficulty with multipactor conditioning after a recent maintenance period during which over half of the LINAC modules were warmed up and opened to air for various minor repairs such as replacement of ion gauge filaments. These problems were in the modules that remained cold (not those opened) suggesting that the lack of pumping for extended periods may have been detrimental. A standard procedure after opening a module is to bake the resonators at 100 deg. C in vacuum for about two days. This treatment invariably allows the resonator field level to be quickly brought up to 3 MV/m.

We also now have very favorable experience with the long term frequency stability of the LINAC, which is certainly less than 100 Hz/month. In fact, on one occasion a module was warmed up to room temperature and upon cooling down again returned immediately to phase lock with no adjustments when the RF power was turned on!!

The one area in which some effort to improve performance has been made in this initial operating period is in the static (quiescent) losses of the helium transfer system. Tests in July showed that losses in this system exceed specified (design) values by over 100 W, so that somewhat more than one-half of the total refrigeration capacity of 400 W was at that time expended before the RF was even turned on. Several changes have already been carried out to improve the situation, and there is now about 225 W available for running losses. One improvement was to provide better (diffusion rather than mechanical) pumping for a known and essentially unrepairable leak in the trunk line of the distribution system. Losses in the helium transfer line bayonnet connections were also reduced from about 3 W per bayonnet to 1 W by installing teflon sleeves to restrict thermal oscillations in the narrow annular neck of the bayonnets. There remain however still somewhat excessive losses of 3 W additional in each module transfer tube which exceed the design value of 1 W.

Acknowledgement

The completion of Stony Brook booster project was possible only with the efforts of a large and dedicated group of individuals (see "Installation of the Stony Brook 20 MV Superconducting LINAC" in Proceedings of the 1982 [Seattle, Washington] SNEAP Symposium). The authors especially wish to thank Chuck Pancake, Al Scholldorf and John Hasstedt for their invaluable contributions to the hardware and software of the LINAC computer control system.

1.5 GHz 铌腔 RF 超导的实验研究[①②]

摘　要

北京大学重离子所自 1988 年开始"超导腔"课题的研究工作,经过三年不懈努力获得了重大进展,受到国内外同行与专家的好评. 本文详述了此课题的实验准备工作及低温超导物理实验过程,总结了 RF(射频)超导实验技术、微波及锁相测量技术、腔体的后处理技术、计算机模拟设计及计算机控制、数据获取与处理等有关工作的进展和成果. 目前,1.5 GHz 铌腔在 CW 模式的低温超导实验中获得了 8.6 MV/m ($Q_0 = 6.5 \times 10^8$)的加速梯度,并在 2 K 温度时,测得 $Q_0 = 8 \times 10^9$,这些实验结果达到了当前国际上的先进水平.

关键词　RF 超导　铌腔　低温恒温器　腔体后处理　锁相技术

一、前　　言

RF 超导技术在加速器领域中的应用已有一段历史了. 1965 年 Stanford 大学首先研制了加速电子的镀铅超导腔,经过 20 多年的努力,超导腔腔形、腔体材料性能、加速梯度都获得了成功的改进[1]. 今天,世界上有 20 多个实验室已使用或正在建造用于重离子、电子的高能超导加速器.

超导腔具有高梯度、大束孔,可以在 CW 模式工作及可采用较低的工作频率等特性[2]. 这些性能使超导腔在高能强流电子加速器的应用中具有较大的优势. 尤其是自由电子激光器对电子束流的品质、流强、平均功率等有较高的要求,选择超导加速器是较好的方案. 80 年代末期,为了提高用于 FEL 或 TEV 直线对撞机的注入器的亮度,国际上提出了超导腔、光阴极高亮度注入器方案[3].

---

①　合作者:赵夔,王光伟,王莉芳,张保澄,宋进虎. 摘自《强激光与粒子束》,Vol. 4, No. 1 (1992) p. 15~31.
②　该项目由国家高技术激光技术领域基金资助.

今天的超导腔具有优异的 RF 性能,实验室单腔的加速梯度可达 25 MV/m ($Q_0 \approx 10^9$). 按照 BCS 理论的预期值,纯铌腔的加速梯度应能达到 50 MV/m[1], 实验值距此还有较大差距. 经过大量的实验, 不断地摸索和探讨,人们认识到腔体的表面性能的优劣至关重要. 多次电子束冶炼,提高 RRR(残留电阻比), 化学清洗,超高真空下的高温处理, 人为制造氧化膜等方法可有效地提高铌材的性能. 但是, 至今场致发射仍然限制加速梯度的提高. 目前世界上一些大实验室(KEK, DESY, CERN, CORNELL)都在寻求新的方法、新的技术, 进行大量的腔体表面改性实验, 以求在提高梯度上有所突破[4~7].

北大重离子所超导腔课题组在腔体的后处理及加速梯度的提高上进行了大量的工作. 在实验过程中逐步建立了一套有效的后处理方法, 使我们在较短的时间里取得了显著的进展, 我们已测得 8.6 MV/m 的加速梯度(CW 模式), 并在 2 K 时获得 8×10^9 的腔体品质因素. 本文重点介绍了围绕 Q 值及加速梯度 E_{acc} 的提高所采用的方法和技术.

二、设计分析,程序组在 PC 机上的运行

1. 设计分析

当铌腔处于超导状态时,其损耗电阻(R_s)为 nΩ 量级, 比在液氦温度下铜腔的 R_s 小约 5 个数量级[1]. 极低的损耗使铌腔具有了独特的性能, 其中重要的一点是: 对超导腔讲, 腔体形状的变化对 RF 功率的转换效率影响很小, 因此在设计腔形时需要考虑的是如何减弱暗弧效应、高阶模(HOM's)的影响等问题. 图 1 是椭球形超导腔, 束管上有耦合器开孔, 束管两端有截止波导. 这种形状的加速腔在腔体部分有圆弧形过渡, 并有大孔径束管, 可以有效地抑制暗弧效应和束流感应场. 我们对 1.5 GHz LE 型椭球形单谐振腔($\beta=1$)的基模、高阶模以及尾场效应分别用 SUPERFISH, URMELT 及 TBCI 等程序进行了计算分析. 有关的结果可参看我们已发表的文章[10~12]. 图 2 是单一聚束脉冲所引起的尾势. 对 1.5 GHz 铌腔用 HP 8753C 网络分析仪及其他手段, 做了 3.0 GHz 以下模式的室温测量. 测试结果与计算分析基本一致, 表 1 是测量值与计算值的比较.

表 1 测量值与计算值的比较

MODE	URMEL/MHz	Meas/MHz
TM010	1470.22	1467.70
TM011	2792.13	2800.40
TM020	2977.62	2975.84
TE111	1889.80	1868.32
TM110	2058.45	2043.54
TM111	2621.95	2585.17
TM112	2955.58	2974.58

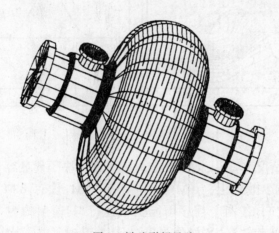

图 1 椭球形超导腔

2. 程序组在 PC 机上的运行

在计算机设计分析过程中,我组取得了一项突出的成果,这就是将 URMELT,TBCI 等程序成功地移植到 386-PC 机上运行. 大程序在微机上运行,首先要解决的是内存问题. URMELT 及 TBCI 程序的本身代码段和巨大的数据段大大超过了 640 K,使得在 DOS 环境下无法直接运行. 解决这个问题过去常用的方法是:对大代码段采用覆盖,对大数据段开临时文件. 但这些方法需仔细分析源程序的每一模块甚至每一句执行语句,工作量是极大的.

我们解决的方法是利用虚拟内存. 该方法将硬盘的一部分划为虚拟的内存,利用 386 微机在保护虚地址方式下运行时,可管理 16 M 实存和 64 T (T=M×M) 虚存容量的便利条件,使大程序在微机上的运行成为

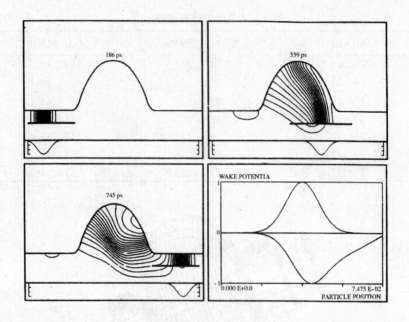

图 2　椭球形腔的尾场计算

可能.在这种情况下,最大能运行多大程序就取决于硬盘容量了.现在在我组的 HP-312 微机上,10 M 内存(实存 2 M,虚存 8 M)的情况下,URMELT 及 TBCI 两个程序均能正常运行,其速度约为 VAX 785 的 1/3~1/2,如果考虑 VAX 机的多用户分时,其速度并不比在 VAX 机上运行慢.图 2 即是 TBCI 程序在 HP-312 微机上运算所得结果.

程序组能在 386-PC 机上运行,为课题组的科研任务带来极大的便利条件,既节省了时间又节省经费.

三、低温恒温器及超高真空系统

1. 低温恒温器

1.5 GHz 超导腔(铌)的实验温度希望在 2 K 至 4.2 K 间进行,为此需要建造一套超低温系统.经过约两年的努力,我们完成了大型恒温器、100 L 液氦杜瓦、两只 100 L 液氮杜瓦、液氦输液管道、氦气回收管道、温度测量及液氦面测量装置的制造与安装工作.这些设备的建成确保了超导腔实验的顺利进行.

100 L 恒温器是实验的主要设备. 研制这种大型低温液氦实验装置在国内还未见报道, 在缺少资料的情况下, 我们将初始设计寄往 DESY (德国), 请国外专家评审. 德国同行就我们的图纸提出了 10 条重大修改意见, 我们根据这些意见重新设计. 新的设计比较简单实用, 突出了安全性及铌腔的超高真空要求. 图 3 是恒温器的示意图. 恒温器约高 175 cm, 最大外径 ϕ 75 cm. 铌腔与恒温器采用了分立的真空系统, 恒温器使用了一台高速分子泵, 系统一般要求 10^{-4} 至 10^{-5} Pa 真空度. 铌腔采用了一套超高真空系统.

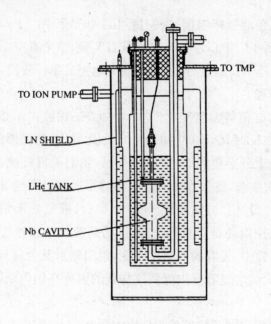

图 3　液氦恒温器

整套低温系统经过约一年的运行检验, 基本达到了设计要求, 恒温器的液氦蒸发量约每小时 1%, 100 L 液氦杜瓦瓶的蒸发率小于 1.5%/天.

2. 超高真空系统

超导腔内表面的洁净度要求极高, 不允许有油污, 否则对 Q 值及

E_{acc} 值影响很大. 1.5 GHz 铌腔竖直安装在恒温器内,以 OVER-HEAD 方式与真空抽气管道连接,这样可以避免污物落入腔内. 铌腔束管两端为铟封接面,铟封法兰内径尺寸要根据束管法兰实际大小设计,精度要求高,封接铟丝直径为 1 mm,这两道封接口径较大,直接影响系统的超高真空的获得,尤其是当液氦处于超流态时,它们是薄弱环节. 两只冷凝吸附泵及一台离子泵为铌腔提供了一套无油机组. 整个系统全部为金属密封,并使用了超高真空金属阀门,RF 馈入窗口是 99 瓷封接的同轴 50 Ω 接头,在真空管道终端安装有超高真空宝石阀,准备用来做氦处理实验的控制阀门.

这套系统在液氮降温前,真空度可达 10^{-6} 至 10^{-7} Pa,加入液氮、液氦后,可以达到 10^{-8} Pa. 经多次液氦降温实验,整个真空系统运行可靠,保证了低温物理实验的进行,使我们顺利地完成第一阶段工作.

3. 减压降温实验

超导腔的 Q 值随温度变化[1],对 1.5 GHz 铌腔在 4.2 K 时,Q 值为 $(2\sim4)\times10^8$,2 K 时,$Q_0=(6\sim9)\times10^9$. 由此看出,超低温的获得对提高梯度的实验是十分必要的. 为获得超低温,我们采用机械泵减压降温的方法,通过多次实验,终于在 100 L 容积的恒温器上做到 2 K 的低温,并测得 8×10^9 的 Q_0 值. 当温度低于 2.12 K 时,液氦处于超流态. 此时对液氦槽内的真空器件是严重的考验,因为超流氦的渗透能力极强. 在多次减压降温实验中,铌腔的真空度不变,说明铟封及金属封是可靠的. 据了解,在这样大的液氦容器内降到这样低的温度在国内是不多见的.

4. 地磁屏蔽

地磁场的存在对超导腔的 RF 性能影响较大,需要在铌腔周围建立起地磁屏蔽. 我们在恒温器外加了多层高 μ 材料薄带,这是一种非晶态金属(Fe-Ni-Si-B),薄带尺寸是 0.04 mm×100 mm(厚×宽),共用了 6 kg. 北京大学地球物理系的专业人员对恒温器内的轴向及水平方向的磁场做了检测,使用的测试仪器是美国的 Digital Magnetometer Model DM 2220,测得结果:在工作区的轴向磁场是地磁的 1/7,水平方向接近 1/20,基本满足了超导实验的要求.

四、铌腔的后处理

超导铌腔的机加工与焊接完成后,还需要做机械打磨抛光、电抛光、化学抛光、钛膜后处理等工艺流程.经过这些处理后,可以获得具有高热导的高质量内表面,这有利于腔体的品质因素 Q 及加速梯度 E_{acc} 的提高.腔体的后处理涉及到材料科学、化学与热处理技术,为了获得理想的 RF 超导条件,我们进行了以下的工作.

1. 铌材分析

用于电子加速器的超导腔目前一般是由高纯铌薄板制成,市场上的商品铌不能直接使用,需要再经过多次电子束冶炼,提高纯度.纯度提高后还要进一步改善铌材的热导,现在通常的办法是在铌的表面蒸发一层钛膜,用其结合的特性带走铌中的杂质.

我们对德国 DESY 实验室提供的高 RRR 铌材及有色院的经过电子束冶炼过的铌材进行了分析对比,表 2 是两种样品的半定量化学分析结果,1# 是 DESY 样品,从杂质含量上看不出优异.图 4 是电子能谱测量结果,可以看出铌的表面有一层几百埃厚的 Nb_2O_5.经过 RF 超导实验的多次验证表明:这层氧化物对于腔体 RF 性能的稳定是有利的.目前 KEK、CEBAF 等实验室在腔体的后处理过程中增加了高质量氧化膜的制备.

表 2　两种铌样品的半定量化学分析结果[a]

1#

Ni	Zr	Rh	Sn	Tl	Ta	Mg	Pt	Ti
0.001	0.003~0.01	0.01	0.001	0.001	0.1	0.001	0.002	0.001~0.01

2#

Hf	Zr	Tl	Al	Pt	Cu	Ti
0.01	0.003	0.001	0.003	0.001	0.001	0.003

a 表内数据均为百分数.

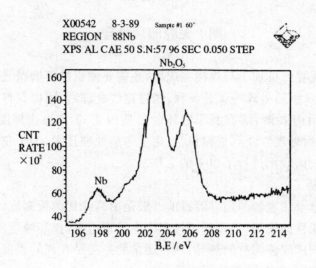

图 4 铌材的电子能谱测量结果

2. 化学清洗

Dornier 公司制造的 1.5 GHz 铌腔的腔体是用 RRR = 300 的高纯铌制成,一般讲 RRR 高的话,热击穿的阈值就高,可获得高的加速梯度 E_{acc}. 但是这种高 RRR 铌却给化学清洗带来较大的困难,在酸洗的过程中,容易将 H^+ 或磷酸化合物引入铌表面,造成腔表面性能的极大下降. 为了避免这种情况的发生,我们进行了多次样品及腔体的化学清洗试验,建立了一套有效的清洗方法.

化学清洗是参照 DESY 及 KEK 等实验室的方法进行的[13]. 图 5 是清洗前后样品的显微照片,从样片上可以看出,清洗后铌表面呈现出明显的晶粒排列. 我们的清洗液是由 HF(40%),HNO_3(65%),H_3PO_4(85%)三种酸按一定比例混合而成,酸洗数次,每次时间要控制好,酸洗中要注意铌腔的冷却. 酸洗后的清洗十分重要,去离子超纯水的电阻率需随时监测,尤其是最后一次冲洗时,其电阻率应在 18 MΩ 左右. 腔体酸洗时安装在专用支架上,可以转动,使酸液与腔壁均匀接触且排放液体方便. 束管两端配有特制的耐酸法兰与阀门. 热处理前的第一次化学清洗要在腔体表面去除 20~30 μ 的厚度. 热处理后要再做一次化学清洗,除掉 5~10 μ 的表层.

(a) 清洗前

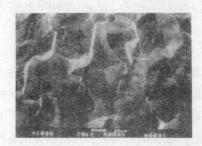 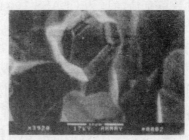

(b) 清洗后

图 5 化学清洗前后的显微照片

3. 高温热处理

我们自 1990 年 10 月开始 RF 超导实验,头两次实验对腔体仅进行了化学清洗,结果不理想. 经过分析并查阅了 KEK DESY,Cornell 等实验室的资料后得到初步结论,腔体需消除应力,这种应力来自机加工和焊接. Cornell 1989 年的文献报道:1.5 GHz 铌腔的 Q_0 值一直提不高,与预期值相差两个数量级,通过热处理消除应力后,这个问题解决了. DESY 实验室在 1989 年会议上提出了同样的问题. 我们参照他们的经验,制定了腔体热处理工艺:铌腔热处理在真空炉中进行,真空度不低于 1×10^{-3} Pa,热处理温度 850℃~900℃,保持一小时,降温时不得关闭真空泵机组. 在热处理过程中,铌腔放置在一钛盒内. 高温时杂质由铌表面蒸发出来,降温时很可能又被铌吸附,有了钛盒后,钛易于吸附杂质离子,可以保护铌腔不会有二次污染.

我们的第三次低温液氦实验所使用的铌腔即是完成热处理的腔,正是在这只腔上我们获得了突破性的进展,测得 10^8 量级的 Q_0 值(4.2 K 时).随着减压降温的实现,又测得了 10^9 量级的 Q_0 值.这些数值与国外的实验值一致,与理论预期值相符.在此同时,我们也注意到,完成热处理的腔在多次实验中重复性好,没出现因升降温循环而引起 Q_0 值的降低现象[15,16].

4. 超净安装

为了防止清洗后的腔受到污染,需要在超净环境中安装.完成腔的清洗后,要用超纯水充满以防止杂质进入腔内,然后立即送到超净室进行安装.安装前再用超纯水将腔冲洗.超净室装有风淋门及空气过滤器,同时装备有洁净度为 CLASS 100 的双人工作台,保证工作台上每立方英尺的空气中大于 0.5 μm 的尘埃不超过 100 个.而空气中一般数目为 21 万个.在安装主耦合器时,将液氮杜瓦蒸发的纯氮气通过加热管道注入腔体,形成腔内正压,以防止腔表面与空气接触,尽量减少杂质进入腔内的机会.完成腔体与主耦合器的安装后,换热氮气从主耦合器方向注入,将这两部分与真空管道相连,在此过程中,一直保持腔体内为正压.

起动离子泵后,我们一般要对整个系统烘烤 20 小时左右,一星期后真空度即可达到要求.

五、微波与锁相测量技术

RF 超导实验的一个关键是微波测量.在超导条件下,1.5 GHz 铌腔的损耗电阻 R_S 仅为数十至数百纳欧,它比铜腔的 R_S 约低 5 个数量级,由此带来了极高的超导腔品质因素($10^8 \sim 10^9$),其谐振曲线的半功率点带宽只有几个 Hz,而场的衰减时间常数达 30～300 ms,这给 RF 参数的测量和谐振点的寻找造成极大的难度.从课题一开始,我们就认识到这样高 Q 值的超导腔的测量是国内微波测量领域的新课题,要求我们应具有相应的测试方法及高精度、高稳定度的测试仪器.

1. 仪器配置

图 6 是用于低温超导实验的微波测量线路,这个线路中所用仪器基

本显示了我们的仪器配置. 合成信号发生器是 HP 8663A, 它的频率稳定度为 10^{-10}/天, 输出信号在 1.5 GHz 附近的分辨率为 0.4 Hz, 小于腔体超导时的半功率带宽. 该信号源有扫频及锁相调频功能. 用扫频法寻找超导点时, 信号源必须具有上述的高分辨率及稳定度.

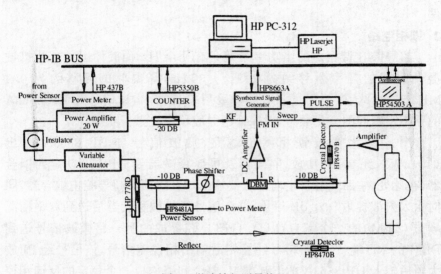

图 6　微波锁相测量线路

数字化示波器 HP 54503A 的带宽为 DC-500 MHz, 有四个通道, 示波器的横向时间轴有 501 个点 (HP-IB 远程获取数据为 1024 点), 纵向电压测量为 8 位, 精度是 $1/2^8=0.4\%$ (HB-IB 远程获取数据为 16 位, 精度 1.5×10^{-5}). 它可自动测量信号的最大值、最小值等, 并在远程控制下对某一给定时间测信号的电压, 也可测信号线过某一给定电压的时间. 此示波器有存储记忆功能.

功率计是 HP 437B, 测量范围: 频率为 100 kHz～50 GHz, 功率为 -70～44 dBm, 精度为 0.1%. 频率计为 HP 5350B, 在 1.5 GHz 附近的精度是 1 Hz, 门时间 200 ms.

HP PC-312 控制器有两个 CPU, 既可做 386 微机又可通过 IEEE-488.2 接口与信号源、示波器、功率计、频率器以及激光打印机相连, 组成一个数据获取、处理、存盘记录并打印输出的全自动系统.

20 W RF 固态放大器是测量系统的末级功放, 它是由机电部 54 所制作的.放大器频率范围是 1.4~1.6 GHz,当铌腔的 Q_0 值为 10^9 量级时,其最大功率输出满足实验要求.

系统中使用了 HP 778D (20 dB)、narda 4242-20、TT3-W (10 dB) 等定向耦合器,HP 8470B 晶体检波器,DMM-6-1500 双平衡混频器等器件.

2. 锁相电路

超导腔在谐振点的半功率带宽只有几个 Hz,而腔体因外部的机械振动引起的频率飘移约为 20 Hz~30 Hz,降温时的频移约为 0.1 MHz/K. 为此,微波测量系统需要采用锁相技术,让信号源 HP 8663A 的频率跟踪腔的变化频率,以保证超导腔稳定在谐振状态.

图 6 即为微波锁相测量线路,微波信号自 HP 8663A 输出 (10 mW),经 20 W 功放、隔离器及可变衰减器到 HP 778 D 定向耦合器,在此处,主路信号自定向耦合器输出口直接馈至超导腔主耦合器,同时由入射波提取 -20 dB 信号,这个信号经机械移相器后送到双平衡混频器(DBM)的一个端口(LO). 自超导腔提取的另一路微波信号送到 DBM 的 RF 端口,两路信号经鉴相后输出相误差信号,该信号经 PI 放大器后输入 HP 8663A 的 FM 端口. 整个回路构成一个稳定的反馈相控环.

3. 主耦合器与提取探针

微波功率经主耦合器馈入超导腔,提取探针自腔内取出微波测量信号.对功率馈入和提取信号讲,它们的外部品质因素 Q_{ext} 的定义为

$$Q_{ext_1} = 2\pi U/W_g \text{(馈入部分)} \quad Q_{ext_2} = 2\pi U/W_p \text{(提取部分)} \quad (1)$$

式中,U 为腔的储能,W_g,W_p 表示一周内在信号源内阻和提取信号电路上分别消耗的能量. 我们知道[1],有载品质因素 Q_1 与 Q_0 的关系为

$$1/Q_1 = 1/Q_0 + 1/Q_{ext_1} + 1/Q_{ext_2} \quad (2)$$

为了表示耦合程度,引进馈入与提取电路的耦合系数分别为 β_1,β_2

$$\beta_1 = Q_0/Q_{ext_1}, \quad \beta_2 = Q_0/Q_{ext_2} \quad (3)$$

把 β_1,β_2 带入(2)式中得到

$$Q_0 = Q_1(1 + \beta_1 + \beta_2) \quad (4)$$

由实验可测得超导腔的有载品质因素 Q_1,腔的无载品质因素 Q_0 则要在测得 β_1,β_2 后才能获得. 图 7 是腔体的主耦合器结构示意图. 超导腔

的 Q_0 值约为 $10^8 \sim 10^9$, 为使微波功率尽可能多的馈入腔内, 要求主耦合器的 Q_{ext_1} 与 Q_0 相当. 提取探针的 Q_{ext_2} 比 Q_0 高两个数量级左右, 这时探针提取的信号大小对 Q_0 测量影响较小, 计算 Q_0 时可忽略提取探针的 β_2.

我们将外 Q 的设计值定为: $Q_{ext_1} = (1 \sim 3) \times 10^9$, $Q_{ext_2} = 1 \times 10^{11}$, 由这两个数值分别确定主耦合器天线及提取探针的长度. 先用计算机估算, 然后用实验测量并调整, 最后在超导时验证, 有问题需重新安装, 这是一件精细的实验工作. 在液氦温度下, 金属的热胀冷缩会使外 Q 值发生变化, 提取探针是用 50 Ω 硬同轴电缆制作的, 当降低温度时, 内外导体间的绝缘层的收缩量比金属的大几十倍, 其结果是液氦温度时提取探针的 Q_{ext_2} 比设计值高出两个量级. 主耦合器也遇到了相似的问题, 主耦合器天线是一段 50 Ω 同轴线, 内外导体间是真空, 其一端经陶瓷封接 50 Ω 同轴接

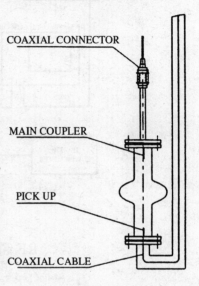

图 7 耦合器示意图

头与功率馈线相连, 这只陶瓷接头可工作在 20 K 温度, 内导体是由金属钼制作的, 天线与内导体用螺纹相连, 初次实验时, 天线选用了铜材, 实验后发现天线几乎脱扣, 仅有一扣螺纹相连了, 这是由于铜的膨胀系数比钼大的原因. 通过实验, 我们修正了设计, 改换了材料, 逐步做到了室温的实验值与低温测量值相符.

用 URMELT 程序计算出腔中的 E_z 电场分布, 根据探针电压、腔电场与探针长度间的关系可以估算出探针长度. 图 8 是实验测定天线及探针长度方法示意图. 在腔的一端固定辅助天线, 用反射波扫频信号调节其长度使 $\beta=1$, 功率计测定信号源的输出功率, 然后测固定在腔的另一端的待用天线的提取功率, 调节待用天线的长短, 以获得合适的提取功率与输出功率比, 这样就可得到要求的外 Q 值. 我们实验中的使用值为: $Q_{ext_1} = 1 \times 10^9$, $Q_{ext_2} = 2 \times 10^{10}$. 后一数值比设计值小了一个量级, 原因是鉴相器需要提取探针供给 5 dBm 以上的信号.

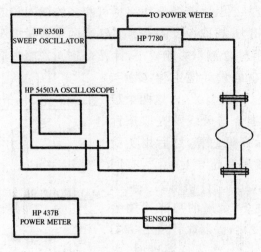

图 8 Q_{ext} 值测量线路

六、RF 超导实验及其结果

自 1990 年 10 月开始 RF 超导实验,首次实验是对整个低温系统进行检验. 经运行测试,低温及真空系统性能良好,工作安全可靠. 系统做到以下状态:降温前铌腔的超高真空系统及恒温器夹层的真空度分别达到了 10^{-6} Pa 和 10^{-4} Pa. 经两天液氮降温后,铌腔温度可达到 160 K 左右. 超导实验时,用液氦输液管将杜瓦的液氦输至恒温器中直至液氦漫过整个铌腔,在此同时,开启回收管道,将氦气送至低温车间回收. 完成输液后,去除输液管,这时铌腔的温度是 4.2 K,液氦面超过束管上端法兰. 温度及液氦面分别用铁铑电阻温度计及液氦面仪监测的. 一台 15 升机械泵经一个三通与恒温器相连,其排气口与回气管道连通,起动机械泵,控制好阀门的流量,就可进行减压降温实验,最低温度可到 2 K.

1. RF 超导实验结果

RF 超导实验是研究 1.5 GHz 铌腔在超导状态下的 RF 性能,这种 L 波段的谐振腔是用在直线加速器上,我们通过实验测量腔的无载 Q 值随加速梯度 E_{acc} 和温度 T 的变化. 实验时,用减压降温方法改变温度,我们测量了 6.5 K,4.2 K 及从 4.2 K 到 2 K 温度间各点所对应的超导腔

的无载 Q 值,这种实验进行多次,重复性好.图 9 给出了表面电阻 R_s 与 T_c/T 间的关系曲线,R_s 与 Q_0 的关系为 $G=R_s×Q_0$,G 为腔的几何常数,对我们的腔,$G=260\ \Omega$. T_c 为临界温度,铌的是 9.2 K.图 9 中的实线是由 BCS 理论推导的 R_s 随温度变化的预期值,小三角是实验值,在大部分区域理论值与实验值符合较好.当温度很低时,实验值偏离理论曲线,渐趋于一固定值,这是正常的,这个数值称为超导腔的残留电阻 R_{res},这也正是 RF 超导与直流超导的不同处.R_{res} 的大小与铌材表面性能相关[1].

图 10 是腔的无载 Q 值与腔的加速梯度 E_{acc} 的实验结果.我们对 4.2 K 与 2.25 K 两个温度点做了功率实验,对应不同的馈入功率,测量腔内的加速梯度及腔的无载 Q 值.由图中的曲线可以看出,随着加速梯度 E_{acc} 的增高,腔的无载 Q 值呈下降的趋势.下降的快慢或出现明显的拐点是衡量一只腔 RF 性能优劣的标准,性能好的腔的 Q_0 与 E_{acc} 曲线应比较平直,在较高加速梯度时 Q_0 出现明显下降或有拐点.我们的实验结果未出现明显的拐点,但 Q_0 值有所下降.我们在初次的功率实验中只获得了 3 MV/m 的加速梯度,经过不断的改善腔体的清洗、热处理、减压降温以及微波锁相测量等技术,我们将腔的加速梯度提高到 8.6 MV/m,这个数值已进入到国际上的先进水平.目前我们正准备作难度较大的腔的氦气锻炼实验,这可以改善我们的 Q_0 与 E_{acc} 曲线,使高加速梯度的 Q_0 值提高.

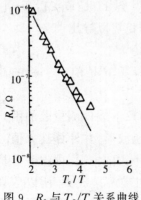

图 9 R_s 与 T_c/T 关系曲线

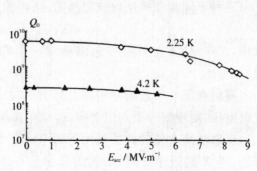

图 10 Q_0 与 E_{acc} 关系曲线

图中 E_{acc} 的数值由下面关系式(5)得到.

$$E_{acc} = (\sqrt{2}/0.1)\sqrt{102 \times Q_0 \times (4\beta/(1+\beta)^2) \times P_i} \qquad (5)$$

式中:0.1 为腔的有效加速长度(m),$(102 \times Q_0)$ 为腔的分路阻抗(Ω),P_i 为 RF 入射功率(W),E_{acc} 单位为(V/m).β 及 Q_0 值的获得见下节详述.

2. 超导状态下的微波测量技术

前面已经提到 1.5 GHz 超导腔的无载品质因素 Q_0 高达 $10^8 \sim 10^9$,半功率带宽只有几个 Hz,而场的衰减时间常数达 30～300 ms,这给 RF 参数的测量和谐振点的寻找都造成了很大困难.而且由于我们的实验腔上缺少频率调谐机构,更增加了确认谐振频率的难度.

(1) 扫频法寻找腔的谐振点

由于 HP 8663A 合成信号发生器具有高的频率稳定度和 0.4 Hz 的扫频分辨率,我们在 1.46～1.47 GHz 频率范围内,对每 100 Hz 的频率区间进行精密扫频,扫频时间是 100 ms/Hz.经过多次的精密扫频便可捕捉到超导时的谐振频率信号.获得这个频率后,将发生器由扫频状态改为 CW 模式,同时调节锁相电路使其工作,在锁相环的控制下,超导腔稳定工作在谐振状态.

(2) 反射法测量 β

测量 β 值的方法是应用谐振电路的瞬态响应过程,脉冲调制的 RF 信号馈入谐振腔,腔内电压有一建场过程及一衰减过程,把瞬态入射电压信号(脉冲)与瞬态腔内电压信号相减,就得到反射电压的瞬态信号,对应不同的 β 值,反射波电压包络有不同形状.图 11 是 $\beta<1, \beta=1, \beta>1$ 三种不同类型的反射波包络图.由传输线理论可以导出[10]

$$\beta = X/(2-X) \qquad (6)$$

式中:$X=b/a$,b 是反射波包络的后一个峰的高度,a 是前一个峰的高度.

我们在图 6 所示的定向耦合器 HP 778D 的反射端口测量腔的瞬态反射电压,对应每一个入射功率,记录下反射电压波形,并计算其 β 值.

反射法测 β 值的方法适用于两个峰高相差不大的情况,由误差分析可知,当 X 趋近于 2 时,X 的测量误差在计算时被严重放大,所以应调节主耦合器的外 Q 值,使其耦合系数 β 尽量接近 1,以便缩小测量带来

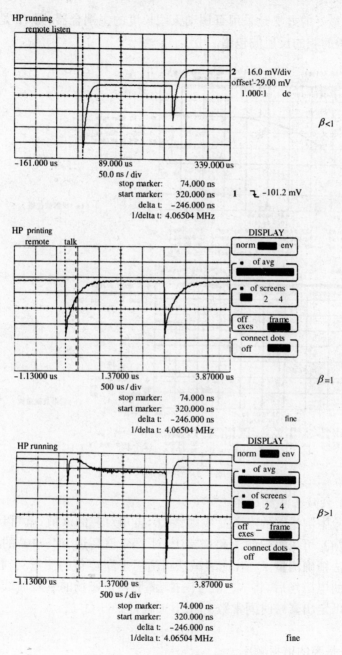

图 11 β值与反射波包络

的误差.最好的办法是采用可调动天线长度的主耦合器.图12是超导实验中获得的腔的反射波包络.

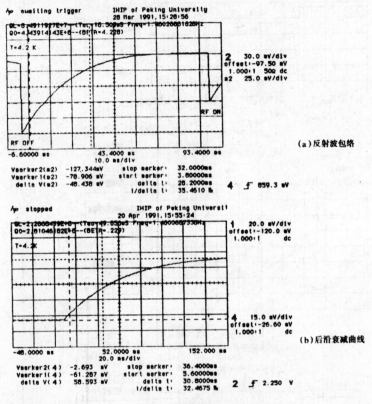

图12 实验测量的两种曲线

(3) 衰减法测量 Q 值

实验中测量腔的有载品质因素 Q_1,再用对应的 β 值,就可计算得到 Q_0.测量 Q_1 可采用扫频法和衰减法.对于 Q 值高达 $10^8 \sim 10^9$ 的情况,扫频法无法精确测量 f_0 值.而衰减法可适用于高 Q 值的测量.当腔内建立起稳定的电磁场后,将 RF 信号断掉,测腔内电磁场的衰减曲线,由衰减曲线可以定出衰减时间常数 τ,用下式就可获得 Q_1 值:

$$Q_1 = \pi \times f_0 \times \tau \tag{7}$$

式中 f_0 是腔的谐振频率.

图12中,另一波形即为腔的电磁场自由衰减信号.

3. 实验仪器的控制和数据采集

在微波测量技术上,我们成功地寻找到高 Q 腔的谐振频率,并用锁相技术,使超导腔稳定工作在谐振状态,这些方法在国内微波测量领域是新的技术和新的发展. 在此同时,我们对图 6 所示的系统实现了微机控制和自动数据处理,利用 HP PC-312 微机的控制性能,编制了程序,控制整个测量系统(信号源、示波器、频率计及功率计),使测量做到准确、直观、快速、方便.

HP PC-312 微机装有一块 HP BASIC 控制板,控制板自带 CPU,控制板上有 2.5 M 字节的 BASIC 专用内存,可以通过 HP-IB 接口对其他测量仪器进行远程控制和获取数据. 我们编制的程序核心是为了测量 RF 系统的 Q 值,其功能有在线计算、屏幕(结果)打印、数据存储、数据再现及数据后处理等. 参数设定及修改均由大光标菜单拾取,操作简单明了. 软件采用自顶而下的模块编成,结构坚固,容错能力强,用户界面友好. 在此程序控制下做到以下几点:

(1) 对 Q 值用扫频法或脉冲衰减法进行测量,结果回送示波器屏幕.

(2) 示波器屏幕送激光打印机打印,任意位置,四种大小. 多幅图可打在同一页上.

(3) 示波器的波形数据存磁盘文件,供以后再现、再计算、后处理.

(4) 存储波形的再现、计算.

图 13 为该系统功能简介图,图 14 为运行开始后的系统主菜单,图 15 为 1.5 GHz 超导铌腔,图 16 为装配好的铌腔,图 17 和图 18 分别为程控微波锁相测量系统和实验恒温器.

图 12 图形上部的数据即为测量时所计算出的腔的有载 Q 值、耦合系数 β,及由此导出的腔的无载 Q 值. 由程序控制的全自动测量及时方便,提高了我们实验的效率和成功率.

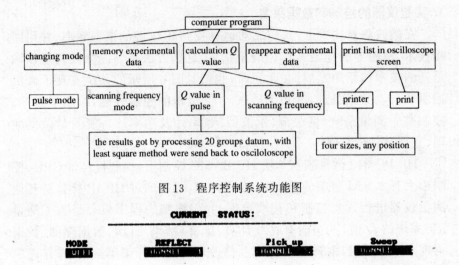

图 13　程序控制系统功能图

图 14　程序运行的主菜单

图 15　1.5 GHz 超导铌腔

图 16　装配好的铌腔

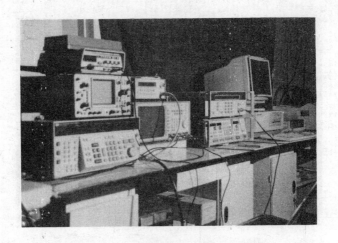

图 17　程控微波锁相测量系统

图 18　实验恒温器

七、结　语

北京大学重离子所的"超导腔"课题组能够以较短的时间在RF超导加速腔研究上取得初步进展,为RF超导腔的应用和今后开展超导型高亮度注入器的研制提供了依据与技术基础.

致谢　首先要感谢有关专家和学校领导的关怀和支持.在这三年的工作中,北大重离子所的沈子林、吕德泉、邬惠民、付晓斌、耿荣礼等同志参加了超导腔的部分工作.北大物理系低温车间的教师和工程人员大力协作,积极支持,保证了液氮、液氦的供应及氦气的回收.刘微浪老师、戴远东副系主任亲临指导输液降温,使实验安全顺利进行.

我们还得到德国DESY国家实验室的D. Proch博士的支持与帮助,以及Cornell大学H. Padamsee博士、Darmstadt的T. Weiland博士、清华大学童得春教授的支持,在此致以衷心的感谢.

参　考　文　献

[1] H. Piel, CERN 89-04, 149(1989).
[2] 王光伟等,RF超导腔及其性能研究,北大重离子所报告SC-91-03 (1991).
[3] L. Serafini et al., EPAC 90, 143(1990).
[4] D. L. Moffat et al., KEK Report 89-21, 445(1990).
[5] D. Proch et al., DESY M-84-10 (1984).
[6] K. Saito et al., KEK Report 89-21, 635(1990).
[7] D. Bloess et al., KEK Report 89-21, 477(1990).
[8] H. Lengeler, CERN 89-04, 197(1989).
[9] K. C. D. Chen, Proceedings of the Beijing FEL Seminar, 172(1988).
[10] 宋进虎,超导谐振腔的实验研究,北大技物系硕士论文(1991).
[11] 赵夔等,高梯度超导谐振腔,加速器会议文集(1988).
[12] Chen Jiaerh et al., RF Superconducting Cavities at IHIP of Peking University, to be published (1990).
[13] Chen Jiaerh et al., IHIP Report SC-91-01 (1991).
[14] Q. S. Shu, KEK Report 89-21, 539(1990).
[15] R. W. Roth et al., EPAC 90, Nice (1990).
[16] R. Byrns et al., EPAC 90, Nice (1990).
[17] K. Asano et al., KEK Report 89-21, 723(1990).

[18] G. Mueller, KEK Report 89-21, 267(1990).
[19] K. W. Shepard, KEK Report 89-21, 139(1990).

RF SUPERCONDUCTING EXPERIMENTS OF 1.5 GHz Nb CAVITY

Abstract

The project of RF superconducting cavities have been started at Peking University since 1988 and the progress has been obtained. The peparation and procedure for RF super-conducting experiments at low temperature are described in detail. RF superconducting experiments, microwave measurements, phase lock technique, the surface treatment of cavity, computer simulation for cavity design, control, data acquire and processing with computer are summaried. So far, an accelerating field of 8.6 MV/m was abtained with CW mode and Q_0 of 6.5×10^8, 2.25 K.

Key Words　RF superconductivity, Nb cavity, cryostat, postpurification, phase lock technique.

THE PROGRESS OF SUPERCONDUCTING CAVITY STUDIES AND R & D FOR LASER DRIVEN RF GUN AT PEKING UNIVERSITY[①]

Introduction

The progress of RF superconducting cavity studies and R & D for high brightness laser driving RF gun are introduced in this paper. So far, single cell of 1.5 GHz Nb cavity has been studied in detail and an accelerating field of 12.6 MV/m was obtained with Q_0 of 2×10^9 at 2.35 K. A single cell cavity of 1.3 GHz is being developed. As an application of RF superconducting cavities, the feasibility study on superconducting photoemission RF gun and the experiment of photo-cathodes are being undertaken.

The Research Progress of Superconducting Cavity

In the past several years, a single cell of 1.5 GHz Nb cavity has been studied with a vertical LHe cryostat in detail. Fig. 1. gives the measured results of Eacc vs. Q_0 at 4.2 K and 2.35 K respectively. The accelerating gradient up to 12.6 MV/m with $Q_0 = 10^9$ was obtained.

---

① Coauthers: Zhao Kui, Zhang Baocheng, Wang Lifang, Yu Jin, Song Jinhu, Geng Rongli, Wu Genfa, Wang Tong, Liu Weilang, Xue Zengquan. Reprinted from the Proc. 5-th Japan-China Joint Symposium on Accelerators for Nuclear Science & their Applications, Oct. 1993, Osaka, Japan, p. 244～246.

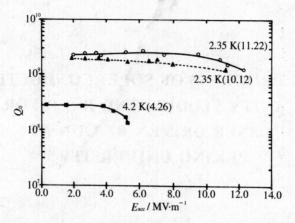

Fig. 1 E_{acc} vs. Q_0 at 4.2 K and 2.35 K

For the further study of superconducting cavities, a DI recirculating water system and a new class 100 room have been installed and operated in our Lab. A 200 W microwave power source of 1.3 GHz is being tested. Considering the specifications of suitable microwave source in the domestic market as well as possible applications of L band superconducting in the future, we select 1.3 GHz as the operating frequency. The design of 1.3 GHz cavity was optimized by computer codes of URMEL-T and SUPERFISH. The designed parameters of 1.3 GHz cavity are shown in the table 1 and illustration Fig. 2.

Table 1 Parameters of 1.3 GHz Cavity

Frequency	1.3 GHz
Beam Aperture	7.0 cm
R/Q	51.06 Ω/m
E_{peak}/E_{acc}	2.16
H_p/E_{acc}	41.09 Gs/MV/·m^{-1}

In order to test the forming and welding procedures of Nb cavity, a set of steel dies, including a female die, two male dies and a hold down plate, have been designed and manufactured. The dies was made of fine steel with chromium plated. The surface roughness of dies is 0.8 μm.

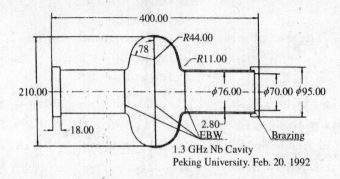

Fig. 2   1.3 GHz Nb cavity with geometrical parameter

Considering the mechanical properties of Cu and Nb are very similar, we selected 2.5 mm Cu sheet to test the forming technology. A 1.3 GHz Cu cavity has been formed successfully by deep drawing with this set of dies. The electron beam welding of 1.3 GHz Cu cavity has been completed and the test of EB welding for Nb material is in progress. The measured frequency of 1.3 GHz Cu cavity is 1.2958 GHz without tuning. In order to manufacture Nb cavity of 1.3 GHz, the production of high purity Nb has been tried in a factory of China. The first batch Nb samples was delivered last year. The measured RRR of the first of Nb samples was only about 60 to 70. Since then, the factory improved the melting technology and new Nb sheets have been delivered recently. Table 2 is the impurity contents of the Nb sheets.

Table 2   Impurity Contents of Nb Sheets

Element	Wt. (%)	Element	Wt. (%)
Si	0.001	Zr	<0.001
W	0.001	Hf	<0.001
Ti	0.001	Cu	<0.001
V	<0.001	O	50 ppm
Al	<0.001	C	20~30 ppm
Fe	0.001	N	50 ppm
M	0.001	H	5 ppm

The Feasibility Study of Photo-Emission RF Gun Using Superconducting Cavity

The idea of high brightness photoemission source using a superconducting cavity was suggested by Prof. H. Piel and Dr. C. K. Sinclair. Some sort of studies and experiments have been done by Dr. A. Michalke at Wuppertal. There are obvious advantages with a laser-driven RF gun. A photocathode located in an RF cavity can deliver very high current densities: more then 400 A/cm^2 while illuminated by a laser pulse. In 1992, a proposal was made by the group of superconducting cavity of Peking University to study the feasibility of S. C. photo-RF gun. As the first step, we decide to develop a 1.3 GHz 1+1/2 superconducting cavity made of high thermal conductivity niobium. The photocathode will be located on the center of the end plate. The design of beam dynamics and the optimization of the first half cell cavity are carried out using ITACA code. Table 3 is the parameters of pre-testing photoemission RF gun with superconducting cavity. The initial values assignment are $E_{cathode}=18$ MV/m, $R_{Laser}=2$ mm, Pulse Length$_{Laser}=10$ ps, Phase$_{in}=54°$ respectively.

Table 3　Parameter of S. C. Photo-RF Gun

Charge of Bunch	nC	0.08	1.3
Energy Gain	MeV	0.9471	1.049
RMS Energy Spread	%	0.289	0.379
I_{peak}	A	12.44	77.12
Norm. Trans. Emmit.	mm·mrad	3.0972	12.083
Norm. Brightness	A/mrad^2	0.82×10^{10}	0.33×10^{10}

Two aspects will be investigated in our photocathode pretest program. The first, various samples of photocathodes will be tested under DC field in order to select a suitable cathode; the second, the field emission and the dark current effects will be studied under RF field without laser irradiation while the photocathode is mounted into the superconducting cavity.

R & D on Photocathodes

One of the important issues for photo-emission RF gun is the requirement of an ideal photocathode which should have high Q_e (quantum efficiency) and reasonable long life time. In order to develop a practical photocathodes which can satisfy these requirements, a series of studies and experiments have been performed in our laboratory. The photocathodes based on the concept of ion beam implantation have been developed and tested since April 1993. In the mean time multi-alkali photocathodes and a special Ag-Ba-O photocathode, whose substrate is BaO embedded with ultrafine silver particles, are being developed in collaborating with Dept. of Radio-Electronics, Peking University.

As the first attempt, low energy cesium ions beam are implanted into four kinds of substrates, namely cooper, aluminum, nickel and niobium. The substrates were mechanically polished and treated with acetone, ethanol and DI water carefully. The vacuum of the target chamber is better than 10^{-6} torr in the period of implantation. The implanted samples were stored and transferred in a vessel filled with nitrogen gas. A test system for photocathodes was established. A high voltage of $1 \sim 6$ kV is applied to extract the electrons from the cathode. Usually, two or three times of heat treatments are necessary to get stable emission. The photoelectric experiments are performed using a Q-switched Nd:YAG laser. The pulse duration of the laser is 6 ns with 30 Hz repetition. 2nd, 3rd and 4th harmonics of the laser can be extracted through different selection of KTP crystals. The sketch of the photoelectric experiment is shown in Fig. 3.

The preliminary results of the ion implanted photocathodes, including the evaluation of Q_e, emission characteristics and the investigation of emission procedure with 2nd, 3rd and 4th harmonics of laser, are obtained. The estimated Q_e of Al:Cs with irradiation of focused laser at the wave-

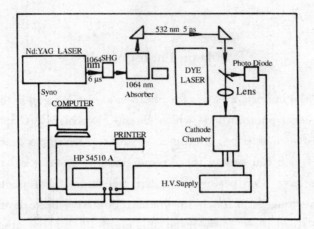

Fig. 3 The sketch of the photoelectric experiment

length of 532 nm is $10^{-5} \sim 10^{-4}$, one order higher than that of pure aluminum. The real time signals are shown in Fig. 4. It has been demonstrated that the Q_e of Ni:Cs is also the same order as that of Al:Cs. Nevertheless, the distortion of the current pulse shape is more critical as shown in Fig. 5. The more serious distortion of Ni:Cs is due to higher temperature increase on the surface because of worse thermoconductivity of Ni comparing with Al. With better thermoconductivity of Cu, Cu:Cs exhibits good photoemission even at 30 MW/cm^2 of laser intensity and 24 nC integrated charge and $>$2000 A/cm^2 current density have been reached. Experiments with 3rd and 4th harmonics without focusing of the laser beam showed excellent current pulse response even better than that of the photo-diode, but the estimated Q_e of Al:Cs is $10^{-7} \sim 10^{-6}$ for 355 nm and 266 nm respectively. Finally, we didn't find that Nb:Cs give a prominent enhanced emission compared to sample of pure Nb. All of the samples implanted Cs ions can be transferred in atmosphere and recycled by heat treatment.

Conclusions

In order to realize the photoemission RF gun, it's necessary to increase the gradient field of Nb cavity. Meanwhile the study of mutual in-

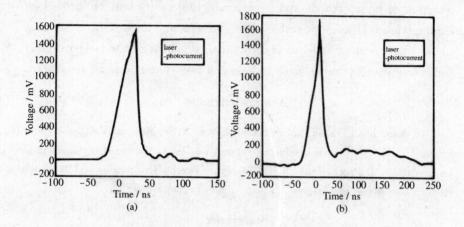

Fig. 4 Time resolved photoemission characteristics of Al:Cs irradiated
by focused laser at wavelength of 532 nm
(a) pure multi-photon effect, laser intensity 1.4 MW/cm^2, integrated charge 1.8 nC.
(b) laser intensity 2.5 MW/cm^2, pulse distortion appears due to thermal effect

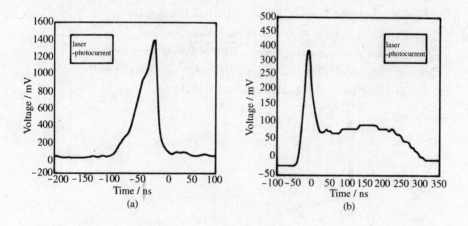

Fig. 5 Time resolved photoemission characteristics of Ni:Cs irradiated by
focused laser at wavelength of 532 nm
(a) pure multi-photon effect, laser intensity 2.3 MW/cm^2, integrated charge 1.6 nC.
(b) laser intensity 2.5 MW/cm^2, pulse distortion appears due to thermal effect

teraction of photocathode and superconducting cavity will be focused on. High quantum efficiency, stable photoemission and long life time are needed for a practical cathode. It is difficult to synchronize the laser to the rf, and to reduce the timing jitter to be much less than one pulse length.

Acknowledgment

We would like to thank Prof. Q. D. Wu, Prof. X. Sh. Zhao and their colleagues of Peking University for their continuous support and help. We are also pleased to acknowledge Dr. D. Proch, Prof. Y. Kojima, Prof. Pagani, Dr. Serafini and Dr. Kneisel for their continuous encouragement, help, and discussion.

Reference

[1] C. Travier, Particle Accelerators, 1991, Vol. 36, pp. 3374.
[2] A. Michalke, Photocathodes Inside Superconducting. Cavities, External Report, WUB DIS 92-1, Jan. 13, 1993.
[3] C. Chia-erh, et al., Proc. of the 5th Workshop on RF Superconductivity, 1991, pp. 102～109.
[4] Geng Rongli, et al., Internal Report, IHIP of Peking Univ.
[5] Yu Jin et al., Internal Report, IHIP of Peking Univ.
[6] Song Jinhu, et al., Internal Report, IHIP of Peking Univ.
[7] Wang Tong, et al., Internal Report, IHIP of Peking Univ.
[8] C. Travier, NIM in Physics Research A304(1991) 285～296.

DESIGN AND CONSTRUCTION OF A DC HIGH-BRIGHTNESS LASER DRIVEN ELECTRON GUN[①]

Abstract

A DC high-brightness laser driven photoemissive electron gun is being developed at Peking University, in order to produce 50 ~ 100 ps electron bunches of high quality. The gun consists of a photocathode preparation chamber and a DC acceleration cavity. Different ways of fabricating photocathodes, such as chemical vapor deposition, ion beam implantation and ion beam enhanced deposition, can be adopted. The acceleration gap is designed with the aid of simulation codes EGUN and POISSON. The laser system is a mode-locked Nd-YAG oscillator proceeded by an amplifier at 10 Hz repetition rate, which can deliver three different wavelengths (1064/532/266 nm). The combination of a superconducting cavity with the photocathode preparation chamber is also discussed in this paper.

1. Introduction

Free electron lasers, high-energy linear colliders, wake field accelerators and new accelerator experiments all require very bright electron beams. During the last 10 years, photo-injectors progressed rapidly [1]. By this means, short pulse, high brightness, intense electron beams have been successfully produced. However the production of such a beam of CW operation mode is still marginal. Superconducting (SC) RF gun has been proposed by some authors considering the advantages of SC cavities

---

① Coauthers: K. Zhao, R. L. Geng, L. F. Wang, B. C. Zhang, J. Yu, T. Wang, G. F. Wu, J. H. Song. Reprinted from the Nucl. Inst. &. Meth. in Phys. Res. A, 375(1996) p. 147~149.

[2]. The preliminary results show that there is no principle barrier to combine photocathode with SC cavity. Nevertheless field emission from photocathode inside SC cavity occurs when the accelerating gradient reaches $2\sim3$ MV/m, which indicates that more experiments should be done to insure the possibility of SC RF gun. Unfortunately very limited recent work has been reported on this topic.

Since 1992 we have began to study the possibility of the SC photoemissive RF gun. The first temptation was concentrated on the photocathodes. A method of ion beam implantation was put forward to make novel photocathodes [3]. The results are very interesting. We have also tested Cs_2Te as intense electron material and we got very good parameters. Meanwhile, we have successfully manufactured two SC niobium cavities with the results of $Q_0 = 1.6 \times 10^9$, $E_{acc} = 10$ MV/m at 2.5 K. To reach the goal of a SC photoemissive RF gun, we will follow the following strategy, as the first step, set up a DC electron gun consists of a photocathode preparation chamber and a DC acceleration chamber, as the second step, substitute the DC chamber with a SC cavity, inserting the determined photocathode directly into the SC cavity. In this paper, we describe the design and construction of the DC photoemissive gun. The combination with the SC cavity is also discussed at the end of the paper.

2. DC High-Brightness Electron Gun

A DC high-brightness laser driven electron gun was designed and constructed by the superconducting cavity group of Peking University. Fig. 1 is the layout of the device. It consists of four main parts, i.e., the photocathode preparation chamber, the DC acceleration chamber, the mode-locked laser and the beam diagnostic system.

The photocathode preparation chamber includes the vacuum chamber and some other items. A metal bellow of 1.5 m length transports the samples from the manufacturing site to the acceleration chamber. Telluride and cesium vaporizers are to provide the necessary materials for making

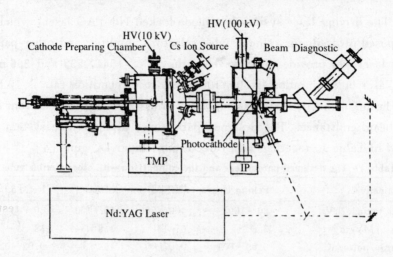

Fig. 1 The layout of the DC high-brightness laser driven electron gun

Cs_2Te. A 30 kV ion source is mounted in front of the vacuum chamber, which can deliver low energy cesium ions to make ion implanted photocathodes. An argon ion gun is located opposite to the sample surface for the cleaning and post-treating of the samples. The sample heating and temperature measuring system is at the rear part of the chamber. An anode fed into the chamber allows monitoring the photocurrent while making the photocathodes. Three different methods can be adopted to make photocathodes, namely chemical vapor deposition, ion beam implantation and ion beam enhanced deposition.

Inside the DC acceleration chamber a focusing electrode is mounted to the inner rear part of the vacuum chamber through a ceramic bellow of 70 mm in length. The photocathode will be located at the centre of the focusing electrode. A 100 kV negative high voltage is fed to the focusing electrode through a 120 mm long ceramic bellow. The anode is grounded and the cathode-anode distance can be adjusted. The laser ports allow the driving laser beam to enter the vacuum chamber through the quartz windows at an angle of 74° to the normal of the cathode. At the exit of the acceleration chamber a magnetic coil is added to insure good transportation of the electron beam in the proceeded drift region.

The driving laser system is a mode-locked Nd-YAG laser, which is composed of oscillator, single pulse selector, amplifier and other parts. This laser can provide three wavelengths, i. e. 1064, 532 and 266 nm. The laser pulse duration is 50~100 ps and the repetition rate is 10 Hz. The beam diagnostic system is supposed to measure the pulse current and the beam emittance. The designed parameters of the whole system are listed in Table 1, together with three other similar DC guns [4].

Table 1 The designed parameters and the comparison with other similar guns

Parameters	Peking Univ.	CEBAF	SLAC	Ref. [5]
High voltage/kV	80~100	400~500	60~120	30~60
$E_{cathode}$[MV/m]	3.6	6~10	1.8	15
Charge/pulse/nC	0.05~0.1	0.07~0.2	16	0.02~0.6
Cathode diameter/mm	6	25	14	—
Emission diameter/mm	4	2~6	14	2
σ_t of laser beam/ps	15	15	500	38
Average current	0.5~1 nA	5 mA	4 μA	0.2~6 nA
Repetition rate	10 Hz	37 MHz	120 Hz	10 Hz

3. The Study of Photocathodes

The most important thing in our experiment is to get a good photocathode for our DC electron gun. We expect the photocathode will have high quantum efficiency, long lifetime and very desirable stability when it is submitted to strong E field. Since 1993, a series of photocathode experiments have been carried out in our group. The development and research of ion implanted photocathodes were finished with some interesting results [3]. This kind of photocathode can be transferred in the air and recovered after post treatment. The quantum efficiency is one order higher than that of pure metal under 1 MV/m electric field.

At the end of last year, Three devices were set to test the Cs_2Te photocathode. The Cs_2Te film was deposited on nickel substrate by chemical vapor deposition method. A nanosecond laser operating at 266 nm in wavelength was used to conduct the photo-electrical experiments. The final result of the experiment is quite desirable. The quantum efficiency reaches 4.3% and the maximum current density is about 100 A/cm^2 at 11 kV extraction high voltage. Fig. 2 illustrates the variation of the quantum efficiency vs. the extraction voltage. It should also be mentioned that the photoemission characteristics of these photocathodes kept unchanged for almost 2 months.

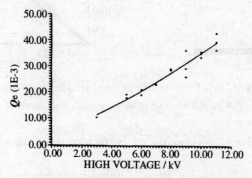

Fig. 2 Quantum efficiency vs. extraction voltage of the Cs_2Te photocathode

4. The Design of DC Accelerating Gap

In the acceleration chamber a 100 kV extraction gap is used. The geometry of the electrodes is optimized with the aid of the simulation codes EGUN and POISSON to get as high as possible bright electron beams. In our case the electron pulse duration is not much less than the transition time of electron between cathode-anode gap; as a result the code EGUN can give a roughly desirable estimate. The simulations with more adequate code such as PARMELA is now being undertaken. Fig. 3 shows the simulation results by EGUN and POISSON, from which we can see that the electron beam is almost laminar and the radial electric field inside the acceleration chamber has quite good linearity which is essential to minimize the spacecharge effect.

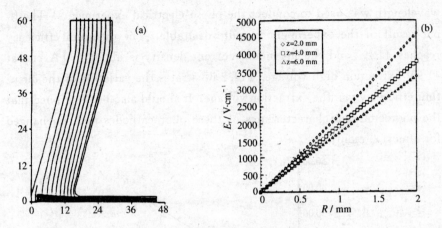

Fig. 3 (a) Electron trajectory inside the electron gun
(b) Radial electric field vs. radial displacement

5. The Considering of Superconducting Cavity

Since DC accelerating voltage is limited to a few hundred kilovolts, it is more appropriate to use RF fields to extract high peak current from the cathode. Our future plan is to replace the DC acceleration gap with a SC cavity. The advantage of a SC cavity is that it can provide a high average electric field in CW mode. Due to the high quality factor, the optimization of cavity shape eliminates the consideration of the cavity shunt impedance. Also, the low temperature provides the photocathode an ultra-clean environment, which is quite desirable to prolong the lifetime of the photocathode. On the other hand, some other drawbacks arise from the introduction of the SC cavity, such as cooling of the cathode-cavity joint, microwave shorting and timing between the laser and the accelerator.

References

[1] C. Travier, Particle Accelerators 36(1991) 33.
[2] H. Chaloupka et al., Nucl. Instr. and Meth. A 285(1989) 327.
[3] Zhao Kui et al., Proc. LINAC 94(1994) Tsukuba, Japan, to be published.
[4] H. Liu et al., Proc. Particle Accelerator Conf., 1995, Dallas, USA, to be published.
[5] A. Aleksandrou et al., Phys. Rev. E51(1995) 1449.

THE HIGH-BRIGHTNESS ULTRA-SHORT PULSED ELECTRON BEAM SOURCE AT PEKING UNIVERSITY[①]

Abstract

In this paper the characteristics of an ongoing new high-brightness ultra-short pulsed electron beam source in Peking University is reported. The results of the experiment on the DC laser-driven electron gun are shown, together with the preparation of the photo-cathode, which has a much improved performance. Based on these achievements the new electron source will adopt a superconducting accelerating section, thus it is expected to obtain electron beams of 1~2 MeV as a first step, and it is expected to reach a higher energy range in the future.

1. Introduction

Thanks to the highly developed science and technology, the performance of photocathode guns such as the increased peak currents and reduced emittances has improved dramatically over the past few years [1~4]. The laser-driven high-brightness electron source is now available in many areas of scientific frontiers. Up to now, the laser-driven electron source has such important characteristics as high repetitive frequency, high-brightness e-beam (electron beam), polarized e-beam, pico- or femto-second e-beam, etc., and it can be synchronized with the laser. As

---

① Coauthers: Kui Zhao, Yin-E Sun, Bao Cheng Zhang, Lifang Wang, Rongli Geng, Jiankui Hao, Yuxing Tang, Xi Yang, Yunchi Zhang, Yu Chen, Zhitao Yang. Reprinted from the Nucl. Inst. & Meth. in Phys. Res. A, 407(1998) p. 322~326.

for the users of the e-beam, a chance of scientific research at a higher level is offered.

Our plan is to set up a photo-RF gun using superconducting cavities. The idea seems desirable for high-brightness e-beam generation. The plan is separated into two steps: The first step is that, using DC extraction, we will add a superconducting accelerating section to enhance the energy of the e-beam after it is extracted. The second step is using the superconducting cavity as the extracting cavity instead of the DC extraction chamber.

2. The Prospects of the High-Brightness Electron Beam Source

Since the 1990s, the reliability and brightness of electron beam generated by photo cathode RF guns is very important for FEL and for the linear collider applications. In fact, it is only at the very beginning of the development that its main applications are in the areas of high energy physics and FEL, and even now this kind of electron source is still very important for FEL. Yet, nowadays, there are various applications of this kind of e-beam such as in surface physics, material physics, solid physics, chemistry, medical and biological areas, etc.

As for our laser-driven high-brightness electron source, the electron beam is available directly after the extraction, which falls in the energy range of 10~200 keV, and after the superconducting accelerating section, the energy will run from 1~20 MeV (see Section 3).

In the energy range of 10~200 keV, it will serve as an unique e-beam source in biology and material science as it can provide the polarized e-beam with ultra-short pulse and at the same time with small emittance. As, for instance, the ultra-short Sweep Electron Microscope (SEM), polarized Sweep Tunnel Microscope (STM) etc., will be developed greatly if the ultra-short e-beam or polarized e-beam is adopted.

As for the energy range 1~20 MeV, it will be an area of particle physics and high energy physics, also the high-brightness electron beam can generate other kinds of radiation connected with the micro bunches, such as the coherent transition radiation, coherent synchrotron radiation, coherent Cherenkov radiation. Smith-Purcell radiation, and diffraction radiation, etc.

3. The New Laser-Driven High-Brightness Ultra-Short Pulsed Electron Source at Peking University

The high-brightness ultra-short electron beam source is a laser-driven electron gun. In this device the cathode is driven by a laser to generate the photo-electron beam, and the electron beam is extracted by a high-gradient field (DC or RF) and is then accelerated by the superconducting section. Thus, it can provide the CW mode or pulsed high-brightness ultra-short electron beam, or polarized electron beam.

Fig. 1 is the schematic of this electron source. It is constructed by the following parts:

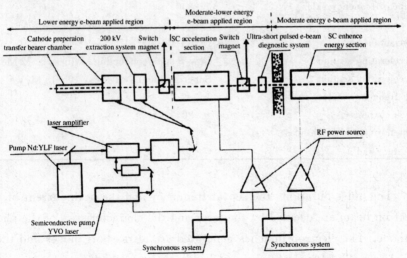

Fig. 1 The schematic of the new electron source

- The pico- and femto-second laser system;
- The photo-cathode preparation chamber. We have three methods

to prepare the cathode, the transport device of the cathode can work in high vacuum condition;

• The DC extracting chamber, providing a DC voltage of 200 kV;

• The superconducting accelerating section, containing the superconducting cavities of 1.5 GHz and a cryostat which can work at 2 K. This section is designed to accelerate the electron beam to 2 MeV;

• The post-accelerating section, which is expected to accelerate the electron beam to 20 MeV.

• The operation platforms for users of the electron beam.

The main parameters of this new electron source are shown in Table 1.

Table 1 The main parameters of the new electron source

Laser	Femtosecond or picosecond laser system
Wavelength (nm)	About 260 500 850
Pulse duration (fs)	About 200 6,000~30,000
Energy (μJ)	2~500
Repeat frequency (kHz)	1~300
Photocathode	Cs_2Te, GaAs, metal ion implantation
Extraction gun	DC, 150~200 kV
Accelerating section 1	Low-β superconducting cavities, 1~2 MeV
Accelerating section 2	Superconducting cavities, 10~15 MeV
RF frequency (GHz)	1.5
Peak current (A)	1~20
The normalized brightness ($A/m^2 \cdot rad^2$)	~10^{10}

The pulse duration, the repeat frequency and the peak current of the electron beam are decided by the laser and the characteristics of the photo cathode. The electron bunches generated are ultra-short pulses and there will be no need for the traditional bunching system, thus, it reduces the energy spread and the beam emittance, and guarantees high brightness. As the RF superconductivity technology is adopted, it can work in CW mode while maintaining the high accelerating gradient, thus it decreases the length of the structure; at the same time the larger beam pipe is possible for superconducting cavities, both are a great help to achieve high brightness and an ultra-short electron beam.

4. The DC Gun Fabricated at PKU

In 1992 the feasibility study of the laser driven superconducting electron gun began at PKU. A laser-driven photo-cathode electron gun has been fabricated and many experiments concerning the preparation of the photo-cathodes have been done.

4.1 The structure of the gun and experiments done on it

This laser-driven electron gun is capable of producing pulsed electron beams of 45~100 keV with a duration of 50 ps. It consists of four main parts: the photo-cathode preparation chamber, the DC accelerating chamber, the mode-locked laser, and the beam diagnostics system.

There are three different schemes for the fabrication of the photo-cathodes: the ion implantation, Chemical Vapor Deposition (CVD), and ion beam enhanced deposition. The result of the experiment shows that the characteristics of the cathode are quite satisfactory. The electron pulse of 2.2 A in peak was extracted at a voltage of 45 kV. The typical charge of the electron beams is measured by a coaxial Faraday cup located downstream of the beam line.

4.2 Electron beam dynamics

It is very important to study the dynamics of the electron beam after it is generated, especially when the beam energy is relatively low, in which case the space charge acts as a main effect of the emittance dilution. In this regard a couple of codes have been supplemented to study the different dynamical behaviour of the beam along the transport line.

POISSON code is used to simulate the electron field of the DC extracting chamber. Then, to study the emittance growth of the electron bunch as it is accelerated through the chamber, the code PARMELA is used. By the result of the calculation, we found that with the laser we used now, which has a FWHM of 50 ps, at the DC extracting voltage of 45 kV, the width of the electron beam pulse is about 160 ps, the relative energy spread is about 3.2%, the beam emittance is about 4.4 mm · mrad; if the voltage is increased to 100 kV without changing other parameters, the width will be 68 ps, the relative energy spread about 1.9%, and the emittance about 4.6 mm · mrad. In Fig. 2 and Fig. 3 the contrast of the beam pulse duration, energy spread and beam emittance at DC extracting voltage of 45 kV and 100 kV is shown.

As for the superconducting cavities, we already have 1.5 GHz niobium cavities using a China made niobium sheet. An unloaded quality factor of 10^9 and an accelerating gradient of 11 MV/m were obtained at 2 K. We are now making researches and experiments on the sputtering cavities.

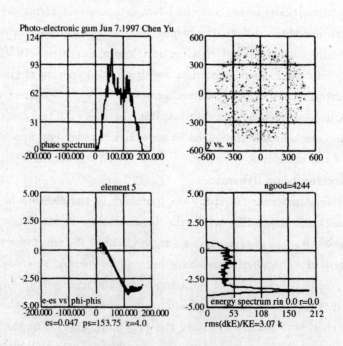

Fig. 2　DC extracting voltage of 45 kV

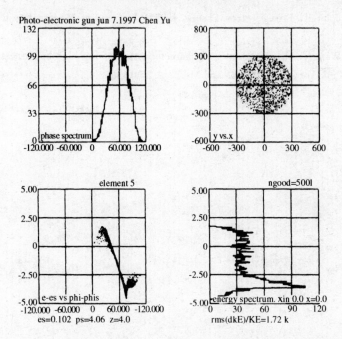

Fig. 3 DC extracting voltage of 100 kV

5. The Plan of the New Electron Beam Source

The new ultra-short e-beam source is expected to generate the sub-picosecond pulsed electron beam. We also expect synchronization between the RF system and the laser. If so, there will be widespread applications in the many areas. Yet there are many issues that need to be discussed for the success of such a device, such as the control of the phase of the electron pulse in the RF cavities, and new laser system, beam diagnostic system, etc. We have already obtained the new cryostat fabricated for the cavities and need funds for the new laser system.

References

[1] K. Zhao et al., Nucl. Instr. and Meth. A 375(1996) 147.
[2] K.-J. Kim et al., Generation of Femtosecond X-Rays by 90 Compton Scattering.

[3] R. H. Pantell, Free-Electron Lasers, Beijing Institute of Modern Physics Series, vol. 2.
[4] X. J. Wang et al., Experimental characterization of the high-brightness electron photo injector, Nucl. Instr. and Meth. A 375(1996) 82.

DESIGN STUDY ON A SUPERCONDUCTING MULTICELL RF ACCELERATING CAVITY FOR USE IN A LINEAR COLLIDER[1][2]

Abstract

A nine-cell superconducting RF accelerating cavity is designed for the TeV electron linear accelerator collider in the next century. The ratio of the maximum surface electric field to the accelerating gradient, E_{pk}/E_{acc}, is reduced to 2.024 and the cell-to-cell coupling remains as high as 1.95%. The distribution of the higher-order mode passbands is reasonable. There is no overlap between these bands, therefore no trapped modes. The circle-straight/line-ellipse-type structure provides good mechanical strength in the accelerating cavity. According to the present state of the art of surface processing techniques of Niobium cavities, it is possible to reach an accelerating gradient of 25~30 MV/m with beam load.

Index Terms—Accelerating gradient, linear collider, multicell Niobium cavity, SRF, trapped modes.

I. Introduction

The maximum beam energy accelerated in existing electron storage ring colliders is limited to under 0.1 TeV due to the synchrotron radiation loss, increasing in proportion to the fourth power of beam energy. A linear collider has no significant radiation loss, and its cost increases linearly with the beam energy. Therefore, to realize TeV range electron-positron

---

[1] Coauther: Donglin Zu. Reprinted from the IEEE. Transactions on Nuclear Science, Vol. 45, No.1, 1998, p.114~118.
[2] This work was supported by the National NSF of China.

collisions, a linear collider is needed. There is widespread consensus among the high-energy accelerator community that an e^+e^- collider with a center-of-mass energy of 2×0.5 TeV and luminosity of a few times 10^{33} cm^{-2}s^{-1} should be the next accelerator[1] after the LHC at CERN. Such a collider would provide a powerful means for the top quark analysis via $t\text{-}\bar{t}$ production and also have the potential for the discovery of new particles.

The advantages of the TeV superconducting linear accelerator (TESLA) collider[2] is to allow low RF peak power (using relatively long pulse length) and average power (using multibunch). However, the key problem is that the accelerating gradient at present is not high enough, as it needs to have a substantial upgrade. The principle cost of the accelerating structure itself also needs to be reduced further. At the anticipated capital cost of the linear collider, for a superconducting RF (SRF) accelerating structure to prevail over a normal RF (NRF) accelerating structure, reliable operating gradients of the SRF cavity must reach 20 MV/m or above. Conventional accelerators have used 5 MV/m for a reliable operating gradient with beam load. The cost of an SRF cavity typically is 200 or 40 k\$/MV. CEBAF uses 360 five-cell cavities. Suppose the accelerating gradient reaches 25 MV/m, the active length of 2×0.5 TeV linear collider will reach 2×20 km. The active length of a 1.3 GHz, nine-cell SRF cavity is about 1 m long. This TESLA collider requires $2\times 20\,000$ cavities of this kind. This paper studies the optimum shape design of the TESLA cavity.

The accelerating gradient of SRF cavities at present is limited by two phenomena, i.e., field emission (FE) and thermal breakdown[3]. The maximum electric field of the internal surface in a Niobium cavity, E_{pk}, is limited to a certain value by FE. The E_{pk} threshold can be upgraded to 100 MV/m through high RF peak power processing[4]. Even when producing a batch of SRF cavities using a Niobium sheet from the same furnace, using the same processing protocol, under the same machine conditions, their E_{pk}'s show variation. Despite using the same RF peak power processing, it is impossible for all the E_{pk} to reach 100 MV/m. It is likely for their low limit to be 75 MV/m. As the gradient without load depends on the E_{pk}/E_{acc} ratio, suppose a good cavity shape with a low E_{pk}/E_{acc} of 2.0 can be round, then the accelerating gradient without beam load could reach 37.5 MV/m. The accelerating gradient with beam load becomes two-thirds of its gradient without beam load, thus the gradient with beam load can reach 25 MV/m. It is difficult to make $E_{pk}/E_{acc} < 2.0$ as E_{pk}/E_{acc} is restricted strongly by cell-to-cell coupling k. A compromise $E_{pk}/E_{acc} - 2$, $k > 1.8\%$ is needed and likely.

Thermal breakdown demands that the ratio of the maximum surface magnetic field to the accelerating gradient H_{pk}/E_{acc} reaches a minimum value. But this parameter has more room and is not a principal limit factor. In general. E_{pk}/E_{acc}. k, and H_{pk}/E_{acc} all depend on the cavity cell shape, among them the FE is the main limiting factor.

II. The Concept and Theory of Multicell Accelerating Cavities

The cells in a multicell cavity behave like weak-coupling oscillators whether there are standing wave modes or traveling wave modes. One mode of a single-cell cavity could split into N modes of a multicell cavity when the single-cell cavity is evolved into the N-cell cavity. The N modes possess slightly different frequencies which form a "passband" with a different phase shift in each cell. In the same passband, if the eigen circular frequency of the qth mode is ω_q and the longitudinal electric field in the nth cell is $E_n(q,t)$, their relation can be expressed with the following formulas:

$$\omega_q = \omega_0[1 + k(1 - \cos\Phi_q)]^{1/2} \quad (q = 1,2,\cdots,N) \quad (1)$$

$$E_n(q,t) = E_0\sin\left(\frac{2n-1}{2}\Phi_q\right)\cos\omega_q t \quad (n,q = 1,2,\cdots,N). \quad (2)$$

Here k is the cell-to-cell coupling, E_0 is the maximum of the longitudinal electric field, and Φ_q is the phase shift in the qth cell.

$$\Phi_q = q \cdot \frac{\pi}{N} \quad (q = 1, 2, \cdots, N) \tag{3}$$

when $q=N$, $\Phi_q=\Phi_N=\pi$, e.g., the oscillation phase difference in adjacent cells is π, and it is called a π mode. The π mode in the fundamental TM010 band is the accelerating mode to be used. When $q=1$ and $\Phi_q=\Phi_1=\pi/N$ in adjacent cells, the oscillation has a phase difference of π/N. When $N\to\infty$, $\Phi_1\to 0$, so it is called "zero mode." Tuning is to make each cell resonate at the same frequency ω_N through mechanical adjustment and to make the peak value of $E_n(N,t)$ in each cell is equal. Cell-to-cell coupling k can be expressed as

$$k = \frac{f_\pi^2 - f_{\pi/N}^2}{2f_{\pi/N}^2 - [1 - \cos(\pi/N)]f_\pi^2}. \tag{4}$$

Here f_π is the mode frequency and $f_{\pi/N}$ is the "zero mode" frequency. The k tends to go down with the increasing cell, such as when an existing four-cell or five-cell cavity is evolved into nine-cell cavity, and its k value will become intolerably small. A more serious issue is that as the mode density in each passband increases, the accompanying number of cells increases, and even passband overlapping appears, thereby resulting in "trapped modes" which will destroy the normal working conditions of SRF accelerating cavity. The trapped modes were analyzed with a computer[5]. The physical mechanisms of generating trapped modes are not clear yet. In the design of SRF cavities, trapped modes are a troublesome problem. So seeking a new cavity shape without trapped modes in an important task for the cavity designer. Another parameter of a multicell cavity $\Delta f/f$ represents the dispersion of passband, and its definition is

$$\frac{\Delta f}{f} = \frac{2(f_\pi - f_{\pi/N})}{f_\pi + f_{\pi/N}}. \tag{5}$$

There is only one mode (π mode) useful for accelerating in the fundamental passband; the rest of the $N-1$ modes are harmful and need to be extracted by the main coupler. Hence the larger the dispersion of the fundamental passband, while the smaller the dispersion of higher order passbands, the better, so as to avoid passbands too wide to be separated. And it is desirable that passbands distribute homogeneously and reasonably.

III. Searching for a New Cavity Shape

To reduce the construction cost of the future linear collider and considering both the operating experience of SRF accelerating cavities at active service and the present state of the art of the surface processing techniques of SRF cavities, the new accelerating structure should satisfy the following conditions.

1) The number of cells per cavity should double to nine cells in order to reduce the number of expensive RF couplers, to save space, and to shorten the total length of the collider.

2) In the fundamental passband, cell-to-cell coupling k must be larger than 1.8%, and the higher, the better.

3) E_{pk}/E_{acc} must be reduced to 2.0 to make the accelerating gradient as high as possible.

4) The higher order modes (HOM) coupler mounted on the beam tube near the cell can efficiently extract the energy of higher-order standing-wave modes to make their external Q_{ext} low enough to be a tolerable level.

5) All traveling HOM's can propagate out of cells; so-called trapped modes are not allowed in cavities.

6) The structure should be reasonable, strong enough, and mechanically tunable.

The structure parameters of a cavity are usually described using one-fourth of the lengthwise section of an SRF cavity with axial symmetry, where the cell is the elemental block of a cavity. Half a cell wall consists of several segments of curves, such as circle-straight/line-circle (CSC) or ellipse-straight/line-ellipse (ESE). The "CSC" type has five independent parameters[6]. i.e., cell outer radius OR, beam tube radius IR, iris radius Rl, wall slope TD, and half-cell length L, respectively. First among them, L depends on the accelerating mode frequency because accelerated electrons must be synchronous with RF oscillation. In consideration of mechanical stability, we take TD=79°, thus only the other three parameters. i.e., OR. IR, and Rl are allowed to vary. Our strategy is to keep right frequency 1.3 GHz resonant to keep a certain OR through adjusting IR and Rl, due to each parameter variation that affects π mode frequency of the fundamental passband.

If 352 MHz. 500 MHz, and 1.5 GHz cavities are all scaled to 1.3 GHz, the OR, IR, and Rl of the LEP four-cell cavity[7] are the minimum (there are no trapped modes in it): in opposition the OR, IR, and Rl of Cornell ten-cell TESLA cavity[8] are the maximum (there appear trapped modes in six-cell cavity of this TESLA shape). We tried to find a good shape in which OR, IR, and Rl were between the two sets of extreme parameters. In the searching process, it is important to keep the resonant frequency of the trial cavity at 1.3 GHz. Otherwise, the E_{pk}/E_{acc} calculated will be incorrect. In addition, E_{pk}/E_{acc} will be incorrect[6] as well if the number of mesh points used by URMEL[9] is less than 25 000.

Using the "CSC" type, although searching for quite some time, we failed to reach the goal. We had to try a new shape, circle-straight/line-ellipse (CSE). At last we found a cell shape shown in Fig. 1, named the Beijing Tesla (BT) shape. The $E_{pk}/E_{acc}=2.024$, $k=1.95\%$ for the accelerating mode of the nine-cell accelerating cavity (shown in Fig. 2) increased with this new cell shape.

Fig. 1 BT cavity shape

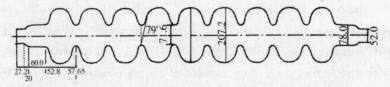

Fig. 2 The geometric structure of SRF nine-cell accelerating cavity for linear collider; the dimensions correspond to 1.3 GHz

IV. The Characteristics of the Accelerating Mode of the BT Cell

In order to command the accuracy and reliability of calculating the BT cell, we repeated the calculation for the Cornell TESLA, KEK TESLA, and non-TESLA LEP fourcell cavities, and the results agree with the data published[8]. For comparison, Table I lists their cell number per cavity, two geometric parameters, the number of mesh points used in running URMEL, four physical parameters, cell type, and trapped modes. Among them there are two key parameters: E_{pk}/E_{acc} and k. The smaller the E_{pk}/E_{acc}, the higher the accelerating gradient will be obtained; the larger the k value, the easier the adjusting, the tuning, and the controlling of the cavities are. However, E_{pk}/E_{acc} and k condition each other strongly. Generally speaking, to keep k larger than 1.8%, one cannot lower the E_{pk}/E_{acc} at will.

Table I Comparison of SRF Accelerating Cavity for Next Linear Collider

	LEP[a][7]	Cornell[8]	KEK1[7]	KEK2[10]	Beijing[12]	Saclay[11]	DESY[11]
N_{cell} per cavity	4	10	9	9	9	9	9
OR (mm)	102.20	109.00	104.15	103.30	103.60	102.20	103.30
IR (mm)	32.67	40.90	40.00	38.00	35.80	32.31	35.00
N_{mesn}	20000	20000	20000	25000	25000	25000	25000
$H_{pk}/E_{acc}((A/m)/(MV/m))$	3129.8	4321.1	3967.7	3429.8	3414.7	3159.2	3318.4
$k(\%)$	2.00	1.80	1.22	2.66	1.95	1.42	1.85
R/Q() per cell	121.48	90.96	112.92	108.00	110.87	124.33	115.22
E_{pk}/E_{acc}	2.359	2.054	2.039	2.220	2.024	2.000	2.070
cell type	CSC	CSC	SCSCS	CSE	CSE	CSC	CSE
trapped modes	no			no	no		

a LEP is not TESLA cavity, introduced as a reference cavity.

V. Monopole HOM Passbands and the Trapped Modes Problem

In 1989, Cornell University was the first to design a ten-cell TESLA shape cavity[8], "CSC" type, with $E_{pk}/E_{acc}=2.10$, $k=1.8\%$. It is necessary to reduce the ratio E_{pk}/E_{acc} through improving the geometry of cell structure so as to raise E_{acc} as high as possible, because raising E_{pk} threshold experimentally is restricted. Usually the maximum E_{pk} appears in the iris area where the electric field lines concentrate, as shown in Fig. 3. If the circle radius Rl at the iris is enlarged, the electric field lines will be diverged out; consequently, the maximum E_{pk} will go down. This is indeed an effective way for the accelerating mode. Cornell/CEBAF cavity's $E_{pk}/E_{acc}=2.56$, but the TESLA cavity E_{pk}/E_{acc} is reduced to 2.10. However, this way results in some difficulty for cell-to-cell coupling and propagation of HOM's. In the six-cell cavity, with this cell shape, under three times

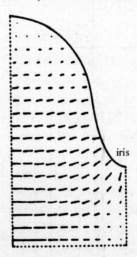

Fig. 3 The electric field distribution of π mode of TM010 band in the BT cell. The arrow indicates the field direction, and the length of arrow is proportional to field strength

the fundamental frequency there already appears to be an overlap of TM021 and TM030 passbands (see Fig. 4), and thereby the probability of trapped modes is formed. This convinced us to change the circle at the iris to an ellipse so that the ellipse at the iris not only diverges the electric field lines and lowers the maximum E_{pk}, but also compresses the physical width of the iris area and facilitates the stagger of HOM passbands.

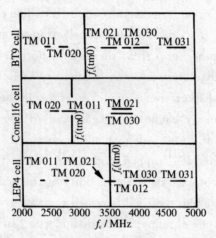

Fig. 4 The monopole HOM passbands distribution of the BT nine-cell cavity, LEP four-cell cavity, and Cornell six-cell cavity. The cutoff frequency of TM0- modes (tm0): f_c is the boundary between standing and travel modes

The "CSE"-type structure is advantageous to cell-to-cell coupling and propagation of HOM's. On the one hand it lowers E_{pk}/E_{acc}. On the other hand it raises cell-to-cell coupling k, as to avoid trapped modes formed. Another disadvantage of the Cornell TESLA is that TD$=70°$ is not firm enough for mechanical stability. Its one-cell cavity collapsed once when it was pumped down. The mechanical stability of both the ESE type of Cornell/CEBAF five-cell cavity and CSC type of LEP four-cell is all verified to be strong enough. Taking LEP four-cell cavity's TD$=79.09°$ as a good reference, in the Beijing CSE-type cavity we have a TD$=79°$. It is believed that mechanical stability of CSE type is good enough. After the CSE cell type was chosen, searching for a good cell shape carefully in a wide range of geometry parameters was conducted, and the BT cell shape was obtained at last. Its HOM passbands distribution is reasonable compared with Cornell TESLA six-cell's and LEP four-cell's. In the Cornell six-cell there appears an overlap of HOM traveling wave modes; frequency degeneration modes from different bands; interaction, drawing each other and energy exchange between modes with the same or near frequencies; and consequently, a result in interference effect. The electromagnetic field enhances in a local area; meanwhile, the electromagnetic field reduces and even eliminates another local area, thereby the normal cell-to-cell coupling and propagation characteristics are changed. When the electromagnetic field of this kind of interference modes in end-cells becomes zero (shown in Fig. 5), the energy cannot propagate out of the cavities into beam tubes. This kind of "propagating" mode stays in a cavity as if in a trap, and thus they are named "trapped modes." A DESY TESLA cavity in design process does not exclude the possibility of trapped modes existing. A BT nine-cell cavity, like the LEP four-cell cavity shown in Fig. 4, has no overlapping passbands of monopole mode under 3.7 times the fundamental frequency. All mode frequencies are discrete. There is no condition of interference, therefore there are no trapped modes existing in it.

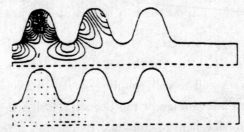

Fig. 5 There appear trapped modes in the Cornell six-cell cavity. The electromagnetic field in end cell nearly equals zero, and the cell's energy cannot propagate out of the cavity

VI. Computer Tuning

In a multicell cavity, to guarantee the end-cell and the in-cell having the same resonance frequency, the geometric dimensions of the half end-cell must be different from the in-cell because the beam tube effect needs to be compensated. We tuned the nine-cell BT cavity for the mode of the fundamental band by changing the end half cell length with URMEL, 25 000 mesh points. The half in-cell length equals 57.65 mm, the half end-cell lengths are 52.7 and 52.8 mm, respectively, corresponding to two sets of tuning curves indicated in Fig. 6(a) and (b). The field flatness

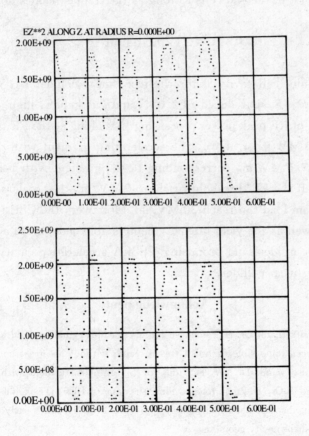

Fig. 6 The computer tuning of nine-cell BT structure with URMEL:
(a) tuning curve when $L=52.7$ mm and (b) tuning curve when $L=52.8$ mm

($\Delta E_z/\overline{E_z}$) is nearly 0.028. The tuning sensitivity is 2880 Hz per micrometer.

In the end cell, a slight mistune to HOM's is possible but not a concern. HOM's always manage to be damped adequately. The higher standing wave modes spread all over the cavity through cell-to-cell coupling and finally are extracted by the HOM coupler. The normal higher travel wave modes can always propagate through cells to the beam tubes in spite of mistune or not in the end cells as long as the trapped modes do not exist.

VII. Conclusion

A BT nine-cell cavity[12] for the next linear collider has no trapped modes, its E_{pk}/E_{acc} is reduced to 2.024, and k is kept as high as 1.95%. Making use of RF peak power processing[4], as long as the FE threshold is raised to 75 MV/m or above, the accelerating gradient with empty load can reach 37.5 MV/m, corresponding to the gradient with beam load of 25 MV/m. If the FE threshold reaches 90 MV/m, the accelerating gradient with beam load can reach 30 MV/m. Moreover, the CSE type is reasonable as well as the yield intensity being stable enough. In a word, the BT shape is an ideal and promising TESLA accelerating cavity candidate for the next linear collider.

Acknowledgment

The authors thank Dr. H. Padamsee for his teaching, guiding, and assistance. It would have been impossible to finish this research without his great and continuous help. They also thank Dr. W. Hartung for his enthusiastic help with this work. Thanks also go to Dr. D. Proch for providing the data for DESY Tesla and Saclay Tesla and stimulating discussions. Finally, they appreciate Prof. Kojima's early interest with nine-cell BT-shape cavity geometry.

References

[1] R. H. Siemann, "Overview of linear collider design," in *Proc. IEEE PAC*. Washington DC, May 1993, vol. 1. p. 532.

[2] H. T. Edwards. "Progress report on the TESLA test facility," in *Proc. IEEE PAC*, Washington DC, May 1993, vol. 1, p. 537.

[3] Q. S. Shu et al., "Reducing field emission in superconducting RF cavities for the next generation of particle accelerators, Cornell Univ., CLNS 90/1020, 1990.

[4] J. Graber et al., "A world record accelerating gradient in a niobium superconducting accelerator cavity," in *Proc. IEEE PAC*. Washington DC, 1993, p. 892.

[5] A. Mosnier, "The trapped modes in a multicell SC cavity," in *Proc. 1st TESLA Wkshp.*, Cornell Univ., Ithaca, NY. July 23~26, 1990.

[6] D. Zu, "Concept design of a new structure of 1.3 GHz single-cell superconducting cavity," in *High Energy Phys. Nucl. Phys.*, vol. 21. no. 1, p. 491~503, 1997.

[7] H. Padamsee, private communication.

[8] H. Padamsee et al., "A new shape Candidate for a multi-cell superconducting cavity for TESLA." Cornell Univ., Rep. CLNS90-985.

[9] T. Weiland, NIM, 216(1983) 329.

[10] E. Kako et al., *Proc. 5th SCRF Wkshp.*, DESY, Hamburg, Germany, 1991, p. 751.

[11] D. Proch, private communication.

[12] D. Zu and J. Chen, *Proc. IEEE PAC*, Washington DC, 1993, vol. 2 p. 1095.

第六部分　粒子加速器综述

THE PROGRESS OF LOW ENERGY PARTICLE ACCELERATORS IN CHINA[①]

Abstract

A brief account on the development of particle accelerators in China is given. Features and performance of typical accelerators are described. R &. D efforts are also included.

Introduction

It was 30 years ago when Prof. Zhao Zhong-yao and his group put China's 1st 2.5 MV Van de Graaff accelerator into operation at the city branch of Institute of Atomic Energy[1]. A little later, 25 MeV Betatrons and 1.2 m cyclotron imported from USSR started working at Peking Univ., Tsinghua Univ. and IAE respectively. Inspired by the needs of basic research and the achievements of nuclear energy, a variety of accelerators were studied and constructed by universities and research institutes in the early period. Among them, the 30 MeV electron linac, developed by Prof. Xie Jia-lin's group, was the highest in energy and was put into operation in 1964[2]. Subsequently the industry produced various types of accelerators e. g. cascade generators, electrostatic accelerators, betatrons and cyclotron etc. By the end of the first decade, the number of accelerators totaled up to about 50 and thus laid down the basis necessary for future development.

The development of accelerator technology in the recent decade turns

---

① Coauther: Zhao Weijing. Reprinted from the Invited talk at the First European Particle Accelerators Conference, Proceeding EPAC Rome, June, 1988, Vol. 1 p. 242.

out to be active and fruitful. New projects for basic research including Beijing Electron Positron Collider, Lanzhou Heavy Ion Research Facility, Hefei Synchrotron Radiation Facility, the 6 MV Tandem accelerator and the 4.5 MV Van de Graaff accelerator etc. all are approaching their completion. Thanks to international collaborations, the design and construction of these accelerators are very facilitated. Meanwhile, a great number of small accelerators aiming at applications in the fields such as radiotherapy, micro electronics, radiation processing and ion beam analysis etc. have emerged to meet the strong needs from medical treatments and various industries. According to recent incomplete statistics, the number of accelerators totals up to 270[3], and about 2000 scientists and engineers are engaged in this field, among them 800 are members of China Society of Particle Accelerators.

The number of various accelerator species is plotted in Fig. 1.

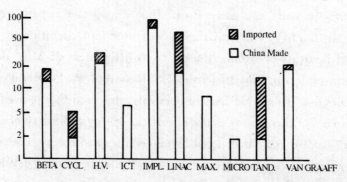

Fig. 1 The number of various accelerator species

The rapid development of ion implantation technology stimulated mass production of high voltage multiplier as implantor. So far 76 sets of 150~600 kV implantors have been produced in China and 22 sets imported from abroad[4]. The medical treatment is also a field where extensive applications of accelerators are absolutely necessary. Up to now, 45 electron linacs and 8 betatrons have been installed in the nation for radiotherapy[5]. To meet the needs of medical isotopes, existing cyclotrons have been producing short-lived isotopes from time to time. Moreover, an imported compact cyclotron CS-30 and a 35 MeV proton linac are dedicated to medical isotope production. Applications of electron accelerators to the field of radiation processing appear to be quite profitable. The manufacturer in Shanghai and Chengdu made million yuan (RMB) profit out of cable irradiation. In addition, more than 20 electron accelerators are used for producing foamed polyethylene, vulcanized silicon rubber and for paint curing and etc.[6]. However, the installed capacity of beam power totaled 182 kW which is obviously too low to meet the requirements from the industry. The number of small tandem Van de Graaff accelerators are growing rapidly because of extensive use of ion beam analysis. The 2 MV tandem developed by Lanzhou Institute of Modern Physics was put into operation in 1986, and the other 12 made by NEC and GIC in U. S. A. are also in operation. They are mainly used for studies in environmental science, archaeology, material science and etc.

Ion Implantors

There are 4 factories and 10 institutes engaged in the manufacture and development of ion implantors. 59 sets of 200 kV implantors have been produced in Beijing, Baoji and Changsha since 1976[4]. main typical products are listed in Tab. 1. All of them are equipped with pre-acceleration analyzer, in general a 90° double focusing magnet with a resolution $M/\delta M = 100$.

Table 1

Type	Year	Voltage /kV	B^+ Yield /μA	Uniformity
J95-200/ZM	1980	20~200	100	<3 %
J59200B/2K	1981	30~200	120	≤1.4%
LC—2B	1984	30~200	150	=5 %
ZLZ—200	1985	30~200	250	≤2 %

Typical accelerating tube developed by Beijing Institute of Automation for Machinery consists of stainless steel electrodes bonded with high boron glass rings. The gradient of a surface break down is 4.5~5 kV/cm and the strength against pulling 61.3 kg/cm. Extensive studies have carried out on ion sources for implantors, e. g. Penning, Duoplasmatron, Calutron, micro-wave source and etc., so as to raise the ion yield, to extend the ion species and to improve the beam quality. The ionizing efficiency of Freeman type source was enhanced at IAE by putting molybdenum screens on both ends of the ionization chamber[7]. For a BF_3 discharge the proportion of B^+ to the total yield = 38.5%, while $I=30$ mA at gas consumption of 3.5 cc/min, and the source life time >10 h. A number of compact Penning source with permanent magnets was developed by Peking University's group[8]. Total yield of a typical side-extraction source is 1~2 mA with the proportion of N^+ 60%~70% and B^+ 40%~50%. Rich content of multiply charged ions has been found.

The implantor SD-400[9], constructed by Shangdong University in 1985, has an energy range of 50~400 keV and a resolution $M/\delta M > 200$. In order to minimize the path length of ions while keeping resolution high they made good use of an unsymmetrical double focusing magnet with a radius of 56 cm for pre-acceleration analysis. An electrostatic quadruple doublet is set in front of the magnet so that the emittance of the injected beam can matched in both directions to form a real image. The path length of ions from source to target is about 6 m.

The C-600 implantor[10] developed by Shanghai Institute of Metallurgy accelerates ions of $A=1$~210 amu. to an energy 200~630 keV. Typical target currents are $Ar^+ \approx 200$ μA, $P^+ \approx 105$ μA, $Mg^+ \approx 754$ μA, and $Ca^+ \approx 65$ μA. The HV terminal is about 7 m high from the ground and is linked by an infrared optical telemetry system transmitting 40 signals with the controlling panel. The heavy ion source model 820 can work with gas, liquid or solid state samples and provide more than 20 species of ions. The accelerating tube is totally 1.76 m long, divided into 4 sections. Each consists of titanium disk shaped electrodes bonded with 95% Alumina ceramic rings. The tube is working under a clean vacuum of 3×10^{-7} tor and can sustain a voltage as high as 850 kV during the "cold" test. The analyzing magnet is set after the accelaration. With a gap of 4 cm and a radius of curvature of 1.2 m the magnet has a resolving power $M/\delta M = 208$. The transmission efficiency from the tube output to target $\approx 80\%$. Five beam lines are available.

Recently, high current implantors for modification of materials have been developed by Sichuan Univ., Beijing Normal Univ., IHEP and etc.

High Voltage Electron Accelerators

Most of the electron HV accelerator are used for radiation Processing. The JJ-2 accelerator produced by Vanguard factory was the basic model in 1960's. It is a 2 MV electrostaic accelerator with 200 μA output current and 40 cm scanning width. However, to make the processing cost effective an average beam power higher than 5 kW is necessary. For this purpose, BIAMI developed series of ICT products including 0.3 MeV/30 mA, 0.6 MeV/30 mA, 1.5 MeV/10 mA and etc., while Vanguard developed a 2 MeV/10 mA model GJ-2 for cable irradiation[11]. It is a Dynamitron providing a scanning width of 80 cm and field flatness 85%. The cable is winded by pullies in a pattern of "8" during the exposure. Besides, they produced a new model of cascade generator providing 500 keV/40 mA beam with a 1 m scanning width and uniformity $> 85\%$. Moreover, an electron curtain accelerator of 200 kV/30 mA will be available commercially for radiation paint curing. It is expected that the production of high power accelerators for industrial irradiation will continue growing, more and more radiation centers will be established in the near future.

Pulsed high current generators have been developed to meet the requirements of flashing radiography[12] as well as the research of ICF and excimers by National Academy of Engineering Physics[13] and IAE[14]. The specifications of basic facilities are listed in table 2.

Table 2

Inst.	Model	V/mV	I/kA	t/ns	Dose	Use
NAEP	FL-1	6~8	75~100	85	950	Flash radiograph
	EPA-1	0.4~0.7	4	90		Microwave FEL
	EPA-2	0.5	120	70		Excimer Studies
IAE	No. 1	1	80	80		Excimer &. Beam
	No. 2	0.65	150	40		Physics

Micro-wave FEL amplifier experiments were carried out on EPA-1[15]. A 100 A, 50 ns pulse beam, 4 mm in diameter with axial momentum spread 0.2% was injected into a cylindrical optical cavity. The Wiggler field generated by 1 m long bifilar helix was a circulary polarized magnetic field with $\lambda_W = 3.45$ cm. At an input level of 20 W, more than 1.4 MW radiation of 34.5 GHz was observed (Fig. 2).

Fig. 2   The FL-1 Accelerator

Heavy Ion Electrostatic Accelerators

The 2 MV tandem at Inst. of Modern Phys. Lanzhou is the first workable tandem Van de Graaff made in China and has run stably since

1986[16]. Its terminal voltage can be adjusted in the range of 0.3~2 MV with a fluctuation $<\pm 1$ kV. The beam current at the analyzing magnet exit is 1~3 μA and the spot size 3 mm. the laddertron developed by IMPL runs at a linear velocity of 12 m/s and can sustain a gradient of 2.06 MV/m. typical charging current is 440 μA. The bakeable titanium-ceramic (95% Alumina) accelerating tube ensures a very clean and high vacuum for acceleration, the gas leakage of the tube is 10 tor^{-1}/s. The tube consists of 6 sections and has an effective length of 1.15 m in total with the mechanical strength against pulling: 500~900 kg/cm. Apart from the above, the 6 MV tandem designed and constructed by Shanghai Institute for Nuclear Research has been installed and commissioned, their laddertron, duoplasmatron source and 60 keV injector have passed the acceptance test[17].

The tandem HI-13 of HVEC was put into operation at IAE, Beijing in Oct. 1986[18]. So far, more than 10 ion species have been accelerated. The highest voltage of 13.4 MV was reached with the acceleration of proton beam. Experiments were carried out smoothly with Q3D and neutron TOF spectrometer and other equipments. The transmission efficiency along 40 m long beam lines appears to be 95%. However, as the beam utilization efficiency of the present buncher is as low as 2/70, so that a higher efficiency beam pulse system has to be considered. In addition an EN-18 tandem of HVEC was transferred from Oxford to Peking University in 1986. It is being commissioned and hopefully it might be able to get the first beam by the end of this year. The other 11 tandems of 1 MV, 1.7 MV, 3 MV produced by NEC and GIC are in operation for ion beam analysis in the nation.

As for single ended Van de Graaff accelerator, the 4.5 MV machine at Peking University has been installed after a long delay due to the building construction[19]. It is expected to get the beam this year. A high efficiency beam pulsing system is equipped in the terminal electrode. the BUE is enhanced by a factor of 1.47 by superimposing the 9th harmonic onto a conventional sine wave chopper[20]. Moreover, a short focused double drift harmonic buncher is installed to ensure the BUE OF bunching to 66%[21]. These compose one of the unique features of the accelerator (Fig. 3).

Fig. 3 The 4.5 MV Van de Graaff Accelerator

Cyclotrons

HIRFL is one of the major accelerator projects in China, which is composed of a SF injector cyclotron ($K=69$) and a Split Sector cyclotron ($K=450$). The SF cyclotron has been in operation since 1987 with typical parameters as following:[22]

Ion	B/kG	F/MHz	V/kV	R_{ex}/cm	E/MeV·A^{-1}	I/eμA
$^{12}C^{4+}$	14	7.1	60	75	5.9	1.1
$^{16}O^{5+}$	15	7.1	60	75	6.0	1.4

Carbon ions are to be accelerated in SSC to 100 MeV/A while Xenon 5 MeV/A with an intensity of $10^{10} \sim 10^{12}$ pps. The main magnet consists of four 52° sector, each weights 2000 tons, with a field up to 16 kG. there are 36 pairs of trim coils mounted inside the gap. When the input RF power is 240 kW, 100~250 kV is generated by each of two 26° sector cavity in two valleys, the operation frequency of which can be adjusted in the range of 6.5~14 MHz. The vacuum chamber contains pole tips, trim and harmonic coils, accelerating cavities, as well as inflection and extraction devices in a high vacuum and thus has a rather peculiar shape with a total volume up to 100 cubic meters. Nevertheless the giant was evacuated to a vacuum as high as 6×10^{-8} tor. last year, which was one of most difficult yet well done tasks during the commissioning. So far they have finished the shimming of the magnet. The field inhomogeneity measured at median radius was 21 gauss and reduced to 1 Gs after careful shimming. However the perturbation due to the inflecting channels can be as high as 17 Gs in the region of injection after incomplete compensation. The center variance of the orbit in the field is about 3 mm. SSC is expected to get its first beam by the end of 1988 for the experiments of nuclear research (Fig. 4).

Fig. 4 The Accelerator SSC

The majority of conventional cyclotrons built in 1960s have been upgraded. For instance, the 1.2 cyclotron of Shanghai Institute for Nuclear Research was reconstructed in 1982 and thus it turned from a fixed energy machine (6.8 MeV proton) to a variable energy isochronous cyclotron providing proton beam in the range of 10~30 MeV, deuteron 10~16 MeV and Alpha beam 10~32 MeV. The energy spread of the extracted beam was minimized to 0.43% by computer optimized orbit programming based on the measured electromagnetic field distribution.

Besides cyclotrons a microtron DHJ-25[23] for dosimetry research was jointly developed by Tsinghua University and BIAMI. It provides electron beam in the energy range of 5~27 MeV, and 18 mA pulsed current at the 27th orbit when the RF power is fed from a 2 MW magnetron. The beam current extracted at 25 MeV, reached 18.9 mA with a spot size 2×2 mm and energy spread $<\pm 1\%$. The dose rate of 25 MeV X-ray is in the range of 400~1000 R/min·m.

Linear Accelerators

The majority of linear accelerators in China are for medical use. Among them BJ-4 and NDZ-20 are two recent products developed by Beijing Institute of Medical Equipments and Department of Physics, Nanking University respectively. The former is a compact, economical 4 MV therapy system[4] suitable for installing in a small treatment room. The length of the entire assembly is only 38 cm. and the gun, sidecoupled cavity, RF window and the target are brazed altogether and sealed off under very clean and high vacuum. As a result it works with high reliability, long life and convenience. Under a peak power of 2.6 MW. BJ-4 provides 4 MeV X-ray with a dose rate of 500 R/min·m.

NDZ-20 is a 20 MeV TW electron linac with a feedback loop[25,5]. It's equipped for X-ray and electron beam therapy (table 3). The feedback of the traveling wave facilitates varying the output energy in a wide range: electron beam 5~20 MeV. and X-ray 5, 15 (or 8, 16) MeV. In this way it is able to provide a variety of depth dose profiles to satisfy various requirements of different treatment. It was found that the system efficiency as well as the beam loading characteristics can be optimized by using a coupler with variable feedback ratio. To limit the extra energy spread of the beam caused by long build-up time associated with the feed-back process, a pair of magnetic chopping coils are used to cutoff the front edge of the beam before entering the waveguide and it works successfully

(Fig. 5).

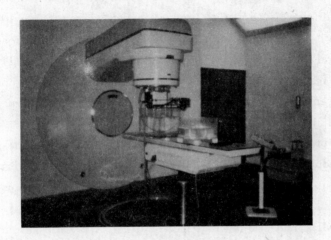

Fig. 5 The NDZ-20 Electron Linac

Table 3

Type	BJ-4	NDZ-20		DZ-10	
		X-ray	e-beam	X-ray	e-beam
Energy/MeV	4	10, 15 (8, 16)	5~12	8	6, 8, 10
SSD/cm	80	100	100	100	100
Dose rate /R·(min·m)$^{-1}$	500	300~500	300~1000	500	500
Spot size/mm	2	3.5		2	
Uniformity (%)	±3	±3	±5	±3	±5
Structure	side coupled	TW feedback		Travel	wave
Frequency /GHz	2.998	2.856		2.998	
RF Power/MW	1.6	4.5		1.8	
Total length /m	0.38	2.58		2.33	

The proton linac of IHEP is also dedicated to medical applications. Proton beam can be accelerated to 35.5 MeV with a peak current of 70 mA and transported down the 35 m beam line to two target halls equipped for short lived isotope production and neutron therapy respectively[26]. With an average beam current of 75 μA, it is able to provide annually several tens Curies of medical isotope like ^{11}C, ^{67}Ga, ^{201}Tl and etc. The first batch of ^{201}Tl was produced in 1987. Moreover, by bombarding proton beam on Be target, 20 MeV neutrons can be produced at a dose rate of 145 rad/min·m.

The preinjector of the linac consists of a 750 kV cascade generator with a duoplasmatron source and a double drift harmonic buncher which is set in front of the Alvarez tank. The total length of the tank is 21.83 m and divided into 6 sections with totally 104 cells. The linac is turned to 201.25 MHz and the average accelerating field varies from 1.65 ~ 2.6 MV/m at 5 MW RF power. To suppress the unwanted TM_{001} and TM_{012} modes two RF feeding ports located at 1/4 and 3/4 tank length are used. LASL type resonant couplers are mounted inside the tank with a periodicity one per two drift tubes so as to get the stabilized field. As a result the mode separation increases from 74 kHz to 139 kHz[26].

Fig. 6 The 35 MeV Proton Linac

Apart from medical applications, there are several linac projects carried out for basic research. As FEL studies arouse great interest in recent years, IHEP decided to upgrade their 30 MeV linac for the first phase of their Compton regime FEL research project[27]. The parameters of the linac are: $E = 10 \sim 30$ MeV, I (bbu) $= 300$ mA, T (e-beam) $= 4$ μs, T (modulator) $= 5$ μs $\delta\gamma/\gamma = 0.5\%$, $\delta f/f = 10^{-6} \sim 10^{-7}$. The longitudinal motion of the electrons in a 3.05 m long constant gradient wave guide has been simulated by computation under the condition of 200 mA beam current and 12 MW RF power input. The longitudinal phase space and factors affecting the output energy spread, e. g. initial phase and energy spread, phase fluctuation of injected electrons and etc. have been carefully studied.

The high brightness preinjector is one of the key tems of linac based FEL project. IHEP is developing a microwave gun for this purpose. It is expected to provide 20 A pulsed current with 4° phase width and an emittance of 30 mm-mrad at an energy of 0.9 ± 0.1 MeV. On the other hand, IAE planned to develop co-axial planner grided gun combined with 1/24 and 1/6 subharmonic buncher so as to reach a peak current of 100 A ($t = 20$ ps) with energy spread $< \pm 1\%$. With their grided pierce gun, 19 A peak current of 2.4 ns has been obtained in a bench test when the pulsing voltage reached 82 kV[28]. The cathode-grid geometry will be studied further and scandate cathode will be tried so as to upgrade the performance of the gun.

The linear induction accelerator is the other important approach to produce energetic high intensity beam. NAEP has completed the first induction accelerator in China. It consists of 7 inducting cells and is capable of accelerating 4 kA beam 90 ns beam to 1.5 MeV[12]. They are going to upgrade the machine and carry out FEL amplifier studies.

As for heavy ion linacs, with the support of NSFC, Peking University's group has developed a 4-rod structure excited by integrated split-ring resonators which can be operated in the range of 14 $\sim$ 100 MHz[29,30]. Typical results of model test are: $F_0 = 27.26$ MHz, $Q =$

1190, Specific resistance $r=135.6$ kΩ. Both the axial and radial field distribution are satisfactory. We are planing to construct a RFQ either for ion implantation or to use as a preinjector for our EN-18 tandem to accelerator ions like C, Be, B, Al and etc. to 300~500 keV for the AMS project supported by NSFC. Apart from the above, the group has also been engaged in developing post acceleration boosters for D.C. accelerators. A variety of low Beta resonators including Helix, Tapperd helices, Spirals, Split-rings designed, constructed and tuned to 28 MHz or 108 MHz. A model cavity containing two 17 cm long helices turned out to be able to accelerate 300 keV proton to 620 keV when the RF input power is 19 kW and frequency 28.8 MHz[31]. The cavity has been in routine operation as a postbuncher providing 1 ns deuteron beam for neutron TOF experiments. In addition, two helical cavities and a split-ring cavity were constructed and installed on the beam line of EN-18 Tandem to provide a boosting voltage of 3 MV[32].

Conclusion Remarks

Particle accelerators in China has progressed smoothly since 1978. While there will be great efforts in the near future to make best uses of these existing facilities, more and more new projects for industrial and medical applications will be founded. On the other hand, new ideas and technologies will be highly encouraged to develop few advanced accelerator projects in China either for basic research or application purpose.

The authors wish to acknowledge: Sun Zuxun, Wang Daji, Yu Juexian, Zeng Naigong, Liang Xiuru, Wang Shuhong, Wang Daming, Zang Enhou, Gu Benguang, Lai Qiji, Sun Quanqing, Chen Maobai, Tao Zucong, Chen Shuze for providing their information and photos.

References

[1] Ye Minghan. *Acta Physica Sinica*, 19(1963) 60.
[2] Xie Jialin. GSI-84-11, *Proc.* 1984. *Linear Accelerator Conference* (1984) p.14.

[3] Liang Xiuru et al., *Policy Studies on the Development of Low Energy Accelerators in China*, p. 3 (1986) (in Chinese).

[4] Zhao Weijiang, *ibid.* p. 113 (1986).

[5] Zhang Zhongzu, *ibid.* p. 35 (1986).

[6] Cui Shan, Wang Daji, "Development of Radiation Accelerators", *Proc. 3rd Japan-China Joint Symposium on Accelerators for Nuclear Science and their Application*, Nov. 1987 (Proc. JCJS), to be published.

[7] Lin zhizhou et al., "A Freeman-Type Ion Source with high B^+ proportion", *Proc. 4th National Electron-Ion-Photon Beams Conference*, Nov. 1986, p. 31 (in Chinese).

[8] Song Zhizhong et al., "Several Mini Ion Source" *Vaccum*, Nol. 136 No. 11~12 (1986).

[9] Wang Keming et al., "SD-400 Ion Implantor" Internal Report, Deparment of Physics, Shandong Univ., Jinan, China (1985).

[10] Jiang Xinyuan, Guan Anming et al., "C-600 keV Heavy Ion Implantor", *Proc. 4th National EIPB Conference*, Nov. 1986, p. 59.

[11] Wan Baofa and Fan Quanle, "Improvements on GJ-2 Accelerator", *Nuclear Techniques*, Vol. 10, No. 11 (1987) 8, (in Chinese).

[12] Tao Zucong, Private Communication (1987).

[13] Huan Sunren et al., "Transport of Intense Electron Beam at EPA Accelerator", *Proc. 3-rd JCJS* (1987).

[14] Zeng Nagong, *Annual Report IEA* (1986). 42 and *Proc. 6th Int. Conf. on High Power Particle Beam* (1986).

[15] Su Yi, Huang Sunren et al., "A Micro Wave FEL Amplifier Experiment", Internal Report, Southwest Institute of Applied Electronics, P. O. Box 527, Chengdu, China.

[16] Zhang Mingchao, "2×2 MV Tandem Accelerrator", *Proc. 3rd JCJS*, 1987, to be published.

[17] Tandem Accelerator Group, "R&D 2×6 MV Tandem Accelerator", Internal Report, Shanghai Institute for Nuclear Research, Shanghai, China (1986).

[18] Yu Juexian "Progress Report on the HI-13 Tandem Accelerator" *N. I. M.* A244 (1986) 39.

[19] Chen Chia-erh et al., "Design of 4.5 MV Electrostatic Accelerator", *Proc. 1st JCJS* (1980) 19.

[20] Xu Guangsheng et al., "A High Efficiency Two Harmonics Beam Chopper", *R.*

S. I. May (1986).

[21] Jiang Xiaoping et al., "A Double Drift Harmonic Buncher for 4.5 MV Van de Graaff", *Proc. 3rd JCJS* (1987), to be published.

[22] Zhang Enhou, "Progress of HIRFL Project", *Proc. 3rd JCJS* (1987), to be published.

[23] Zhu Xian et al., "DHJ-25 Microtron", Interenal Report, BIAMI, Beijing, June 1985.

[24] Gu Benguang, "A Compact Medical Standing Wave Electrton Linac" (1986), Private Communication.

[25] Wang Yanling et al., "A linac with Feedback under Optical Coupling", *Proc. 3rd JCJS* (1987), to be published.

[26] Zhou Qingyi et al., "Design and Beam Test of Beijing 35 MV Proton Linear Accelerator", *Proc. 3rd JCJS* (1987), to be published.

[27] Xie Jinlin et al., "Design Considerations of the Beijing FEL Project", *Proc. 9th International Free-Electron Laser Conference*, Williamsberg (1987), to be published.

[28] Guan Zhexin et al., "A NS Pulse High Current Electron Beam Source", *Proc. 3rd JCJS* (1987), to be published.

[29] Chen Chiaerh et al., "A RFQ Injector for EN Tandem Linac Heavy Ion Accelerator", *ibid*.

[30] Fang Jiaxun et al., "A Integral Splitring ReEsonator Loaded with Drift Tube and RFQ", *IEEE Trans. NS*-32 5(1985) 2891.

[31] Chen Chiaerh et al., "Bunching Characteristics of a Buncher Using Helix Resonator", *IEEE Trans. NS*-30 2 (1983), 1254.

[32] Li Kun et al., "Experimental Test of Helix Loaded Cavity for Heavy Ion Accelerator", *Proc. 2nd JCJS* (1983) 127.

RECENT ION SOURCE DEVELOPMENT IN CHINA[①]

Abstract

The recent development of various types of ion sources and their application in China is reviewed. Emphasis is given to new improvements of the electron cyclotron resonance ion source, MEVVA ion source, electron beam evaporation metal ion source, compact multicusp ion source, as well as compact negative ion sources with permanent magnets. Some of the new proposals are also presented.

I. Introduction

The general status on the R&D of ion sources in China was reported in a previous conference.[1] The development of ion sources since then has been pushed mainly by the requirements of material science, especially by various needs of thin film formation, surface modification, and ion implantation. Only a few ion sources were developed or upgraded to meet the needs of particle accelerators. We shall present in the following the progress of these sources according to their structure and functions.

II. Ion Sources for Particle Accelerators

A. HIRFL-ECR ion source

At the Institute of Modern Physics, Lanzhou, the Caprice type ECR1 source from Ceng, France, has been in full operation for the heavy ion cy-

---

① Coauther: Weijiang Zhao. Reprinted from the Rev. Sci. Instrum. 67(3), March 1996.

clotron HIRFL at all times. Recently, a new ECR2 ion source, similar to that of ECR1 in structure (Fig. 1) was constructed to provide for HIRFL higher intensity beams of higher charge state ions, and also to separately conduct research of high excitation state atomic physics. Tests on the materials of the plasma cavity and the effect of the cavity wall coatings showed that a copper cavity is better than a stainless steel one, the perme-

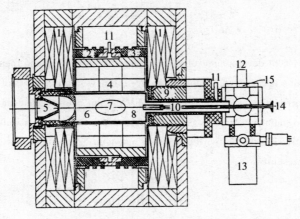

Fig. 1 HIRFL-ECR2 ion source

1. solenoids, 2. insulator in extraction, 3. insulator in injection, 4. hexapole, 5. puller, 6. plasma cavity, 7. ECR surface, 8. quartz tube, 9. conical ring, 10. coaxial, 11. entrance and exit of cooling water, 12. wave guide, 13. TM pump, 14. entrance of working gas, 15. microwave window.

Table 1 The beam current of ECR2. The extraction voltage is 15 keV, the diameter of the extraction hole $\phi = 7$ mm

Charge state	Beam current (eμA)			
	O	Ne	Ar	Xe
6+	200	150		
7+	44	55		
8+		20	245	
9+			125	
14+				28

ability of which is not low enough. It is very interesting to notice that an Al coating is quite effective for obtaining a fraction of highly charged ions. Three kinds of thin tubes, including Ta, Zr, and stainless steel tubes were also tested as the liner of the cavity. It turned out that a Zr tube of 0.1 mm in thickness is most favorable in terms of beam intensity and stability. In addition, an Al plasma electrode always gives better results independent of the coating material. The ECR2 source has worked smoothly since last year and four species of beams, i.e., N, O, Ne, Ar, have been delivered for the physical experiments. The preliminary result of the beam current for some gaseous ions is given in Table 1.[2] Tests on a new extraction geometry and new hexapole magnet are well underway.

B. ECR2 ion source with lasers

The acceleration of highly stripped ions of solid elements is a very demanding issue for HIFRL. However, technical difficulties along with a very compact ECR source like the 10 GHz ECR2, excludes the use of an external oven or a specially designed internal oven developed in Ganil[3] and Grenoble.[4] Therefore a design study for incorporating a laser system into the ECR2 ion source was recently carried out and completed.[5] Based on the preliminary results of the laser-plasma and laser-oven experiments,[6] two lasers are to be employed. A pulsed power infrared Nd: YAG laser of 15 MW will be used to generate the primary plasma while a continuous 50 W CO_2 laser is to be used as an oven heater for evaporating solid elements. A thin optical window of 4 mm in thickness functions both as a vacuum seal and a laser lens with a focal length of about 50 cm. Through the lens, the gap of the coaxial guide, and the plasma chamber, the Nd: YAG laser beam is directed onto the target, which is an inserted part of the stainless steel plasma electrode, as shown in Fig. 1. With a special structure the target can be changed to Al, Cu, or Mo, Ta, etc. The laser-generated plasma, with a relatively low mean charge state (Fig. 2), will expand rapidly into an ECR plasma with highly charged ions. Moreover a laser oven is put in the quartz tube near the first stage of the ECR source, where it can be heated by the CO_2 laser from the back. Both variations of the laser-oven temperature and of the laser output power with the laser working current (i.e., the CO_2 gaseous discharge current) are shown in Fig. 3. The volatile ions such as Ca, Zn, Ag, Mg, Cr, and Pb can be produced easily with a remarkable stability. The experimental results from the ECR2 source with lasers will be obtained before long.

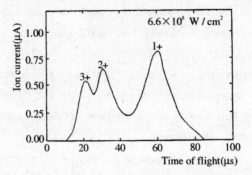

Fig. 2 A time-of-flight spectra of copper ions produced by a Nd: YAG laser

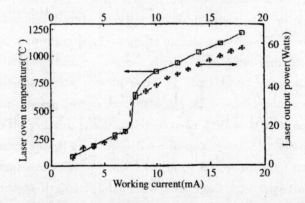

Fig. 3 Dependence of laser output power and laser-oven temperature

C. Multicusp negative hydrogen ion source

A multicusp volume-produced H^- ion source for the cyclotron has been developed at the China Institute of Atomic Energy.[7] The source chamber ($\phi 12 \times 18$ cm) is surrounded externally by ten Nd-Fe-B magnet columns to form a longitudinal line-cusp configuration for primary electron and plasma confinement. The maximum magnetic field at 1 cm from the wall is about 0.08 T. Four iron pipes which are connected with externally changeable permanent magnets, are located at 1 cm from the exit plate to form a magnet filter for the source. The filter plays an essential role in suppressing the electron current. An H⁻ beam of 4.4 mA was obtained at an arc power of 3.2 kW. The emittance of the extracted beam has recently been determined by a device consisting of a multislit screen (slit width 0.35 mm at a spacing of 2 mm) and a movable probe (with a slit of 0.3 × 5 mm). The emittance measured for a total beam current of 2~3 mA and an energy of 25~30 keV is about 20 πmm·mrad for 50% of the beam (Fig. 4).

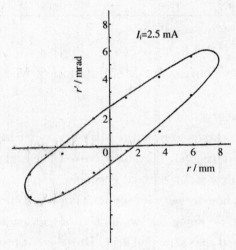

Fig. 4 Emittance of H⁻ beam extracted from a CIAE multicusp ion source

D. Pocket size negative PIG ion source

Compact PIG ion sources with permanent magnets were developed at Peking University to produce metallic as well as gaseous ions, mainly for Van De Graaff and other positive ion accelerators. However, recent experiments with these pocket-size sources showed that quite a number of negative ion species with relatively high electron affinity can be extracted directly from these sources too. For extracting more than 1 mA of an O⁻ beam, the power consumption for the arc discharge can be as low as

50 W. The working parameters of the side-extraction and end-extraction types of PIG sources for negative ions are presented in Table 2.[8] Further tests for the applications to accelerators are to be carried out.

Table 2 Parameters of a PIG source for negative ions

Type	Ion	I_{ext} /μA	V_{ext} /keV	Working gas	I_{arc} /mA	P_{dis} /W
Side extraction	O^-	130	15	O_2	250	112
	F^-	75	15	BF_3	200	280
End extraction	H^-	54	20	H_2+O_2	80	40
	O^-	1497	25	$O_2+C_2H_2$	100	50
	F^-	1800	20	BF_3	100	75

III. Ion Sources for Modification of Materials

A. MEVVA ion source

The ion beam extracted from the MEVVA source has a spatial Gaussian distribution, in general. Obviously only the central part of the beam is good for ion implantation without scanning. Hence, it is important to improve the beam homogeneity for ion implantation over a large area.[9,10] A simple approach for doing that would be using a cusp magnetic field between the anode and the extraction grid. However, it turned out not to be very efficient. It could be much better if the cusp field were moved to the cathode area. Recently, a new structure of the electrode system has been developed to enlarge the homogeneous area of the extracted beam from the MEVVA source at Beijing Normal University. A resistance of about 300 Ω is put between the anode tube (without anode grid) and the extraction grid, and an axial magnetic field is formed near the cathode by permanent magnets outside the cathode mask. Both the schematic diagram of a new anode structure and the measured radial distribution of the extracted beam are shown in Fig. 5. In fact, between the cathode and the anode, there exists such vacuum arc routes that increase the plasma density at the border. By choosing both the position of the extraction electrode and the value of the shunt resistance, a more homogeneous ion beam with a large diameter could be obtained.

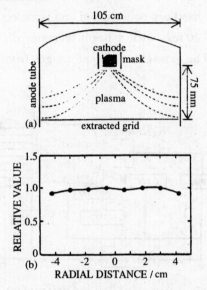

Fig. 5 (a) Schematic diagram of a new anode structure for a MEVVA ion source and (b) the measured radial distribution of the extracted beam

Based on the concept of MEVVA and TITAN,[11] a new multicomponent gas and metal vapor vacuum arc ion source (MUGMA) is being developed at Dalin University of Technology[12] to extend the ion species so as to form metal oxides or nitrides on the material surface. This ion source is composed of four cathode components and a common hollow anode with an accelerating grid system. Each cathode component consists of a metal vacuum arc cathode with a trigger and a pair of cold cathodes where a magnetic field is applied by Sm-Co magnets to form Penning gas discharge. A specially designed compression channel is located between the arc cathode and the cold cathodes linking together the gaseous and arc discharge chambers. An electric circuit diagram of the source is shown in Fig. 6. The source can deliver either gaseous ion beams or metal ion beams separately, when switching on the power supplies C. Trig plus C. Arc or

V. Trig plus V. Arc, respectively. When switching on C. Tric, C. Arc, and V. Arc simultaneously, both gas and metal ions can be produced, and the gas ions also act as a plasma trigger for the metal vapor arc. The MUGMA ion source has been constructed and is expected to extract an average ion current of 20 mA with a beam diameter of 15 cm and a homogeneity of ±20%. The extraction voltage ranges from 20 to 100 kV.

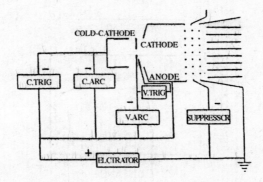

Fig. 6 Electric circuit diagram for a MUGMA ion source

B. Electron beam evaporation broad beam metal ion source

The electron beam evaporation broad beam metal ion source (EBE) has been developed at the Center for Space Science and Applied Research, Academia Sinica (CSSAR).[13] The discharge of the source occurs between a filament and an anode. The electrons extracted from the discharge form a high current density beam by magnetic fields, and then bombard the crucible that contains the metal. Hence a very high temperature area can be obtained on the crucible center; this causes the formation of a high density metal vapor and the discharge can be maintained even without supporting gas. New results of extracted ion beams are shown in Table 3. The extraction area of the ion beam was 10 cm^2. The ion beam current versus the extraction voltage for Ti and C is shown in Fig. 7. Some thin films on solid substrates have been obtained using this ion source.

Table 3 The extraction beam current for EBE at an extraction voltage of 4 kV.

Element	B	C	N	Al	Ar	Ti	Cr	Ni	Cu	Mo	Ta	W
I_{ext}/mA	50	50	30[a]	50	30[a]	90	50	50	20[a]	48	50	44

[a] Nonmaximum extractable beam current.

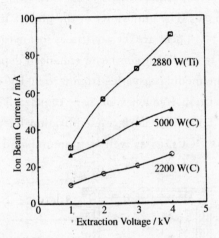

Fig. 7 Dependence of ion beam current on the extraction voltage for EBE

C. Microwave ion source

In order to study the formation of buried oxide layers by oxygen ion implantation on a LC-2F ion implanter, a small microwave ion source is being developed at Peking University.[14] The overall dimension of the source is about 140 mm × 100 mm. The microwave plasma is excited by a microwave power of 2.45 GHz in a stainless steel chamber of about 48 × 80 mm where a magnet field is produced by both an electromagnetic coil with a ferromagnetic loop and a pair of ring-shape permanent magnets. With a seven-hole accel-decel extractor an ion beam of 5 mA has been extracted and further tests are scheduled.

D. Plasma source for ion implantation

A middle sized plasma source for ion implantation (PSII-EX) was constructed at the Southwestern Institute of Physics and has been in operation since October 1993.[5] It is shown schematically in Fig. 8. The size of the vacuum chamber is $\phi 53 \times 100$ cm, on the wall of which there is a linear magnetic cusp field with a strength of more than 0.01 T. The plasma is generated by the distributed heater-cathode discharge in the chamber, and can also be excited by rf power. The pressure of nitrogen ranges from 2×10^{-3} to 5×10^{-2} Pa. The plasma density is $10^8 \sim 10^{10}/\mathrm{cm}^3$ while the electron temperature is $1 \sim 6$ eV. The uniformity of the plasma in a diameter of $\phi 40$ cm is better than 3%. There are Ti vaporizers located inside the chamber near the wall so as to study the ion beam enhanced deposition with PSII. The amplitude of the modulator varies from 0 to 50 kV with a pulse width of 5 μs, and a repetition frequency of $50 \sim 100$ Hz. The target arm can sustain a weight of 5 kg. The surface modification for materials, such as Ti6A14V, Cr12MoV, GCr15, as well as titaniumbased biomaterial, were studied.

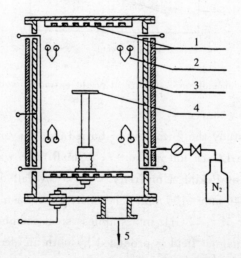

Fig. 8 Schematic diagram of the SWIP PSII-EX
1. permanent magnets, 2. filaments,
3. Ti vaporizers, 4. target, 5. vacuum pump.

E. High power density plasma generator

A new technique based on the interaction between a high power density pulsed plasma and the material surface is proposed for the surface modification of various materials. A pulsed plasma generator, similar to the plasma focusing device, has been developed at the Institute of Physics, Academic Sinica.[16] A schematic presentation of this device is given in Fig. 9. The working principle can be described as follows: the gas is puffed by a fast electromagnetic valve and is then ionized in a coaxial plasma gun. Under the axial force $j \times B$, the plasma is accelerated and injected from the gun into the target chamber where the pulsed plasma bombards the samples at a speed as high as $10 \sim 50$ km/s. The measured plasma density ranges from 10^{14} to 10^{16}/cm^3 while the plasma temperature is $10 \sim 100$ eV and the plasma pulse width is about 60 μs. The kind of film to be deposited on the substrate depends on the supporting gas and the materials of the inner and outer electrodes of the plasma gun. When it is working with nitrogen gas, and the inner electrode and outer electrode consist of titanium and carbon, respectively, the polymorphous film of TiN on the steel GCr15 is obtained, and the film thickness measured after a single pulse of bombardment is 1.5 μm.

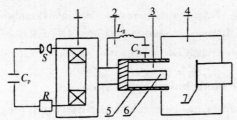

Fig. 9 Schematic presentation of high power density plasma generator
1. fast electromagnetic valve, 2. charge-discharge circuit, 3. coaxial plasma gun, 4. target chamber, 5. outer electrode of the gun, 6. inner electrode of the gun, 7. sample.

IV. Ion Source for Special Application

A. Broad beam electron source

A new type of broad beam electron source based on the Kaufman ion source has been developed at CSSAR[17] for the simulation of the magnetosphere substorm in space. The scheme is shown in Fig. 10. In the two-grid extraction system, the screen grid potential is negative to the extraction electrode which is at the ground potential. It is featured by a wide range of electron energies, from 5 to 100 keV, and a low beam current density on the order of nA/cm^2 with a large beam diameter of 40 cm in the vacuum environment of $1.5 \times 10^{-4} \sim 1.0 \times 10^{-3}$ Pa. It seems to be an advantage of the Kaufman plasma source which enables one to get a broad electron beam with uniformity and stability.

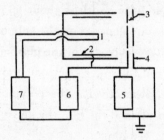

Fig. 10 Schematic diagram of broad beam electron source
1. cathode, 2. anode, 3. screen electrode, 4. acceleration electrode,
5. extraction power supply, 6. anode power supply, 7. cathode power supply.

B. Ion source for biotechnical application

The ion implanter has been taken as a basic tool for ion beam biotechnique studies in the Institute of Plasma Physics in Hefei, China. A Duo-Penning ion source was developed to meet the requirements of massive processing of rice seeds.[18] The cathode of the source consists of a LaB_6 electron emitter (0.8 cm^2 in area) and two tungsten spiral filaments ($\phi 1.5$ mm in diameter) which are wound opposite to each other so that the magnetic field can be compensated for.

There are four shielding layers around the heated cathode so as to save the filament power and to enhance its lifetime. As the emitter is not exposed directly to the arc discharge, the electron emission is fairly stable and the noise of the plasma is also low. In order to optimize the geometrical compression parameter, the cross section at the "throat" of the intermediate electrode is made adjustable by using ring liners of different inner diameter. The multicusp magnetic field is applied to the anode chamber so as to confine the plasma in the anode chamber, where an anode plasma with a density of about $10^{12} \sim 10^{13}/cm^3$ will be formed. In front of the anode, a ring of soft iron is mounted to suppress the escaping electrons. The ions are extracted by an electrode with seven holes. The diameter of the central hole is $\phi=7$ mm and $\phi=5.5$ mm for the other six. With an arc current of 10 A and an extraction voltage of 50 kV, the source can provide 200 mA of N^+ or 300 mA of H^+. The size of the beam spot is 200 mm in diameter with a spatial uniformity better than 80%.

V. Summary

Typical structures and features of various kinds of ion sources developed recently in China have been presented. Considerable progress has been made on the basis of application-oriented studies. The spectrum of ion sources in China was extended from traditional ones to include new species such as a plasma generator and the Kaufman sourcebased broad beam electron gun, etc., so as to meet practical needs. The outcome of this progress has not only had a positive impact on the development of science and the performance of accelerators in China but also on the development of related industries and agriculture. On the other hand, it appears to be an urgent issue for China to put substantial efforts into basic research on the source itself, if the level for the performance of ion sources is to be remarkably advanced. However, it depends heavily on the funding status in China.

Acknowledgments

We wish to acknowledge all who provided us with valuable references and the latest results.

References

[1] H. Z. Zhang, C. -E. Chen, and W. Zhao, Rev. Sci. Instrum. **65**, 383 (1994).

[2] Z. W. Lin, W. Zhang, X. Z. Zhang, X. H. Gho, P. Yuan, S. X. Zhou, and B. W. Wei, Proceedings of 12th International Workshop on ECR ion source, Riken, Japan 25~27 April 1995 (unpublished).

[3] M. P. Bourgarel, M. Bisch, P. Leherisser, T. Y. Pacquel, and J. P. Rataud, Rev. Sci. Instrum. **63**. 2854 (1992).

[4] D. Hitz, G. Melin, M. Pontonnier, and T. K. Ngugen, Proceedings of the 11th International Workshop on ECR Ion Source, Groningenl, 1990 (unpublished), p. 91.

[5] P. Yuan, Z. W. Lin, X. Z. Zhang, and Z. H. Guo, in Ref. 2.

[6] P. Yuan, Annual Report, IMP&NLHIA, Lanzhou, 1993, p. 85.

[7] W. Jiang, B. Cui, and H. Li, Rev. Sci. Instrum, **65**, 1242 (1994).

[8] J. X. Yu, Z. Z. Song, X. T. Ren, and R. X. Li, Rev. Sci. Instrum. **65**, 1337 (1994); **67** (1996), these proceedings.

[9] F. S. Zhou, X. Y. Wu, F. Zhou, H. X. Zhang, and X. J. Zhang, Rev. Sci. Instrum. **65**, 1263 (1994).

[10] A. J. Ryabchikov *et al.*, Proceedings of the Beijing Workshop on MEVVA Ion Source and Applications, 1993 (unpublished), p. 59.

[11] S. P. Bugaev, E. M. Oks, P. M. Schanim, and Y. Yushkov, Rev. Sci. Instrum. **63**, 2422 (1992).

[12] W. Shi, F. Zheng, Z. Cong, and K. Yan, Internal Report (in Chinese), Dalian University of Technology, Dalian 116023, China, 1995.

[13] Y. C. Feng, D. W. You, and Y. Z. Kuang, Rev. Sci. Instrum. **65/4**, 1034 (1994).

[14] Z. Song, D. Jiang, and J. Yu, Rev. Sci. Instrum. **67** (1996), these proceedings.

[15] Z. K. Shang, M. Geng, E. Y. Wang, Y. R. Chen and X. H. Liu, International Conference on Plasma Science and Technology '94, Chengdu, China, 16~20 June 1994 (unpublished).

[16] B. X. Yan, S. Z. Yang, B. Lee, and X. S. Chen, Chin. Sci. Bull. **39**, 17(1994).

[17] L. Xia, Master thesis, CSSAR, 1994.

[18] C. Hu, J. He, Z. Yu, S. Hu, and S. Wang, Internal Report (in Chinese), Institute of Plasma Physics, Academic Sinica, Hefei, 230031, China, 1995.

超导重离子直线加速器[①]

摘　要

本文回顾了近年来高频超导体的状况,叙述并讨论了 SUNYLAC 和 AT-LAS 两台超导重离子直线加速器的结构和性能.在此基础上讨论了超导直线加速器取得的成就、遇到的问题和发展前景.

一　前　言

60年代初期人们发现低温超导体的高频表面电阻只有铜在室温下的十万分之一[1,2].这个发现展示了建造高效率的超导直线加速器的光明前景.从此人们开始了对于超导直线加速器的奋力探索.二十多年来尽管高频超导技术的开发还远未达到理想的境地,但已经出现了诸如超导电子直线加速器、超导高能粒子分离器以及超导重离子直线加速器等卓有成效的应用.

近年来美国阿贡国家实验室和纽约州立大学石溪分校各建成了一台超导重离子加速器 ATLAS[3,4] 和 SUNYLAC[5,6].它们充分证实了超导直线加速器的优越性.这两台加速器都以 FN 型串级静电加速器为注入器,在约 3 MV/m 的加速场强下以 100% 的负载周期加速各种中等质量的重离子.它们的终能量超过一个 25 MV 的串级静电加速器(图1),在束流性能上则保持了串级静电加速器所具有的优异品质.重要的是这样一套静电—超导直线加速器系统的总投资,包括 FN 注入器在内,大大低于建一个 25 MV 串级静电加速器所需的费用.同时超导直线加速器运行的总功耗,包括低温液氦系统在内,又只为同类型室温直线后加速器的五分之一左右.以上这些事实吸引着国际上其它许多从事核物理研究的单位,他们纷纷投身于建造和研制超导重离子直线加速器的行

[①] 摘自《核物理动态》,Vol.3, No.2 (1986),p.19～26.

列. 目前正在建造或已决定建造超导直线加速器的试验室有法国的萨克雷核子中心[7]、英国的牛津大学核物理试验室[8]、美国的佛罗里达州立大学[9]和华盛顿大学、印度的达达研究所以及以色列的维兹曼研究所[8]等,美国加州还开设了一个"超导应用公司"[10],专门供应超导腔及有关设备. 超导重离子直线加速器所取得的成就,还进一步引起人们在高能电子储存环、高能直线对撞机方面开发和应用超高频超导技术的兴趣. 一种新的场强高达 80 MV/m,而造价 1 百万美元/GeV 的超导加速结构正在成为下一步的行动目标!

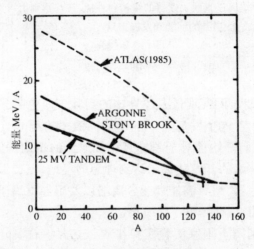

图 1 超导直线加速器的能量

为了具体说明超导重离子直线加速器的成就,它所遇到的困难以及发展动向等,本文下面分别介绍和讨论(1)高频超导体的基本特性;(2)加速腔的结构与性能;(3)超导加速器运行中的几个技术问题等.

二 高频超导体的基本特性[11]

超导体的高频表面电阻和高频临界磁场是超导技术在加速器方面应用的两个最重要的参量,前者决定着加速腔的功耗,后者规定了能量梯度的上限.

高频场下超导体的性能与直流情况下不同. 当温度降至某个临界温

度 T_c 后,超导体的电阻不为零,而等于某个称为"高频表面电阻"的小量 R_s. 根据 BCS 理论,常用超导材料 Nb 的表面电阻随温度 T、频率 f 变化的情况可表述为[13]

$$R_{BCS} = \frac{2.4 \times 10^{-21}}{T} \cdot f^{1.8} \cdot \exp\left(-1.8\frac{T_c}{T}\right)(\Omega)$$

式中 f 以 Hz 为单位,T 以 K 为单位. 实际测量表明:R_s 在随温度降至 $10^{-8}\Omega$ 后,就偏离理论规律而趋于所谓"剩余"电阻 R_{res},所以实际上 $R_s = R_{BCS} + R_{res}$. 式中 R_{res} 的大小与温度无关但与超导体的表面处理状况密切相关,它反映了超导表面不完善的程度. 表面电阻 R_s 随温度和磁场变化的实验曲线如图 2 所示. Pb, Nb 和 Nb$_3$Sn 三种超导材料, 在低场下的典型数据列在表 1 上.

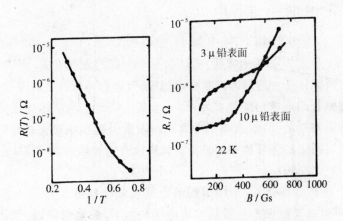

图 2 (a) 表面电阻随温度变化 (b) 表面电阻随磁场变化

超导体的高频临界磁场规定了谐振腔的上限. 常用的 Nb 和 Pb 的临界场决定于所谓"超热临界场"(Super Heating Critical Field)[11] H_{sh}, 而略高于直流临界场 H_c. 高频磁场一旦超过 H_{sh}, 超导体就立即转成普通导体. Nb$_3$Sn 的临界场也等于 H_{sh}, 但略低于 H_c. 表 2 上列出了三种典型材料的直流和超热临界场的理论值, 还列出了假定腔内单位加速电场的磁场峰值 $H_p/E_a = 100$ Oe/MV·m^{-1} 时, 允许的加速场强.

表1　$T=4.2\text{ K}$，$f=100\text{ MHz}$ 时的表面电阻

材　料	类　型	T_c/K	R_{BCS}/Ω	R_{res}/Ω
Pb	Ⅰ	7.2	5×10^{-9}	2×10^{-8}
Nb	Ⅱ	9.2	2.5×10^{-9}	10^{-9}
Nb$_3$Sn	Ⅲ	18.2	4×10^{-11}	$\sim 10^{-9}$

(*100 MHz，300 K 时，铜的表面电阻为 $2.5\times10^{-3}\ \Omega$)

表2　临界磁场的理论值

材　料	T_c/K	$H_c^*(T=0)$ [奥]	H_{sh} [奥]	$E_a/\text{MV}\cdot\text{m}^{-1}$
Pb	7.2	804	1125	11
Nb	9.2	2000	2400	24
Nb$_3$Sn	18.2	5400	4000	40

$^*H_c(T)=H_c(0)(1-(T/T_c)^2)$

实验上测得的临界场明显的低于理论值，Pb 为 450 Gs，Nb 为 1600 Gs[11]，其原因不完全清楚，估计仍与不完善的表面有关. 据认为：材料的表面上可能存在一些局部的非超导的"核"，它们不断地对周围加热，最后导致超导状态的破坏（称"热磁破坏"）. 事实上局部结晶缺陷或成分不纯引起的当地临界场的降低、表面不平等几何因素引起当地磁场的局部升高以及表面残留的尘埃等异物引起的介质加热或欧姆加热等都可成为形成上述非超导"核"的原因.

除了临界磁场之外，电场引起的电子负载包括电子共振负载（Multipactor）和场致发射也都会限制加速电场的水平. 经验表明，在氦气中长时间的进行高频锻炼，有助于克服这类电子负载.

在已知的上百种具有超导性能的元素、合金或化合物材料中，以 Pb、Nb 和 Nb$_3$Sn 三种对在加速器方面的应用最具实用价值. 三种之中又以 Nb$_3$Sn 的 T_c 和 H_{sh} 最高，因而应用的潜力最大. 可惜目前还没有找到一套成熟的工艺，能用以得到表面光洁而平滑的 Nb$_3$Sn. 这不能不暂时限制了它在高频技术方面的应用. 反过来，Pb 的 T_c 和 H_{sh} 都较低，然而它价格便宜，且易于镀于铜的表面形成导热良好的超导层，因此有不少实验室仍用 Pb 来制备高频加速腔.

三 加速腔的结构与性能

适于加速重离子的超导腔,大体有三种结构.美国的 ATLAS 和 SUNYLAC 等采用分离环(SLR)的结构,西德卡尔斯鲁厄与法国萨克莱核中心合作采用锥形螺旋线(Tapered Helix Resonator)的结构,还有以色列和美国合作搞的 1/4 波长共振线(QWR)结构(图3).这些结构的特点与室温的直线后加速器有不少类似之处.然而,同样一个结构参量其含义对于超导和常温往往是不同的.下面就几个主要参量包括谐振频率、电长度、几何尺寸稳定性等问题进行讨论.

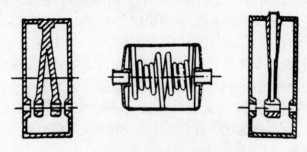

图3 三种超导加速腔结构

超导腔谐振频率的确定是多种相矛盾的因素综合平衡的结果.一方面速度低的重离子要求用频率低的 RF 场来加速以达到较高的能量增益并获得高亮度的优质束.然而另一方面,谐振频率低意味着腔的机械尺寸大,结果机械加工和表面处理的难度和成本都增高了,制冷的耗费亦增大了,更重要的是尺寸大的腔储能高,会给难度较大的频率、相位稳定控制问题增加困难(详见下一节).作为一种折衷,谐振频率 f_0 通常取在 90~150 MHz 之间.有些设计,如 ATLAS,比较注重腔的能量增益和保持注入束的优良品质,就选取 $f_0 = 97$ MHz.另一些,如石溪的 SUNY-LAC 和萨克莱的 Helix 结构,更强调解决相位控制方面的困难,他们分别将 f_0 定为 150 MHz 和 135 MHz.

现有超导腔的电长度都只有 $1 \sim 1\frac{1}{2}$ 个驻波波长,都属于所谓"短结构".这种结构的优点一般说是可以在大的范围内改变加速离子的品种

和能量而不影响有效的加速效率,以电长度为 $1.5\beta\lambda$ 的 SLR 和 THR 结构为例,它们对于能量或质量数在 5∶1 范围内变化的离子(例如 A=16～80)的加速效率都在 70% 以上,而对于电长度为 1 个 $\beta\lambda$ 的 QWR 结构,允许改变的范围就更宽,达 10∶1.(参见图 4. 图上的加速效率用渡越时间因子 $T(\beta)$ 来表示. $T(\beta)=\int E(Z)\cos(WZ/\beta_c)dZ/|\int E(Z)dZ|$,速度等于相束 β_0 的同步离子的效率为 $T(\beta_0)$). 短结构在这方面的优点,无论对超导或常温的重离子加速器都具有相同的意义,然而对于工艺过程更为复杂,且成熟程度还不高的超导腔来说,采用短结构还有更多的好处. 一是短腔易于制备和调运. 即使局部有一点点小毛病也不会像长腔那样让许多仍然有用的加速单元跟着报废. 其次,加速器上腔列的相速分布仅为二、三个预定的常数,不需连续变化.(例如 SUNYLAC 上有 40 个腔它们的相速只有二个;前面 16 个腔的 $\beta_0=0.055$,后面 24 个腔的 $\beta_0=0.10$)这种分布非常有利于腔的互换. 在调试中一旦发现某个腔有毛病可以方便地换上备用腔. 最后,在加速器运行中如遇有一个或几个腔发生故障时只需将它们暂时关掉就是了,整个加速器仍可继续运行. 因为经验表明,全年的运行计划中只有很少的几次是需要在加速器的最高能量下运行的.

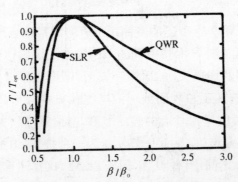

图 4　短结构腔的渡越时间因子(对同步粒子规一化后的值)

　　超导加速腔的几何参量需按稳定运行的要求最佳化. 这里有三方面的要求. 一是要求单位加速电场下表面电场和磁场的峰值趋于极小,前者是为了避免场致发射引起的电子负载;后者是为了使超导体的工作点远离临界场 H_{sh},防止热磁破坏,两者都是为了尽可能地提高加速场强,发挥超导腔的优越性;二是要求储能 U/E_a^2 趋于最小,以利于相位的稳定控制. ATLAS 的腔由于谐振频率低,储能的问题更为突出,为此他们采取了限制腔径向尺寸的措施(如让漂浮管与腔壳同轴等),尽量降低储能. 最后还要求腔的结构尺寸具有高度的稳定性. 由于超导腔的 Q 值高达约 10^8,频带极窄(约 1 Hz),结构尺寸的稳定问题就特别突出. 实际上在一些敏感的部位上,如分离环腔的漂浮管电极,仅十几 Å 的位移就可使谐振频率移出一个带宽! 为此,要特别注意让这些部件具有足够高的机械强度并处于电场辐射压力的平衡之中,使它们不致在高电场下变形或为实验室的机械噪音激发起过量的微振动. 石溪的分离环腔以 ϕ 20 mm 的镀有约 10 μm 纯铅的铜管制成环臂以保证必要的机械强度. 该腔固有振动的基频 $\nu=45$ Hz 左右,高于一般实验室噪音基频. 环端漂浮管电极中央电隙的间距约为两侧间距的二倍以求得电场的平衡. 石溪的 QWR 腔以更粗的锥形管($\phi_{max}=40$ mm, $\phi_{min}=20$ mm)作为电极,其固有基波频率 $\nu=190$ Hz. 鉴于实验室噪音的谱线密度与频率的高次幂成反比,且通常强迫振动的振幅 $\propto 1/\nu^2$,显然 QWR 的动态稳定性要比 SLR 高得多. 此外,QWR 是轴对称结构,电磁压力是自然平衡的. 现有超导腔的一些典型数据列在表 3 上.

表 3 超导腔的性能参数

	分离环 (SLR)		锥形螺旋线	λ/4 共振线
	ATLAS	CUNYLAC	THR	QWR
材料	Nb	Pb	Nb	Nb
相速 β_0	0.105	0.1	0.06	0.085
频率 f_0/MHz	97	150	135	159
加速电场 E_a/MV·m^{-1}	3	3	2.25	3
Q_0	2×10^8	6×10^7	5×10^8	1.5×10^8
能量增益/MeV	1.07	0.67	0.42	0.54
E_p/E_a	4.7	5.5	7	4.2
E_p/E_a	182	105	280	54
U/J ($E=3$ MV·m^{-1})	1.32	0.36		0.52
ν		45		190

超导腔的 Q_0 与表面电阻间存在着简单的关系:$Q_0=G/R_s$,其中 G 是腔的几何因子,与导体的电阻及所加的电场等无关.对于 SLR 腔,典型 $G=20\,\Omega$.由此 Q 值随着加速电场的变化,反映了超导的性能状态的变化.有关的实验曲线画在图 5 上. Q 值在高场区的迅速下降反映了场致发射引起的电子负载的效应.

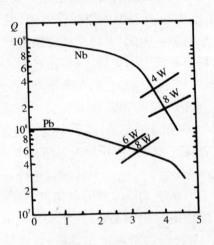

图 5 Q 值随加速电场变化的情况

四 超导加速器运行中的几个技术问题

这里我们将介绍和讨论现有超导加速器所遇到的有关 RF 相位稳定控制、低温冷却以及束流动力学方面的问题.

1. 频率和相位的稳定控制

现有超导重离子直线加速器都由几十个相位可以独立调节短结构谐振腔所构成.由于腔的频带极窄,每个腔都用一个自激回路来激励.自激振荡的频率是不稳定的,随腔体结构的瞬时参量飘动;而它的振幅则是稳定的,不随频率而变.然而从重离子加速的要求来说,必需有一个统一的频率和确定的相位分布.这就提出了超导加速器频、相稳定控制的课题.事实上超导腔谐振频率的偏差有两种成分,一种是所谓"静态的".例如结构相同的腔,即使室温下有相同的频率但冷至 4.5 K 时,因材料或加工上的不均匀或其它差异,谐振频率的差别可达 10 kHz.这类偏差

可依靠速度较慢的自动调谐器来调整.如ATLAS上的腔用步进电机控制的"活塞"作慢调谐器,SUNYLAC则用使腔壁微量变形的"压力杠杆".另一种偏离是因实验室机械噪音引起的快振动,这需要用锁相反馈的方法来调整,即利用腔和标准讯号间的相位误差信息同时控制回路的瞬时频率和振幅,迫使腔的频率锁在加速器的标准频率上而同时保持腔的振幅不变.为了锁频,SUNYLAC用一个与回路原有相位正交的功率放大器,亦称调谐放大器(Tunning Power Amplifier)调整自激回路的相位和频率(图6).计算表明调谐所需的功率$P_{TUN}\approx U(\omega_a-\omega_s)$[12],其中$U$系储能,$\omega_a$加速器的频率,$\omega_s$腔的瞬时频率.石溪SLR腔的储能$U=0.36$(J),频飘$\frac{\omega_a-\omega_s}{2\pi}\approx\pm 30$ Hz,$P_{TUN}=68$ W.用这样的方法可使腔和标准讯号间的相差锁到±0.1°之内.为了使锁相回路稳定工作,需增强自激回路与腔间的耦合,以使有效带宽与频率误差的振幅相当,即$Q_1=Q_0/1+\beta\approx\frac{\omega_a}{|\omega_a-\omega_s|}$.对于体积大、储能更高的ATLAS上的腔,因所需的P_{TUN}相当大,上述方法不一定方便,故改用与腔强耦合的"压控电抗"即"VCR"(Voltage Controlled Reactance)来调整回路的频率."VCR"由十多个受PIN二极管控制的固定电抗构成,通过改变开关管的数量就可改变与腔强耦合的电感量,这样的方式可以在kHz的范围内调整频率,使相位误差<0.1°.

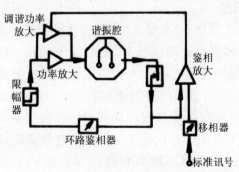

图6 锁相自激振荡回路

2. 低温冷却

现有加速器上,超导腔以3～6个为一组,分别安装在十来个低温罐中冷至 4.5 K. 低温罐的内部结构相当复杂. 大体可分为三层. 中心部分是由液氦冷却的腔体, 外边是一层充液氮的热屏蔽层. 其外至容器壁间是由绝缘和辐射屏蔽材料构成的绝缘层. 最外边的容器壁是真空密闭的. 工作时罐内真空保持在 $10^{-7} \sim 10^{-8}$ Torr[①]. 腔和罐壁之间通过一些绝缘性能好的构件联结或支撑. 冷却用的液氦、液氮以及各种信号控制电缆分别通过罐顶板上的接口输入至罐内各有关部分.

阿贡和石溪的腔在冷却方式上各有特点. 石溪的腔是倒挂着的, 输入的液氦经一个贮氦池, 靠重力由腔顶灌入分离环管及漂浮管电极. 由于腔的体积小, 涂的铅层又很薄, 热传导比较好, 所以在液氦的自然蒸发下就可使 Pb 表面冷至 4.5 K. 这样的冷却方式效率较高, 且可避免因压力输氦引起的机械振动. 不过阿贡的超导腔因体积大, 还需采用压力输氦液的方式来冷却. 同样, 流体阻抗较大的 Helix 也不得不采用强迫冷却的方式.

在 $3\sim4.5$ MV/m 场强下工作的超导腔, 典型功耗为 $4\sim8$ W. 可见一个有 40 个腔的加速器, 就需在 4.5 K 下冷却 $200\sim300$ W 的功率, 加上液氦输运线等上的热损耗等, 冷却功率达 400 W 左右. 现有液氦机的电效率, 高的达 10^{-2}, 低的仅 10^{-3}, 但价钱便宜. 据此, 加速器的总功耗, 根据不同的液氦机的条件在 40 kW 到 400 kW 之间. 除了液氦外, 还有液氮的消耗, 石溪每天消耗液氮约 240 L.

3. 束流动力学问题

超导直线加速器具有"高保真"地加速优质束流的能力. 这首先是因为它能同时在高梯度和连续波 (CW) 的状态下运行, 这可以减少许多畸变. 同时作为直线加速器它的注入和引出过程是自然的, 没有圆形加速器所遇到的那些困难. 然而超导加速器的这一优势得以发挥的根本条件是: 束流的相空间分布必需落在加速器的线性接收空间之内. 对于纵向运动来说, 这意味着注入束的相宽 $\Delta\phi < 6°$, 能散 $\Delta E/E \lesssim 10^{-3}$[13] 为了达到这个条件, 需要在束流的群聚和脉冲束的输运方面作许多努力. 以 SUNYLAC 为例, 在注入器 (FN 串级加速器) 之前设置了高效率的双飘移谐波聚束器, 它在 FN 加速器之后形成宽约为 1 ns 的脉冲束, 束流利

---

① 1 Torr=133.322 4 Pa.

用率可达 60%~70%[14]. 高的束流效率有助于弥补 FN 上电子剥离器所造成的束流损失. 在超导加速器之前,还设有一个超导的分离环腔作后聚束器,它能充分利用束流纵向相空间经 FN 加速之后,充分收缩的特点,进一步将 ns 级束流压缩至 100 ps 左右[14]. 此外,还有一个后切割器它与 FN 加速器的分析磁铁组成一个纵向光阑,切除一切不需要的粒子本底. 与群聚系统同样重要的是脉冲束的后输运系统. 限于实验室的空间,石溪不得不在后聚束之后再将束偏转 180°以注入直线加速器. 为了防止偏转中引起的时间弥散和束流损失,他们采用了由两个 90 磁铁和一个三元四极透镜组成的消色散、等时性输运系统,"高保真"地将束注入直线加速器. ATLAS 加速器采用了类似的注入系统[3],但他们用了一个四个谐波的谐振腔作前聚束器[15],它的电压波形接近于理想的锯齿波形,束流效率可达 70%. 迄今 ATLAS 的前级超导加速器已运行了 10^4 小时,获得了横向相面积 1~2 mm·mrad、纵向相面积 20~80 keV·ns 的各种离子束 50~100 nA,很好地满足了核物理研究的各项要求,他们还打算利用散束器,将脉宽扩展至 100 ns 而使 $\Delta E/E \sim 10^{-4}$[3].

五 结 束 语

归纳已有实践,超导重离子直线加速器所取得的主要成就有以下几个方面:

1. ATLAS 前级超导加速器的上万小时的运行实践充分肯定了高频超导体的实用性. 他们的经验证实:油污、暴露大气、辐照等等都不足以对超导体造成基本的损害,个别的缺陷完全可以通过淋洗—抛光或充氦下的锻炼等迅速恢复.

2. 在能量梯度方面已大大超过高压型加速器和连续运行(CW)的常温直线加速器. 在结构投资和运行费用方面也都显示了它的明显优势.

3. 已运行的 ATLAS 和 SUNYLAC 两台超导加速器都未配备专职开机人员. 一旦机器启动即交由核物理试验人员操作,运行中遇有需要变化束流参量(例如改变终能量)时,可由试验人员通过简便操作,在一分钟之内迅速实现,这充分说明超导加速器运行的稳定和可靠性.

4. 两台加速器加速出的重离子束的优良性能,证实了超导直线加速器所具有的"高保真"地加速优质粒子束的能力.

然而在实践中也暴露了超导加速技术的一些不足.例如由于低温罐的升温和冷却所需时间过长.因为检修一个安装在束流线上的超导腔往往要花上七至十天时间,这不能不影响加速器的开机率.看来今后必需备有随时可换到束流线上工作的备用加速段,以保证加速器在最高能量下的开机率.此外,还有一些更基本的问题,例如因工艺过程不成熟,同样的结构和处理方法往往可以得出相当不同的结果;而且实验上达到的表面电阻和临界磁场总比理论值差得多.这说明高频超导体的应用仍还处于初期阶段,许多基本问题尚待认识,更好的材料尚待开发,而一旦人们对高频超导体的规律有了进一步的掌握,超导直线加速器就将以更高的加速场强,更低的建造费用在高能领域作出新的贡献!图7上画着最佳化梯度下加速器设备投资、运行功耗随着参量 $f \cdot Q_0$ 的变化[11],其上 cl 是单位长度的建造费用(包括土建等基本设施、加速结构、真空束流输运、诊断等)以 M\$/m 为单位,η 是液氦机的机械效率.由图可见 Nb 腔预期的最高梯度达 50 MV/m 而 Nb_3Sn 腔可达 80 MV/m!这无疑将是超导直线加速技术的下一个目标.

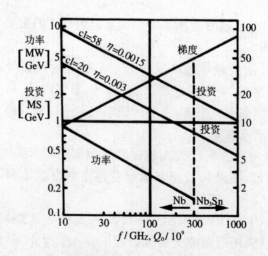

图7 超导直线加速器单位能量的投资、功率以及最佳能量梯度随 $f \cdot Q_0$ 变化情况

参 考 文 献

[1] Fairoank, W. M., et al., Proc. 8th Intern. Conf. Low Temperature Physics, London, 1962, p. 324.
[2] Pierce, J. M. et al., Proc. 9th Conf. Low Temperature Physics, Oiho, 1964, p. 396.
[3] Bollinger, L. M. IEEE Trans. Vol. NS-30 No. 4 (1983) 2065.
[4] Bollinger, L. M. IEEE Trans. Vol. NS-24, No. 3 (1977) 1076.
[5] Brennan, J. M., Chen, C. E. et al., Bulletin of American Physical Society Vol. 28, 2(1983) 91.
[6] Ben-Zvi, I IEEE Trans. Vol. NS-28, No. 3 (1981) 2488.
[7] Compte Rendu d'active 1981~1982, Division De La Physique CEN Saclay ISSN 0750~6678.
[8] Lapostalle, P., IEEE Trans. Vol. NS-30, 4(1983) 1957.
[9] Chapman, K. H., N. I. M. 184(1981) 239.
[10] "Applied Superconducting, Inc." 707 West Woodbury Road, Altadena, California, 91001.
[11] Tigner, M. et al., AIP Conf. Proceedings No. 105 p. 801.
[12] Paul, P. et al., Comments Nucl. Part. Phys. Vol. 11, 5(1983) 217.
[13] Bollinger, L, M. et al., IEEE Trans. NS-22, 3(1975) 1148.
[14] Brennan, J. M. Chen, C. E. et al., IEEE Trans. NS-30, 4(1083) 2798.
[15] Lynch, F. J. et al., N. I. M. 159(1979). 245.

PROGRESS OF RFQ AND SUPERCONDUCTING ACCELERATORS IN CHINA[①②]

Abstract

Efforts for developing high current proton and heavy ion RFQ accelerators in China are presented. Progress in Integrated Split-ring Resonator based heavy ion RFQ accelerator is reviewed. The conceptual design of a proton RFQ for the proposal of "Accelerator Driven Radiological Clean Nuclear Power System" is outlined.

R&D activities on superconducting cavities in China are reported. Progress in L-band superconducting cavities and a photo-cathode electron gun for generating high brightness electron beam is presented. The superconducting quarter wave resonator for the heavy ion post-accelerator is also described.

1. Introduction

Efforts for developing high current RFQ accelerators in China were initiated in 1984 to meet the needs of low energy ion implantation[1]. An Integrated Split-ring Resonator (ISR) RFQ with water-cooled mini vane electrodes was developed for this purpose at Peking University. N^+, O^+ and O^- ions have been accelerated to more than 300 keV. The beam transmission efficiency of the RFQ reached more than 84% with an average current of $\sim 38.4\ \mu A$. Feasibility study of accelerating both O^+ and O^- ion beams simultaneously in the same RFQ was also successfully per-

---

① Coauthers: Fang Jia-Xun, Guan Xia-Ling, Zhao Kui. Reprinted from the Invited talk at International Symposium on Frontiers of Modern Physics, Kuala Lumpur, Malaysia, Oct. 1998.
② Work supported by National Natural Science Foundation of China.

formed. An ISR RFQ for accelerating oxygen ion beam up to 1 MeV was then designed and constructed[2~4].

RFQ for accelerating protons was developed at the Institute of High Energy Physics (IHEP) since 1993 as an injector for Beijing Proton Linac (BPL)[5]. However, feasibility studies on high power proton RFQ are intensified by the proposal of AD-RCNPS[6]. Scientists from IHEP, CIAE and Peking University work together for this purpose. Preliminary parameters of RFQ injector have been worked out and a test model is to be constructed[7].

With the support of the State High Tech Program, studies on RF superconductivity have been active in China since 1988. Two SC cavities with China made niobium were successfully developed by the RF Superconductivity group at PKU[8]. A DC photo cathode electron gun with a 2 MeV superconducting booster was then installed[9]. A Cu-Nb sputtering system was constructed at PKU[10] and SC cavities with Cu-Nb sputtering technology are being developed. The feasibility of a heavy ion superconducting booster with Nb sputtered QWR cavities are studied jointly by PKU and CIAE for the proposal of Beijing Radioactive Nuclear Beam Facility (BRNBF)[11]. RF SC labs are to be set up by IHEP and China Academy of Engineering Physics (CAEP) for the upgrading of BEPC and CW FEL respectively[12,13]. Feasibility studies on superconducting sections for AD-RCNPS is also to be carried out jointly by PKU, CIAE and IHEP[7].

2. Heavy Ion RFQ

The RFQ group at Peking University has been engaged in developing ISR RFQ cavities for ion implantation. A 300 keV cavity was developed and studied extensively with a series of beam tests. A 1 MeV cavity was designed and constructed. The main parameters of these two cavities are shown on the first two columns in Table 1.

Table 1 Main parameters of RFQ accelerators

Type	ISR	ISR	4-rod	4-vane
Ions	$N^+O^+O^-$	$N^+O^+O^-$	P	P
F_0/MHz	26	26	201.25	350
W_{in}/keV	20	22	40	75
W_f/keV	300	1000	750	7000
$I_{av}/\mu A$	38.4*	≈100	≈34	3000
I_p/mA	≈1*	5	60	60~120
L/cm	90	250	118.7	800
D_{out}/cm	50	70	≈25	30
V_0/kV	75	70	130	≈90
Duty Factor	16.7%	16.7%	0.23%	5%

*Measured figure, limited by input beam current

2.1 The Beam Test of a 300 keV Cavity

The structure of an ISR RFQ resonator operating at 26 MHz[1~2] is shown in Fig. 1. High duty factor is ensured by the cooling water flowing through the spiral tubes, supporting rings and quadruple mini-vane electrodes so as to have high average beam current. The end section of the electrodes was specially shaped so that the transverse beam quality is improved. The RF power is fed from a linear power amplifier (XFD-D5) with maximum power output of 30 kW (CW) or 50 kW (pulsed) through a water-cooled loop. The amplitude of the field gradient in the RFQ cavity is stabilized by a feedback loop with a Double Balanced Mixer.

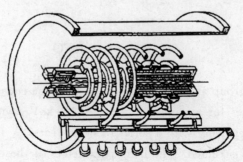

Fig. 1 View of a 26 MHz 300 keV ISR RFQ

2.2 The acceleration of the N^+, O^+ and O^- beam

The layout of the beam test is shown in Fig. 2. The positive and negative ion sources are located at $\pm 45°$ with respect to the beam axis. The ion beams extracted from both sources are focused respectively by the Einzel lenses (EL) next to them. The ions can be bent separately or funneled simultaneously on to the beam axis by a combining magnet (CM) and then focused by a matching Einzel lens (MEL) in front of the RFQ. Two beam monitors (BM) are mounted at the RFQ entrance and exit respectively to measure the input and output beam intensity. The output beam can also be measured with two off-axis cups (FC) with a small magnet (DM). The energy spectrum is measured by the analyzing magnet (AM).

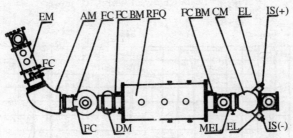

Fig. 2 Schematic layout of the beam test

Cold cathode PIG ion sources with permanent magnets are used for producing O^+ and O^- ions. The input current of O^+ ranges from $200 \sim 490$ μA under an extraction voltage of $17 \sim 20$ kV. The output beam energy determined by AM is shown in Fig. 3 versus input RF power. The highest energy gain of 306 keV was reached at a RF power of 30 kW. The beam transmission efficiency raises with vane voltage until it reaches 43% for O^+, 48% for O^- and 78% for N^+ beams[4].

Acceleration of O^+ Beam was also performed in a pulsed ion source mode. The arc voltage of the source is triggered and synchronized with the RF modulation with a frequency of 166 Hz and the duration of the pulse can be adjusted by the trigger signal. The waveforms of the input and output beam current are recorded on a digital oscilloscope as shown in Fig. 4.

The peak currents I_{in} and I_{out} are 330 μA and 280 μA respectively at an input power of 45 kW, which indicates a beam efficiency of 84.3%. The merit of a pulsed operation lies both in enhancing peak arc power as well as in eliminating background of non-accelerated ion[4].

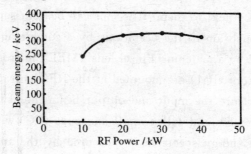

Fig. 3 Beam energy of O^+ vs RF power

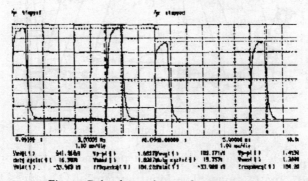

Fig. 4 Pulse form of input & output beam

2.3 Simultaneous Acceleration of O^+ and O^-

Simultaneous acceleration of both O^+ and O^- ions in the same RFQ was performed. The ion sources and the focusing lenses related should be set according to their own characteristics. However, the set-up of the operating voltage of the matching lens has to be a kind of trade off and 13 kV was chosen for the first trial. The simultaneous acceleration was performed at various RF power levels with the average current output varied from 0.1 to 21 μA under a duty factor of 1/6. The current ratio of O^+ to O^- varied from 0.1 to 5. The sum of O^+ and O^- current for simultaneous acceleration I_{sim} is about the same as separated acceleration I_{sep} in all current ranges as can be seen from Fig. 5. This means that both positive and negative half periods of the RF cycle can be used to accelerate corresponding sign of ions at the same time. So far as the micro-peak current is of the order of ~ 1 mA, the interactions between negative and positive ion bunches are negligible, as it was expected. It is also worth mentioning that the result implies that dual species of ions of different amount can be implanted at once with deliberate ratio by using the present set-up[3,4].

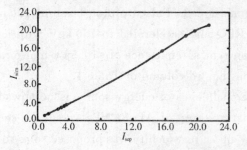

Fig. 5　I_{sim} vs. I_{sep}

A 1 MeV heavy ion RFQ is being built based on the above experiences. The main parameters of which are listed in Table 1. The accelerating cavity has been designed and constructed. The mini-vanes and its supporting arms are mounted on the bottom plate with a movable cover so that they can be accessed and set up easily as shown in Fig. 6. The water-cooled vanes are made of Cr-copper so as to reinforce its rigidity. The vacuum of the cavity tank has already reached 2.5×10^{-6} Torr. The high power and

Fig. 6　The 1 MeV RFQ cavity

beam tests are expected to be carried out by the end of 1998.

3. High Power Proton RFQ

A four-rod type RFQ, BPL-RFQ, was developed at the IHEP to replace the present bulky Cockroft generator[5]. Comparing with the 4-vane structure, it has the merit of lower cost and broader mode separation. As the result of design studies and cold model tests, a cavity was constructed and is being commissioned. Pulsed proton beam of 100 mA/40 keV is to be input to the RFQ and accelerated to 750 keV with an output current of >60 mA and an r.m.s. emittance of <0.56 π-mm-mrad. Main parameters are listed on the 3rd column of Table 1.

To meet the challenges of energy source issue nowadays in China, it was proposed that a set-up of AD-RCNPS should be considered seriously[6]. As a follow-up, a test facility was proposed so as to examine the key items of physics and technology[7]. Scientists from CIAE, IHEP and PKU have been working together since 1995 on the feasibility study of such facility. As a preliminary conceptual design, it consists of a 150 MeV high power proton accelerator, a swimming pool reactor and some basic research facility for material and target assembly test. The conceptual layout of the test facility is shown on Fig. 7, where one of the key elements is the 150 MeV/3 mA high power proton accelerator (HPPA), which consists of an ECR proton source, a low energy beam transport system (LEBT), an RFQ injector, a cavity coupling drift tube linac (CCDTL) and superconducting accelerator sections. It is expected that the HPPA will provide useful information on a number of critical issues, such as beam instabilities, tolerance on beam loss, feasibility of high duty factor operation or even CW operation and etc. It should be used not only for driving sub-critical reactor but also for neutron physics measurements.

The four-vane structure RFQ is preferred in this case because of strict demands on high average beam power with high reliability. It is designed to accelerate more than 60 mA peak current of proton beam (3 mA in av-

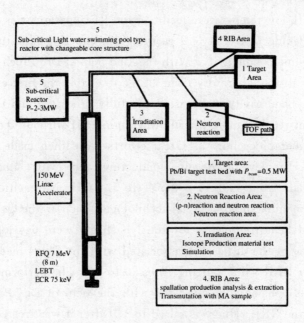

Fig. 7 The layout of the test facility

erage) to 7 MeV, with an input energy of 75 keV. The structure of which is similar to that of Los Alamos Laboratory (APT). In our case, the RFQ will be composed of four segments joining together with three coupling plates, which separate the segments and reduce the magnetic coupling. The RFQ linac is 8 meters overall in length. Each segment is 2 meters long. The specifications of the proposed RFQ are listed in the last column of Table 1. Since the tolerance on beam loss for an intense proton beam is very strict, about as low as 10^{-7}, many efforts are made to study the beam instabilities, such as beam halos caused by mismatched particles[14,15]. To reduce the transient beam loading, the simultaneous acceleration of both positive and negative ions is again helpful.

4. Progress on RF Superconducting Accelerators

4.1 The R & D on single cell RF Superconducting Cavity

Thanks to the support of the State High Tech Program, a laboratory equipped with low temperature facilities and RF apparatus was established in 1988 in Peking University for carrying out experimental studies in the field of RF superconductivity. With the kind help of Dr. Proch, a Nb cavity was sent to the laboratory from DESY to enable the first experiment. It was an L-band cavity and was successfully processed and tested. The field gradient reached more than 12 MV/m in 1991 without Q degradation after a number of cycling[16]. Great efforts were then made by the SC group of PKU to develop the superconducting cavity with China made niobium materials from 1992 to 1994. Two 1.5 GHz Nb cavities were designed and manufactured. Special technologies such as the electron beam welding and heat treatments for the Nb sheets were developed at this stage[8]. One of the difficulties associated with the China made Nb sheet was the low RRR value ranging from 50 to 60. The performance of the material was considerably upgraded with the help of Dr. P. Kneisel of TJNL that the RRR value was raised to 470 after it was treated with high temperature ($\approx$1400 ℃). Based on the above experience, one of the cavities was then treated in KEK with the help of Dr. T. Furuya, and Dr. K. Saito in Sept. 1994. The RRR of this cavity reached more than 270 and the field gradient increased from 4.5 MV/m to more than 10 MV/m, while the Q_0 value from 3×10^8 to more than 10^9, as shown in the low temperature tests. Main parameters of the cavity are shown in Table 2, and the photo of the cavity in Fig. 8, where as Fig. 9 shows the Q versus E_{peak} value verified under 1.8 K by Dr. Kneisel.

Fig. 8　Photo of the Nb cavity

Table 2 Main parameters of the SC cavity

F_0	R/Q	G	H_{sp}/E_{acc}	E_p/E_{acc}
1.487 GHz	102	286	41	2.1

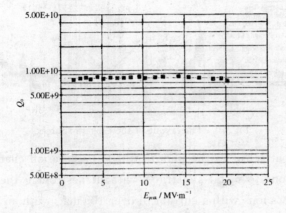

Fig. 9 Q vs E_{peak} of the Nb cavity

4.2 Photo-cathode electron gun with a 2 MeV SC booster

During the last 10 years, photo-cathode injectors developed rapidly to meet the demands of high brightness e-beam for Free Electron Lasers, Linear Colliders, wake field accelerators and etc. A High Brightness Short Pulse Electron Beam Source was developed by the SC Group at PKU. The electron beam is generated by a laser driven photo-cathode and extracted by a high-gradient DC field. An SC accelerating section is followed to accelerate the beam to ≈ 2 MeV. In this way, it was expected to provide Pico-second or Femto-second high brightness electron beam either in CW or pulsed mode. Fig. 10 is the layout of the whole system. It consists of five main parts, the photo-cathode preparation chamber, the DC acceleration chamber, the mode-locked laser, SC energy booster and the beam diagnostic system.

The unique feature of the photo-cathode preparation chamber is its capability of fabricating photo-cathodes by three different schemes, including ion implantation, CVD, and ion beam enhanced deposition. A Cs

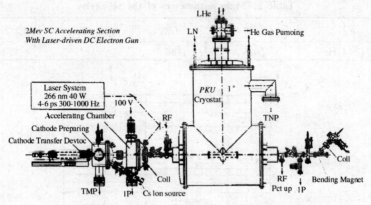

Fig. 10   The layout of the e-beam source

ion source is attached to the photo-cathode preparation chamber for implanting Cs ions to metal substrates without destroying the vacuum. It can generate Cs ions with a current of over 100 mA, with an implantation energy of 25 keV at maximum. The photo-cathode thus processed will be transferred from the preparation site to the DC acceleration chamber located 60 cm downstream via the photo-cathode load-lock system under high vacuum. This can avoid the contamination of the cathode surface and ensure good efficiency. A number of cesium telluride photo-cathodes have been successfully fabricated with CVD scheme inside the preparation chamber. The feasibility of Mg and Mo is to be studied as the candidate substrates, so as to improve the quantum efficiency and acquire with a longer lifetime. Fig. 11 shows the photo of the cathode preparation chamber attached with a Cs source and the booster cryostat.

The mode-locked Nd-YAG laser system is able to provide laser pulses of 10 μJ at a wavelength of 266 nm and a pulse duration of 30~100 ps with 1~10 Hz repetition frequency. The electrons emerging from the photo-cathode are to be extracted at a voltage of $\approx$100 kV and then accelerated up to about 2 MeV by 1.5 GHz Nb superconducting cavity. Preliminary beam tests were performed with an extraction voltage of 45 kV applied to the cathode. Beam pulses of about 35 ps of $\approx$0.05 nC were measured by a coaxial Faraday cup right after the extraction. The results were

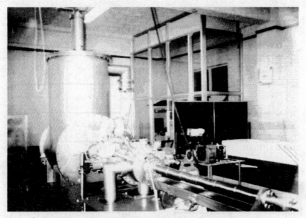

Fig. 11 The cathode chamber & the cryostat

encouraging, the emittance of the electron beam was in the range of 0.5~2 πmm · mrad, and the brightness was estimated to be 5×10^{10} A/m² · rad² correspondingly.

To study the beam dynamics of the DC electron gun, the program PAEMELA is used to simulate the transportation of the beam and to get the pulse duration, energy spread and emittance of the beam after the acceleration. Fig. 12 shows the pulse duration, energy spread and emittance at 100 kV of accelerating voltage, the total charge under consideration is 150 pC. The simulation shows that when the accelerating voltage is raised from 45 kV to 100 kV, the pulse duration will change from 160 ps to 70 ps. For the next step, we will develop an RF microwave photo-cathode electron gun of one and half cell-cavity to obtain pulsed electron beam with a duration of 2~6 ps, peak current of over 100 A and an emittance of better than 1 πmm · mrad. This requires a high performance laser generator, which can produce 100 fs~4 ps pulse at peak energy of 20 μJ with a repetition rate of 360 Hz.

4.3 SC for the Beijing Radioactive Nuclear Beam Facility (BRNBF)

The research and application of radioactive nuclear beams is one of the main issues of nuclear physics in China. CIAE has proposed to build the BRNBF based on the HI-13 tandem accelerator and a superconducting heavy ion post-accelerator[11]. Fig. 13 is the layout of the BRNBF. The

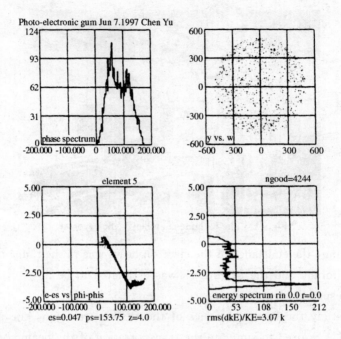

Fig. 12 The result of a simulation

study groups of CIAE and Peking University have worked out jointly the conceptual design and the layout of the SC booster LINAC for the proposed BRNBF project. The niobium-sputtered copper quarter wave resonator (QWR) is the first choice for the accelerating structure, since it has good RF superconducting performance and low construction and running cost. The geometry of the QWR (Fig. 14) is optimized with the program MAFIA by the joint efforts of PKU and CIAE.

Great progress has been made in the technology of the niobium-sputtered resonator in Europe. Niobium has good mechanical properties but low thermal conductivity, which might induce thermal instability of the resonator and thus increase the danger of quenching. One effective way to tackle with the problem is to sputter a layer of high pure niobium, about several microns thick, on to the surface of the OFHC copper cavity, so as to increase the thermal conductivity by about ten times at low temperature.

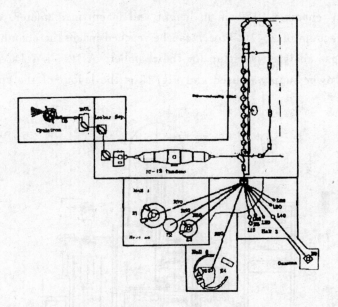

Fig. 13 The layout of the BRNBF

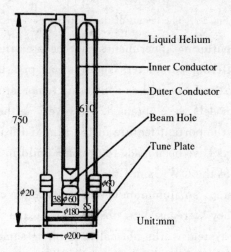

Fig. 14 The structure of the QWR

The SC group of PKU started exploring Cu-Nb sputtering QWR technology at the end of 1995. A DC diode sputtering system were installed for this purpose in the middle of 1997. The main part of the system is a UHV chamber, 120 cm in height and 60 cm in diameter. A background vacuum of 2×10^{-9} torr. can be reached inside the chamber and a residual gas analyzer is set up for the chamber. A 12.5 kW DC constant current power supply is used and Fig. 15 is the layout of the sputtering system.

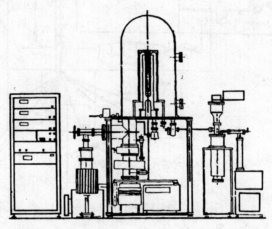

Fig. 15 The layout of the sputtering system

A series of sputtering experiments have been carried out to examine the effect of sputtering parameters such as gas pressure P, sputtering voltage U and current I, as well as sputtering temperature T. Some good results on QWR models were obtained. However, as the thickness homogeneity is the most important feature of the film and it is very difficult to attain, great efforts have been made to obtain a uniform sputtering film on the inner surface of the QWR since then.

As the first step, an aluminum QWR cavity and a copper target were manufactured. They were used to observe firstly the state of discharge and to test if the system works normally. Probing samples of glass were then placed on the various parts of the outer wall and the inner conductor of the QWR to test the properties of the film versus positions. By sputtering copper to the surface of the Al resonator under various conditions, we managed to define the geometry and dimensions of the target as well as the sputtering parameters. After a series of trials and tests, the ratio of the thickness of the films on the inner conductor and the outer wall is quite close to 1.5.

Since Al cavity and Cu target cannot sustain the high temperature, stainless steel QWR and target were used in the second step. It might reflect the Nb sputtering process more realistically. The film thickness and the structure of the contents were measured with glass probes under different parameters for many times. Finally the geometry and dimensions of the target as well as the sputtering parameters were settled. The film thickness of the inner and outer conductor is approximately 1 : 1.5; the thickness homogeneity of the film appears to be quite satisfactory.

Experiments with stainless steel cavity and Nb target were carried out as the third step. It is found that the microstructure of the Nb crystal changed greatly if sputtered with or without bias voltage. RRR of the Nb samples was measured and was also verified by INFN at Legnaro. The result is shown in Fig. 16.

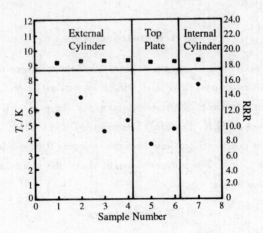

Fig. 16 The RRR of samples

At the same time, the copper resonator and the niobium target were fabricated. The RRR of the niobium is to be measured by a system including a 50-liter volume liquid helium cryostat. Finally the Cu-Nb sputtered QWR will be examined with appropriate beam tests.

4.4 SC cavity for Beijing Electron Positron Collider (BEPC)

BEPC of IHEP is to be upgraded to BEPC-II in the future. With the same mono-ring tube, the luminosity range of $5\times10^{31}\sim1\times10^{32}\,\mathrm{cm}^{-2}\cdot\mathrm{s}^{-1}$ is anticipated for a 150 mA single bunch beam at the energy range of 1.5~2.5 GeV[12]. Two 500 MHz superconducting cavities, with an operating voltage of 1.5 MV, will be installed in this machine, and the Q of HOM will be damped to less than 100. Four superconducting magnets will be used in the area of collision for BEPC-II, so that the β function in the vertical direction will be 1.0~1.5 cm at the colliding point.

5. Conclusion

Considerable progress in the development of RFQ and superconducting accelerators was made in China. Once the proposals on BRNBF or AD-RCNPS are approved, both RFQ and SC accelerators will step into a new stage in China.

6. Acknowledgements

The authors would like to give their thanks to all the members in the RFQ and SC groups of Peking University, CIAE and IHEP, as well as Professors Ding Dazhao, Fang Shouxian for providing all the information for the paper. We also appreciate greatly to colleagues at KEK, Frankfurt University, DESY, TJNL as well as Cornell University for their generosity and kind help in developing RFQ accelerator and RF superconductivity in China. The authors wish to thank the National Natural Science Foundation of China for the support.

7. References

[1] J. X. Fang, C. E. Chen IEEE NS-32 (1985) 2981.
[2] C. E. Chen, J. X. Fang et al., The Progress in the Natural Science, Vol. 4, No. 3 (1994) 271.
[3] J. X. Yu C. E. Chen et al., Trends in Nuclear Physics Vol. 13, No. 2 (1996) 34.
[4] C. E. Chen, J. X. Fang et al., Proceedings 1st Asian Particle Accelerator Conference KEK, Tokyo, Japan (1998) (to be published).

[5] Deng-Ming Kong, Zi-Hua Luo, Shu-Hong Wang, et al. LINAC98, Chicago (1998) (to be published).
[6] Ding Dazhao, The 3rd workshop on neutron science project, March 17~18, 1998 JAERI Tokai-Muka, Japan (to be published).
[7] Guan Xialing, Luo Zhanglin, Proceedings of OECD/NEA workshop on "Utilization and Reliability of High Power Proton Accelerator", 13~15 Oct. 1998, JAERI, Japan.
[8] Wang Lifang, Zhang Baocheng, et al., High Power Laser and Particle Beams, Vol. 8, No. 3 (1996) 349.
[9] Zhao Kui, Geng Rongli, et al., Nuclear Instrument and Methods in Physics Research A 375, (1996) 147.
[10] Zhao Kui, Zhang Baocheng, et al. Internal Report IHIP, Peking University (1997).
[11] Guan Xialing, Proposal for a Radioactive Nuclear Beam Facility at CIAE, Nucl. Inst. And Meth. A382 (1996) 241~245.
[12] Wang Shu-hong et al., Internal Report IHEP, IHEP (1996).
[13] Xiang-wan Du, Ke-song Hu, private communication (1997).
[14] P. K Liu, J. X Fang, C. E. Chen, Proc. 6th particle accelerator physics symposium, (1997) 279.
[15] S. H. Wang, S. N. Fu, S. H. Fang, ibid page 273(1997).
[16] Chia-erh Chen, Zhao Kui, et al., High Power Laser and Particle Beams, Vol. 4, No. 1 (1992) 15.

发表的论文和学术报告总目录

（一）静电加速器与基于串列静电的加速器质谱计

1. Chen Chia-erh, Zhang Yinxia and et al. Design of 4.5 MV Electrostatic Accelerator. Proceedingss 1-st Japan-China Joint Symposium on Accelerators for Nuclear Science & their Applications, University of Tokyo, Japan, 1980: p.19

2. 陈佳洱（以静电加速器小组名义发表）. 4.5 兆伏静电加器. 全国加速器技术交流论文选编. 1980: p.43

3. Li Kun, Liu Hongtao, Chen Chia-erh and et al. The EN Tandem Accelerator Mass Spectrometry at Peking University. Bulletin of the Amer. Phys. Soc., Vol.33, No 9, 1988

4. Li Kun, Chen Chia-erh, and et al. A Project of Tandem Accelerator Mass Spectrometry Facility at Peking University. Proceedings of the 1-st European Particle Accelerator Conference, Rome, June 1988, World Scientific(1989), p.1471

5. 陈佳洱, 李坤, 郭之虞, 严声清, 刘洪涛, 张如菊等. 北京大学加速器质谱计. 第四次全国粒子加速器学术年会论文集, 1988: p.621

6. Chen Chia-erh, Liu Kexin, Guo Zhiyu, Li Kun. Tandem Accelerator Based AMS System and the Project at Peking University. Nuclear Science and Techniques, Vol.1, No. 1-2 (1990)14

7. Chen Chia-erh, Li Kun, Chen Tiemei, Guo Zhiyu, and et al. Status of the Tandem Accelerator Mass Spectrometry Facility at Peking University. Nuclear Instruments and Methods in Physics Research B 52 (1990)306

8. 陈佳洱, 郭之虞等. 北京大学加速器质谱计进展. 第五次全国

^{14}C 学术讨论会论文集，青岛，1991

9. Chen Chia-erh, Guo Zhiyu, Li Kun, Yan Shengqing, Chen Tiemei, Zhang Zhengfang, Yuan Sixun, Lu Xiangyang and Li Bin. Recent Progress of the Peking University Accelerator Mass Spectrometry Facility. Abstracts of XIII INQUA (International Congress, Inter. Union for Quaternary Research), August, 2-9, 1991: p. 48

10. Chen Chia-erh, Li Kun, Guo Zhiyu and et al. The Test Run of AMS System at Peking University. Conference Record of the 1991 IEEE Particle Accelerator Conf., May 6-9 1991, San Francisco, Vol. 4, World Scientific, 1992: p. 2616

11. 陈佳洱，郭之虞等. 北京大学加速器质谱计及^{14}C与^{10}Be的测量. 第五次全国粒子加速器学术年会论文集，北京，1992：p. 579

12. Chen Chia-erh, Guo Zhiyu, Yan Shengqing, Li Renxing, Xiao Min, Li Kun, Liu Kexin, Liu Hongtao, Zhang Ruju, Lu Xiangyang, Li Bin. Beam Characteristics and Preliminary Applications of the Tandem Based AMS at Peking University. Nuclear Instrument and Methods in Physics Research B79(1993) 624

13. Chen Jiaer, Zhang Ying Xia, Wang Jianyong, Gong Linghua. Progress and Application of the 4.5 MV Electrostatic Accelerator. Proceedingss 5-th Japan-China Joint Symposium on Accelerators for Nuclear Science & their Applications, University of Osaka, 1993, Japan, p. 174

14. Liu Kexin, Li Kun, Liu Hongtao, Li Bin, Lu Xiangyang, Guo Zhiyu, Chen Chia-erh and et al. The Progress of AMS at Peking University and CIAE. Proc. Fifth Japan-China Joint Symposium on Accelerators for Nuclear Science & their Applications, Oct. 1993, Osaka, Japan, p. 26

15. 陈佳洱，于金祥，李认兴，韦伦存，巩玲华等. 北京大学 6 MV 串列静电加速器的运行及其应用. 原子能科学技术，Vol. 27 5(1993) 401

16. Lu Xiangyang, Li Kun, Guo Zhiyu, Chen Chia-erh. The Fast

Switching Control and Data Acquisation System of AMS Facility at Peking University. Proc. Beijing International Workshop on AMS, August 15 1990, Beijing, China, Collected Oceanic Works, Vol. 16, No. 1(1993)14

17. Chen Chia-erh, Guo Zhiyu, Yan Shengqing, Li Renxing, Xiao Min, Li Kun, Liu Hongtao, Liu Kexin, Wang Jianjun, Li Bin, Lu Xiangyang, Yuan Sixun, Chen Tiemei, Gao Shijun, Zheng Shuhui, Chen Chengye, Liu Yan. Accelerator Mass Spectrometry at Peking University: Experiments and Progress. Nucl. Instr. & meth. in Phys. Research B 92(1994)47

18. Guo Zhiyu, Liu Kexin, Li Kun, Wang Jianjun, Li Bin, Lu Xianyang, Chen Chia-erh, Chen Tiemei, Yuan Sixun, Gao Shijun. Improvements and Applications of AMS Radiocarbon Measurement at Peking University. Radiocarbon, 37(1995)705

19. 郭之虞，严声清，肖敏，张征芳，杨凤玲，陈佳洱，刘克新，鲁向阳，李斌，钱伟述，袁敬琳，李坤. 北京大学加速器质谱计注入系统. 原子能科学技术，Vol. 28(5) (1994) 390

20. 郭之虞，汪建军，陈佳洱，李坤，刘克新，鲁向阳，李斌，袁敬琳，钱伟述，陈铁梅，原思训，高世君. 北京大学加速器质谱计研究与应用进展. 第六次全国^{14}C 会议报告，1994.11.9～13，四川，巫山

21. Li Kun, Chen Chia-erh, Accelerator Mass Spectrometry at Peking University, Science Foundation in China, Vol. 2(2)(1994)64

22. Li Xinsong, Liu Yuanfang, Shi Jinyuan, Wang Xiangyun, Liu Kexin, Guo Zhiyu, Li Kun, Chen Chia-erh and Yuan Sixun, Genotoxicity of Nicotine Studied by Accelerator Mass Spectrometry. Chinese Chemical Letters, 5(1994)1043

23. Li Xinsong, Liu Yuanfang, Shi Jinyuan, Wang Xiangyun, Liu Kexin, Guo Zhiyu, Li Kun, Chen Chia-erh and Yuan Sixun, A Study on DNA Adduction with Nicotine-derived Nitrosamine by Accelerator Mass Spectrometry. Chinese Chemical Letters, 5(1994)873

24. 郭之虞，李坤，刘克新，鲁向阳，陈佳洱等. 北京大学加速器质

谱计研究与应用进展. 自然科学进展, Vol. 5, 5(1995)513

25. 郭之虞, 鲁向阳, 刘克新, 汪建军, 李坤, 李斌, 袁敬琳, 刘洪涛, 陈佳洱. 北京大学加速器质谱计. 核物理动态, 1996, 13(2): 23~25

26. 郭之虞, 鲁向阳, 李斌, 刘克新, 汪建军, 赵镪, 袁敬琳, 李坤, 陈佳洱. 北京大学质谱计的研制、改进与应用. 中国粒子加速器第6次学术讨论会文集, 1996, 10月27-30日, 四川, 成都, p. 63

27. Zhao Qiang, Guo Zhiyu, Chen Chia-erh, et. al. Recent Progress of the PKUAMS Facility. Proceedings of the 6-th China-Japan Joint Symposium on Accelerator for Nuclear Science and Their Applications, Chengdu, Sichuan, Oct. 21~25, 1996

28. Guo Zhiyu, Wang Jianjun, Zhao Qiang, Liu Kexin, Lu Xiangyang, Li Bin, Yuan Jinglin, Li Kun, Chen Chia-erh. AMS at Peking University. The 14th International Conference on the Application of Accelerator in Research and Industry, Nov. 6~9, 1996, Denton, Texas, USA. CP392 APplication of Accelerators in Research and Industry, 1997, p. 391~393

29. 郭之虞, 赵镪, 刘克新, 鲁向阳, 李斌, 汪建军, 原思训, 陈铁梅, 高世君, 袁敬琳, 李坤, 陈佳洱. 北京大学加速器质谱计装置及^{14}C测量. 质谱学报, 1997, Vol. 18(2): 1~6

30. 郭之虞, 鲁向阳, 任晓堂, 张征芳, 赵镪, 汪建军, 李斌, 张桂筠, 袁敬琳, 巩玲华, 于金祥, 宋广华, 于茂林, 马宏骥, 周广辉, 刘克新, 李坤, 陈佳洱. 夏商周断代工程:北京大学加速器质谱计改造. 高压型加速器技术及应用学术交流会论文集, 1997年9月, 安徽, p. 50~53

31. 赵镪, 刘克新, 郭之虞, 陈佳洱. 古代人骨的加速器质谱计^{14}C测量. 核技术 20(10)(1997)595

32. Zhao Qiang, Lu Xiangyang, Guo Zhiyu, Shi Zhaomin, Wang Jianjun, Liu Kexin, Li Bin, Li Kun, Chen Chia-erh, Lu Hanlin. Measurement of the ^{27}Al(n, 2n)^{26}Al Cross Section Using Accelerator Mass Spectrometry. Chinese Physics Letters, 1998, Vol. 15(1): 8

33. Li Kun, Guo Zhiyu, Lu Xiangyang, Wang Jianjun, Li Bin, Xia Songjiang, Zhao Qiang, Yuan Jinglin, Ma hongji, Zhou Guanghui, Liu Kexin, Chen Chia-erh. Upgrade of the Accelerator Mass Spectrometry in Peking University. APAC98(the 1st Asian Particle Accelerator Conference), 23～27 March 1998, Tsukuba Japan.

（二）射频离子直线加速器

1. 陈佳洱. 螺旋波导直线重离子加速器.《自然科学通讯》北京大学汉中分校. No.3, 1973: p.1

2. 陆善坤, 陈佳洱. 常温螺旋线谐振腔增能器.《粒子加速器及其应用》科学技术文献出版社重庆分社, 1980: p.125

3. 梁仲鑫, 林揆训, 李伟国, 陈佳洱等. 螺旋波导加速腔的不载束高功率试验. 高能物理与核物理, Vol.7, 4(1983)515

4. 梁仲鑫, 李坤, 陈佳洱等. 螺旋波导腔载束全功率增能试验. 1982年全国直线加速器会议报告论文集.（北京大学加速器研究报告 BDJN/82/03）

5. Helical Post-accelerator Group, Peking University. Experimental Test of Helix Loaded Resonator for Heavy Ion Accelerator. Proc. 2-nd China Japan Joint Symposium on Accelerators for Nuclear Science and their Applications, Lanzhou, China, 1983: p.127

6. Fang Jia-xun, Chen Chia-erh. An Integral Split-ring Resonator Loaded with Drift Tube and RFQ. IEEE. Trans., NS-32, 5 (1985) 2891

7. Zhang Qingyun, Chen Chia-erh et al. Experimental Study of Helical Accelerating Structures. Proc. 3-rd China Japan Joint Symposium on Accelerators for Nuclear Science and their Applications, Riken, Saitama, 1987: p.155

8. Chen Chia-erh, Fang Jiaxun et al. A RFQ Injector for EN Tandem LINAC Heavy Ion Accelerator. Proc. 3-rd China Japan Joint Symposium on Accelerators for Nuclear Science and their Applications,

Riken, Saitama, Japan, 1987: p. 116

9. Fang Jiaxun, Chen Chia-erh. An Integral Split Ring Resonator RFQ Structure. Proc. 3-rd China Japan Joint Symposium on Accelerators for Nuclear Science and their Applications, Riken, Saitama, Japan, (1987), p. 120

10. 方家驯, 潘欧嘉, 陈佳洱. 重离子整体分离环 RFQ 与离子的加速. 重离子物理学术讨论会论文集, 兰州, 1987: p. 164

11. 陈佳洱, 方家驯, 潘欧嘉等. 适宜於离子注入的重离子 RFQ 的研究. 第四次全国粒子加速器学术年会论文集, 1988: p. 161

12. Fang Jiaxun, Chen Chia-erh, A. Schempp. Integral Split Ring RFQ Structure for Heavy Ion Accelerator. 1989 Particle Accelerator Conference, March 20 1989, Chicago, Proc. IEEE. 89CH, 2669-0, 1989: p. 1725

13. 陈佳洱, 方家驯. 高功率重离子 RFQ 加速腔的研究. 粒子加速器十年发展报告会文集, 中国粒子加速器学会, 1990, 11, 西安, 1990: p. 294

14. Chen Chia-erh, Fang Jiaxun, Li Weiguo, A. Schempp. Investigation of the Integral Split Ring RFQ Structure. Proceedingss of the 2-nd European Particle Accelerator Conf. Nice, June 12~16, 1990, Editions Frontiers: p. 1225

15. Fang Jiaxun, Li Weiguo, Chen Chia-erh et al. An Integral Split-ring Prototype RFQ Resonator. Proc. 4-th Japan-China Joint Symposium on Accelerators for Nuclear Science & their Applications, Oct. 15 1990, Beijing, China, p. 86

16. Chen Chia-erh, Fang Jia-xun et al. Layout and High Power Test of a 26 MHz Spiral RFQ. Proc. EPAC 92, Editions Frontiers Vol. 2, 1992: p. 1328

17. 陈佳洱, 方家驯等. 整体分离环 RFQ 的高功率试验. 中国粒子加速器第五次学术年会论文集, 1992

18. Chen Chia-erh, Fang Jiaxun, Li Weiguo. A 26 MHz Prototype Integrated Split-ring RFQ Resonator and the Full Power Test. Pro-

ceedingss 5-th Japan-China Joint Symposium on Accelerators for Nuclear Science & their Applications, 1993, University of Osaka, Japan, 1993: p. 52

19. 陈佳洱, 方家驯, 李纬国, 潘欧嘉, 李德山. 重离子整体分离环高频四极场(RFQ)加速结构的研究. 自然科学进展, Vol. 4, 3(1994) 271

20. 李纬国, 方家驯, 于金祥, 陆元荣, 李德山, 袁敬琳, 王忠义, 陈佳洱. 26MHz ISR RFQ 加速器高功率及束流实验. 中国粒子加速器学术讨论会文集, 合肥, 1994 年 8 月, p. 132~136

21. Chen Chia-erh, Fang Jia-xun, Li Weiguo, Yu Jinxiang, Lu Yuanrong, Li Deshan, Yuan Jinglin, Wang Zhongyi. The Beam Test of a 26 MHz ISR RFQ at Peking University. Proceedings of the 4th European Particle Accelerator Conferance, London, July 1, 1994

22. 李纬国, 于金祥, 陆元荣, 袁敬琳, 方家驯, 陈佳洱. 北京大学 RFQ 高能离子注入机的研究. 第八届全国电子束离子束光子束学术年会论文集, 1995 年 5 月, 桂林, p. 21~24

23. 于金祥, 陈佳洱, 李纬国, 方家驯, 任晓堂. MV 级可同时加速正负氧离子的 RFQ 型注入机的设想. 中国电工技术学会第六届电子束离子束学术年会论文集, 1995 年 8 月, 宜昌, p. 284~287

24. 方家驯, 李纬国, 于金祥, 陆元荣, 袁敬琳, 郭菊芳, 陈佳洱. RFQ 加速器在高能离子注入中重要的应用. 中国电工技术学会第六届电子束离子束学术年会论文集, 1995 年 8 月, 宜昌, p. 288~292

25. 于金祥, 陈佳洱, 李纬国, 任晓堂, 方家驯. RFQ 加速器同时加速同 Q/M 正负离子的设想. 核物理动态 Vol. 23, No. 2(1996) 34

26. 李纬国, 陆元荣, 李德山, 袁敬琳, 潘欧嘉, 方家驯, 于金祥, 郭菊芳, 陈佳洱. 重离子整体分离型 RFQ 加速器的研究. 核物理动态, 1996, 23(2): 31~33

27. C. E. Chen, W. G. Li, J. X. Yu, Y. R. Lu, X. T. Ren, Y. Wu, P. K. Liu, J. F. Guo, D. S. Li, W. L. Du, Z. Y. Wan, J. X. Fang. Study of ISR Heavy Ion RFQ Accelerators at Peking University. '96 中日加速器学术讨论会, 1996, 10 月, 成都

28. C. E. Chen, J. X. Fang, J. F. Guo, W. G. Li, D. S. Li, X. T. Ren, J. X. Yu. Experimental Studies on the Acceleration of Positive & Negative Ions with a Heavy Ion ISR RFQ. Proc. 5th European Particle Accelerator Conf., June 1996, Sitges (Barcelona), IOP Publishing, Bristol & Philadelphia, p. 2702

29. 刘濮鲲，方家驯，陈佳洱. 强流束传输中束晕与混沌的 Poincare 图象与 Lyapunov 指数分析. 第六届全国加速器物理学术交流会论文集. 1997: p. 279

30. 陆元荣，李纬国，郭菊芳，于茂林，方家驯，陈佳洱. 离子注入用 RFQ 加速器的研究. '97 全国荷电粒子源粒子束学术会议论文集, 1997 年 10 月, p. 151~154, 湖南，索溪屿

31. 陈佳洱，方家驯. 26 MHz 分离环 RFQ 加速器的研究. 第二届全国加速器技术学术交流会论文集, 1998.7: p. 98

32. 陆元荣，李纬国，郭菊芳，任晓堂，于金祥，方家驯，陈佳洱. 整体分离环 RFQ 加速器的研究. 原子核物理评论，Vol. 15(3)(1998) 144

33. Chen Chia-erh, Fang Jiaxun, Yu Jinxiang, Zhao Kui. Activities on Heavy Ion RFQ and Superconducting Cavities at Peking University. Proc. of The 1st Asian Particle Accelerator Conference, 23~27, March 1998, KEK, Japan

34. Jufang Guo, Weiguo Li, Jinxiang Yu, Yuanrong Lu, Xiaotang Ren, Yu Wu, Zhongyi Wang, Jiaxun Fang, Chia-erh Chen. Investigation of 26 MHz RFQ Accelerators. Proc. of the 1st Asian Particle Accelerator Conference, 23~27 March 1998, KEK, Japan

35. 方家驯，陆元荣，李纬国，郭菊芳，于金祥，吴瑜，于茂林，陈佳洱. 26MHz 整体分离环 RFQ 加速器的研究. 第二届全国加速器学术交流会论文集，宁波, p. 98~101, 1998.7.10~15

36. 陆元荣，W. Gutovski, 李坤，方家驯，李纬国，张保澄，陈佳洱. 加速器控制用模拟压控移相器的研制. 第二届全国加速器技术学术交流会论文集，宁波, p. 352~355, 1998.7.10~15

37. Yuanrong Lu, Jufang Guo, Weiguo Li, Jinxiang Yu, Xiaotang

Ren, Yu Wu, Zhongyi Wang, Jiaxun Fang, Chia-erh Chen. Investigation of 26 MHz RFQ accelerators. Nuclear Instruments & Methods in Physics Research A 420(1999)1~5

（三）射频超导直线加速器

1. J. M. Brennan, Chen Chia-erh et al. First Operation of the Stony Brook Superconducting LINAC. Proc. Symposium of Northeastern Accelerator Personnel, Oct. 1983, University of Rochester, U.S.A.

2. 赵夔，陈佳洱，赵渭江，张保澄等．高梯度超导谐振腔 第四次全国粒子加速器学术年会论文集，1988：p.364

3. Chen Chia-erh, Zhao Kui, et al. R & D Efforts on Nb Superconducting RF Cavity at IHIP Peking University. Proc. 4-th Workshop on RF Superconducting Cavity, August 1989, KEK Tsukuba, Japan. p.859

4. 陈佳洱，赵夔，张保澄，王莉芳等．北京大学重离子所 RF 超导腔研究进展．粒子加速器十年发展科技报告会论文集，西安，1990，p.276

5. Chen Chia-erh, Zhao Kui, Zhang Baocheng, Wang Lifang, Wang Guangwei, Song Jinhu. RF Superconducting Cavities at IHIP of Peking University. The 4th China-Japan Joint Symposium on Accelerators for Nuclear Science and Their Applications, p.176~178, 1990

6. Chen Chia-erh, Zhao Kui, Zhang Baocheng, et al. Superconducting RF Activities at Peking University. Proceed. 5th Workshop on RF Superconductivity, Hamburg, Germany, August, 1991, p.102~109

7. 张保澄，赵夔，陈佳洱，王莉芳等．射频超导腔实验研究．全国电子加速器会议，南京，1991.11

8. Chen Chia-erh, Zhao Kui, Zhang Baocheng. The Status of RF Superconducting Activities at Peking University. Proceedingss of the

3rd European Particle Accelerator Conference, March 1992, Editions Frontiers, p. 1257

9. 赵夔, 王光伟, 王莉芳, 陈佳洱等. 1.5 GHz 铌腔 RF 超导的实验研究. 强激光与粒子束. 第四卷 第一期, 1992 年 2 月, p. 15-32

10. 赵夔, 陈佳洱, 张保澄. 射频超导腔及其研究进展. 全国第六届粒子加速器会议, 北京, 1992 年 10 月

11. Chen Chia-erh, Zhao kui et al. A Proposal of Superconducting Photo-emission RF Gun. IHIP Report SC-01-93, R&D On RF Superconducting Techniques, 1993: p. 1～8

12. Chen Chia-erh, Zhao kui, Zhang Baocheng, Wang Lifang, Geng Rongli, Xu Zengquan. The Progress of Superconducting Cavity Studies and R&D for Laser Driven RF Gun at Peking University. Proc. 5-th Japan-China Joint Symposium on Accelerators for Nuclear Science & their Applications, Oct. 1993, Osaka, Japan, p. 244

13. Chen Chia-erh, Zhaokui, Zhang Baocheng, Wang Lifang, Yu Jin, Song Jinhu, Geng Rongli, Wang Qingyuan, Wu Genfa, Wang Tong. A Proposal of Superconducting Photoemission RF Gun. The First Symposium of Asian Free Electron Laser, Beijing, 1993. 6

14. Zu Donglin, Chen Chia-erh, Study on TESLA Cavity Shape. Particle Accelerator Conference, May 17～21, 1993, Wangshington, D.C., USA

15. K. Zhao, R. L. Geng, L. F. Wang, B. C. Zhang, J. Yu, T. Wang, G. F. Wu, J. H. Song and J. E. Chen. Design and Construction of a DC High Brightness Laser Driven Electron Gun. Nucl. Inst. & Meth. in Phys. Res. A, 375(1996) 147

16. R. L. Geng, J. H. Song, K. Zhao, C. E. Chen. A Hybrid Laser-driven E-beam Injector Using Photo-cathode Electron Gun and Superconducting Cavity. Proceed. 5th European Particle Accelerator Conf., June 1996, Sitges (Barcelona), IOP Publishing, Bristol and Philadelphia, p. 1547

17. 俎栋林, 陈佳洱. RF 超导腔低温测量原理. 高能物理与核物

理. Vol. 20, 7 (1996) 663

18. 赵夔，王莉芳，张保澄，于进，王彤，吴根法，耿荣礼，宋进虎，陈佳洱. 国产铌材超导腔的研制和实验. 核物理动态, 1996, 23(2): p. 26

19. 赵夔，耿荣礼，王莉芳，张保澄，于进，吴根法，王彤，宋进虎，陈佳洱. 激光驱动高亮度电子束源. 核物理动态, 1996, 23(2): p. 28

20. 俎栋林，陈佳洱. 直线对撞机用的 RF 超导多胞加速腔和陷阱模研究. 高能物理与核物理 Vol. 21, 2(1997)180

21. Chen Chia-erh, Zhao Kui, Zhang Baocheng, Wang Lifang, Geng Rongli, Sun Yin-E, Hao Jiankui, Tang Yuxing, Yang Xi, Zhang Yunchi, Chen Yu, Yang Zhitao. The R&D of the RF Superconductivity in IHIP of PKU. 第一届亚太物理国际会议, 1997 年 8 月, 北京

22. Kui Zhao, Yin-E Sun, Bao-cheng Zhang, Lifang Wang, Rongli Geng, Jiaer Chen et al. The High Brightness Ultra-short Pulsed Electron Beam Source at Peking University. Nucl. Inst. & Meth. in Phys. Res. A, 407(1998)322

23. 陈佳洱，赵夔. 射频超导技术的研究进展. 第二届全国加速器技术学术交流会论文集 (1998.7)p. 41

24. Donglin Zu, Jiaer Chen. Design Study on a Superconducting Multicell RF Accelerating Cavity for Use in a Linear Collider. IEEE. Transactions on Nuclear Science, Vol. 45, No 1, 1998: p. 114～118

25. Jiankui Hao, Kui Zhao, Baocheng Zhang, Dalin Xie, Yin-e Sun, Lifang Wang, Zilin Shen, Zhitao Yang, Chia-erh Chen. R&D of Niobium-Sputtered Copper QWR for Heavy Ion Linac in Peking University. The 1-st Asian Particle Accelerator Conference(APAC98), March 1998, KEK, Japan

26. Zhitao Yang, Kui Zhao, Genfa Wu, Lifang Wang, Jiankui Hao, Baocheng Zhang, Chia-erh Chen. Research on Microwave Property of High-T_c Superconductor. The 1-st Asian Particle Accelerator Conference(APAC98), March 1998, KEK, Japan

（四）回旋加速器

1. 徐同仁，顾彪，陈佳洱等. 强聚焦回旋加速器. 全国第一次和平利用原子能会议报告，No. 0117，1960 年 5 月，北京

2. C. E. Chen. A Note on the Magnetic Focusing Wedge. NIRNS Internal Design Note，CDN 500/40/03 September 1964，Rutherford High Energy Laboratory，U. K.

3. C. E. Chen T. C. Randle. A Collection of Data on Wedge. NIRNS Internal Design Note，CDN/500/40/90 Rutherford High Energy Laboratory，U. K.

4. C. E. Chen. An RF Buncher for Axially Injected Cyclotron Beam. Internal Report of Birmingham University，December，4，1965，Department of Physics，Birmingham University，U. K.

5. C. E. Chen，P. S. Rogers. Experimental Studies of the Central Region of the Cyclotron. RHEL/R 116 (1965)，U. K.

6. C. E. Chen，P. S. Rogers. On the Centering of Cyclotron Orbits by Means of Harmonic Coils. NIRL/ M /76，(1965)，U. K.

7. C. E. Chen，P. S. Rogers. Experimental Comparison of Fundamental Mode and 3-rd Harmonic Mode in the Central Region of the Cyclotron. NIRL/M/75，March (1965)，U. K.

（五）电子感应加速器

1. 陈佳洱，高尔布诺夫，阿那尼也夫，徐同仁，李认兴等. 30 MeV 电子感应加速器. 全国电子感应加速器研讨会邀请报告. 1959 年 11 月，北京

2. 陈佳洱，顾彪等. 30 MeV 电子感应加速器磁铁平面模型. 全国电子感应加速器研讨会邀请报告. 1959 年 11 月，北京

（六）束流物理与脉冲化技术

1. C. E. Chen. Resolving Power and Acceptance of the Oxford Injector Magnet. January/1964, Ref. 143/64, Nuclear Physics Laboratory, Oxford University, U. K.

2. 陈佳洱. 静电加速器脉冲化装置及有关特性. 全国加速器技术交流论文选编. 1980: p. 75 （北京大学内部研究报告 BDJ-B-1/1978）

3. 陈佳洱. 离子在高频切割器中的运动. 高能物理与核物理. Vol. 4, 3(1980)401

4. 陈佳洱，郭之虞，赵夔. 螺旋波导后聚束器的群聚特性，1982年全国直线加速器会议报告论文集，南京，1982年10月，（北京大学加速器研究报告 BDJ-B-16/1982）

5. Chen Chia-erh, J. M. Brennan. The Beam Dynamics of a Sine Wave Sweeper. Stony Brook Tandem LINAC report. Jan. 1983, Stoney Brook

6. Chen Chia-erh, Guo Zhiyu, Zao Kui. Bunching Characteristics of a Buncher Using Helical Resonators. IEEE. Trans. NS-30, 2(1983)1254

7. 严声清，沈定予，赵渭江，傅惠清，陈佳洱. 束流相图仪的标定. 核技术. 1, 1983: p. 39

8. J. M. Brennan, Chen Chia-erh and et al. Bunching System for the Stony Brook Tandem LINAC Heavy Ion Accelerator. IEEE. Trans. NS-30, 4(1983)2798

9. G. S. Xu, Z. B. Qian, C. E. Chen. High Efficiency Two Harmonics Beam Chopper. Rev. Sci. Instrum. Vol. 57, 5 (1986)795

10. 郭之虞，张征芳，陈佳洱等. 毫微秒脉冲可变栅距同轴靶. 原子能科学技术，No. 20, 1986: p. 6

11. Jiang xiaoping, Qian Zubao, Chen Chia-erh. A Double Drift Buncher for 4.5 MV Van de Graaff Accelerator. Proc. 3rd Japan-China Joint Symposium on Accelerators for Nuclear Science &. their Applica-

tions, Riken, Siatama, Japan, 1987: p. 212

12. 吕建钦, 谢大林, 张胜群, 陈佳洱. 600 keV 强流毫微秒脉冲加速器中的前切割与后聚束系统. 核技术, Vol. 20 9(1997)555~560

（七）粒子加速器综述

1. 陈佳洱. 重离子加速器技术发展概况. 全国第一届低能核物理讨论会资料选编. 1974 年, 北京. 原子能出版社出版

2. 陈佳洱. 医用带电粒子加速器. 同位素在生物医药中的应用, 第七章, 科学出版社, 1977, 北京

3. 陈佳洱. 超导重离子直线加速器. 核物理动态. 3 (1986)19

4. 陈佳洱. 粒子加速器. 中国大百科全书. 物理卷, 1987: p.733

5. 陈佳洱. 超导加速器. 中国大百科全书. 物理卷, 1987: p.78

6. Chen Chia-erh, Zhao Weijiang. Progress of Low Energy Particle Accelerators in China. Invited talk at the First European Particle Accelerators Conference. Proceedings EPAC 88, Rome, June, 1988, Vol. 1 p. 242

7. 郭之虞, 李坤, 陈佳洱等. 加速器质谱计的原理、技术及其进展. 原子能科学技术. Vol. 23, No. 6(1989)p. 76

8. 陈铁梅, 李坤, 刘东生, 陈佳洱等. 加速器质谱计在地球科学中的应用. 海洋地质与第四纪地质, Vol. 9, 1(1989)103

9. 陈佳洱, 赵渭江, 梁岫如. 我国低能加速器的发展. 物理, 18 卷(1989) 581

10. 陈佳洱, 卢希廷, 包尚联等. 北京大学加速器和核物理状况及研究进展. 核物理动态, Vol. 6, 1(1989)11

11. 郭之虞, 李坤, 陈佳洱等. 加速器质谱计——^{14}C 测年的新手段. 第四纪冰川与第四纪地质论文集, 第六集（碳十四专集）, 地质出版社, 1990: p. 4

12. Chen Chia-erh, Li Kun. Progress of the Tandem Accelerator Based AMS Facilities in China. Proc. 4-th China-Japan Joint Symposium on Accelerators for Nuclear Science & their Applications

(CJJSANSA), Beijing, China, Oct. 15 1990: p. 131

13. 陈佳洱,张仲木. 我国离子静电加速器的发展. 粒子加速器十年发展报告会文集. 中国粒子加速器学会邀请报告. 1990, 11: p. 61

14. 陈佳洱,梁岫如. 我国粒子加速器在技术和应用方面的发展. 中国核学会十周年学术报告会邀请报告. 1990 年 12 月,北京

15. Li Kun, Chen Chia-erh. Status of tandem based AMS facilities in China. AMS Requirements in Canada, Edited by D. B. Carlisle, Natural Sciences & Engineering Research Council, Burlington, Ontario, Canada, 1991: p. 47~50

16. Wang Qingyuan, Zhao Kui, Chen Chia-erh. High-power Multi-electron Beam Cherenkov Free-electron Lasers at mm Wavelengths. Nucl. Instr. & Meth., A, 337(1993)224

17. Zhang Hua-sheng, Chen Chia-erh, Zhao Weijiang. The Progress of Ion Sources in China. Rev. Scient. Instrum. Vol. 65, 4 (1994)1383

18. 赵渭江,陈佳洱. 北京离子源会议评述. 国际学术动态, 4 (1994) 38

19. 陈铁梅,原思训,高世君,胡艳秋,李坤,刘克新,刘洪涛,鲁向阳,李斌,郭之虞,汪建军,陈佳洱,刘嘉麟,宋春郁. 黄土剖面的 AMS 和常规 ^{14}C 初步测年研究. 第六次全国 ^{14}C 会议报告, 1994. 11. 9~13,四川,巫山

20. Chia-erh Chen, Weijiang Zhao. Recent ion source development in China (invited). Rev. Sci. Instrum., 1996, 67(3): 1399-1403

21. Chen Chia-erh, Zhao Kui. The Status of RF Superconductivity R and D in China. Invited talk. Proc. 8th Workshop on RF Superconductivity, October 6-8, 1997, Legnaro

22. C. E. Chen, J. X. Fang, J. X. Yu, K. Zhao. Activities on Heavy Ion RFQ and RF Superconducting Cavities at Peking University. Invited talk at the First Asian Particle Accelerator Conference March 1998, Tsukuba, Japan

23. 关遐令,陈佳洱. 低能核物理中的加速器技术. 第二届全国加

速器技术学术交流会论文集. 1998.7：p.1

24. Chen Chia-erh, Fang Jia-Xun, Guan Xia-Ling, Zhao Kui. Progress of RFQ and Superconducting Accelerators in China. Invited talk at International Symposium on Frontiers of Modern Physics, Kuala Lumpur, Malaysia, Oct. 1998

（八）其他

1. 哈宽富，陈佳洱. X 线粉末照相正中调整新法. 东北人民大学（吉林大学）学报，自然科学版，第一期. 1955 年 3 月

2. C.E. Chen. Measurement of Alternate Magnetic Fields from a Single Turn Coil with a Shield. NIRNS Internal Design Note, CDN 500/14/072, 1964, U.K.

（九）编著的书籍

1. 加速器物理基础，原子能出版社，1993（陈佳洱主编，合作者：方家驯，李国树，裴元吉，郭之虞.）

2. Free Electron Lasers, Beijing Institute of Modern Physics Series 2, World Scientific Publishing Co. 1988. (with Xie Jialin, Du Xiangwan, Zhao Kui)

3. Beijing International Workshop on Accelerator Mass Spectrometry, Beijing, 1990. Collected Oceanic Works, 1993. (with Liu Dongsheng, Li Kun, Chen Tiemei, Guo Zhiyu)

4. Proceedings of the 5-th International Conference on Ion Sources, Review Scientific Instruments. Vol. 65, No. 4 (with Zhao Weijiang)

合计：147

自　　述

　　我生长在上海一个知识分子的家庭。父亲陈伯吹毕生致力于教育和儿童文学创作。母亲吴鸿志是师范学校的钢琴教师。在家里我是一个独子，父母把他们的爱心倾注在对我的培育上。小时候，妈妈经常给我讲故事，教我唱歌、弹钢琴，爸爸以他收藏的儿童图书吸引我学习。可惜好景不长，在日本军的侵略下上海沦陷，爸爸为抗日去重庆北碚国立编译馆工作，妈妈被日本宪兵逼得吐血卧床并被关在医院里，我只得寄居在外婆家，在一家教会小学里上学。在那苦难的日子里哪敢有半点对未来的奢望。

　　抗日战争胜利后，父亲送我进上海市位育中学就读。校长李楚材先生聘请了一批高水平的老师任教。如教数理化的陈安英、周昌寿、李玉廉先生和教语文的朱家泽先生等，使我有机会获得良好的中等教育。在中学的课程中，起初我对英文最有兴趣。爸爸还鼓励我作课余翻译。在他的帮助下，我的第一篇翻译习作《森林中的红人》发表在上海《华美晚报》上。那时候我很希望继承父业成为一个作家。但是，有一次在学校的校庆科技展览上，我看到高年级同学表演用自制的无线电发射机把校庆的消息广播出去，觉得这实在太神妙了，于是决心要掌握这一技术。回来后，就和同学王洪等一起组织了一个课余无线电小组，从单管收音机一直搞到功率放大器，到了废寝忘食的地步。为了在学校里普及无线电知识，我们还搞了一个名叫"创造"的油印刊物，连续出版了一年。上海《大公报》曾专门报道我们的事绩，以资鼓励。从这时候起，我开始立志从事科学。

　　1950年我考进大连大学，1952年又因院系调整转到吉林大学物理系学习。大学的教育为我日后的事业打下了关键性的基础。我有幸得到了许多造诣高深的名师指导，其中对我影响最深的是王大珩、朱光亚、吴式枢和余瑞璜先生。王先生当时是大连大学物理系系主任，为了培养学生的实验能力，亲自带大学普通物理实验课。他对学生要求非常严格，每次实验完毕都要把实验记录交他批阅同意后才能回去写报告，遇有不合格的数据就要打回来重测。要是某同学实验做得好,王先生给了一个 5 分,某同学就得给同伴请客吃花生米,因为这实在太难得了。那一学期我请了 3 次客,至今我仍引以为荣!到了吉林大学后,朱光亚和吴式枢先生分别给我们上原子物理和量子力学课。他们讲的课物理图象非常清楚,逻辑十分严密,而且重点突出,层次分明。每听一课都是一次享受。学了这两门课后,不仅引起了我对近代物理的浓厚兴趣,还帮助我打下很好的理论基础。朱先生还指导我的毕业论文,以"盖革计数管的研究"为题。在他的指导下,我顺利地制成了第一只薄窗型 β 放射线计数管。

1955 年国家决定在北京大学建立第一个培养原子能人才的基地——北京大学物理研究室(现在的技术物理系)。我被调来参加筹建工作。物研室的第一任主任是胡济民先生,参加领导的还有虞福春、卢鹤绂和朱光亚先生。在他们的领导下,我负责建设我国第一个核物理教学实验室。按照实验大纲,需要一大批计数管。当时无处可买,虞先生便让我自制。他把他在美国用过的一整套吹玻璃用的工具都送给了我,让我从学吹玻璃干起。我日夜加班,花了整整三个月的时间在物理系玻璃工的帮助下建起了一个小车间,又经过半年多不眠之夜,终于和同事们一起制成了一批批带薄窗的计数管,其中包括国内首次制成的 α 计数管和康普顿电子符合计数管等,保证了第一批实验的顺利开出。

1963 年学校派我去英国进修,从事等时性回旋加速器中央区的研究。指导我工作的是牛津大学的 D. H. Wilkinson 和卢瑟福研究所的 J. D. Lawson。两位都是英国皇家学会的会员,前者是著名的多道分析器的发明人,后者是著名的等离子体聚变判据的提出者。当时 Lawson 先生向我提出一个问题:"为什么由离子源引出的束流经过中央区加速后会发生大幅度衰减,那些离子丢到哪里去了?"这是一个十分复杂、不易回答的难题,他要我设法解决。我用微分探针逐一确定各种离子成分的轨道并将实验诊断与计算机模拟、理论分析相结合,逐圈分析各轨道的动力学特性。终于弄清束流衰减的基本规律,使束流的传输效率提高了 3 倍以上,还找到了证实越隙共振存在的实验判据。过去我在研究中常常追求严格推导和理论上的完整,通过这一轮研究我更加重视形象思维和结合实际的理论分析。

1982 年我在杨振宁先生的安排下去美国石溪大学访问,承担该校超导直线加速器上脉冲化系统的研制任务。早在70年代我就曾将束流光学的理论和方法拓展到束流的群聚、切割和输运过程,这一次便将有关成果应用于该系统上,结果非常成功。64 MeV的硫离子被压缩到100皮秒,束流利用效率高达60%以上,为当时国际先进水平。我又将有关成果制成软件,通过计算机直接控制加速器各项设备,运行起来既方便又精确,这样的运行方式由1984年一直使用至今。石溪的同行们高兴地把它称为"陈氏模式"。

70年代末,我与教研室的同事们赴上海先锋电机厂研制4.5 MV静电加速器。我们首次摆脱了传统的仿制模式,从优化物理参数着手,自行设计、建造。这台加速器于1983年加工完毕,但因"文化大革命"中原有实验室被占,不能安装。好容易等到1986年新实验楼建成,刚要开始安装,却因基建工人施工不慎,新楼起火,把加速器的高压电极、均压环等主要部件都烧坏了。这对我和我的同事们无疑是一个晴天霹雳!面对这场灾难我们谁也没有退却,大家只是加倍努力,把烧坏的部分一一整修、擦亮,逐一进行安装,经过4年的努力终于调试出来。这台加速器的建成填补了我国单色中子5~7 MeV能区的空白,并为北京大学中子与裂变物理实验室的建设奠定了基础。去年这台加速器不仅为北京大学、清华大学、四川大学等高校和有关国防单位提供了束流,还接待了来自美国伯克利大学、俄国杜布纳研究所和镭学研究所的访问科学家进行实验。我们为之奋斗了半辈子的实验基地终于开始运行,这对我是个莫大安慰!

科学事物,必须不断研究,认真实验,得寸进尺地深入、扩展,通过韧性的战斗,才有可能获取光辉的成就。这是我从事核物理工作40年来的一点感想。

难忘的游戏——代后记

　　那是一个夏日雷天的傍晚。窗外的浓浓乌云使整个天空变得黑沉沉的,真叫人害怕。我不敢独自一个人在房里呆着,便去找爸爸。

　　跨进爸爸的房门,只见他正在聚精会神地伏案写作着。我不愿打扰他,便静悄悄地坐在一旁,呆呆地看着窗外。

　　突然间,闪电像一条金蛇在天空飞舞,耀眼的亮光,刺得我睁不开眼睛。紧跟着"啪啦啦…轰隆隆"震耳欲聋的巨响,把年方6岁的我吓得"哇……"的一声哭了起来。爸爸转过身来,把我搂到他的怀里,安慰我,叫我别怕。他慈祥地看着我,问我:"你知道天上为什么会打雷吗?"

　　"是雷公发火,要劈不孝的人啊!"

　　"是吗?!是谁告诉你的?"爸爸惊奇地看着我说。

　　"隔壁老奶奶讲的呀!"我认真地告诉他。

　　爸爸听了后,"噗嗤!"一声笑了起来。他摇摇头说:"打雷,不是雷公发火。这是阴电和阳电相遇时,放电的结果,懂吗?"我摇了摇头。

　　他见我不懂,便耐心地向我解释说:"你看,我的左手上有阴电,右手上是阳电。两种电相互作用就像两个巴掌拍在一起一样。"他边说着,边"啪!"的一声,把两个巴掌拍响了。爸爸的表演引起了我极大的好奇心。

　　"那电是从哪里来的呢?为什么它们会放电呀?"我睁大着眼睛问道。

　　"电可以通过物体间的摩擦产生,人们常讲'摩擦生电'就是这个意思。夏天,天热,云层飘动时与周围空气摩擦,就产生了电。有的云块带阳电,有的云块带阴电。它们相互吸引,当它们碰在一起时就发生放电,发出火光和声音!"

　　爸爸的解释十分形象、生动,但我还没有弄懂,随口就问道:"摩擦真能生电吗?"

　　爸爸见我有兴趣,就请妈妈来一起做个游戏给我看。他要妈妈找来一块玻璃板,还请她剪一个小小的纸人儿。于是,他用两本书把玻璃板架在桌面上,将纸人儿放在玻璃下。接着他就用擦眼镜的绸布,包在我平时玩的一块长方形积木块的外边,快速地在玻璃板上擦动着。一种有趣的现象竟然在我的眼前发生了。只见随着绸布在玻璃板上的擦动,板下的小纸人儿在桌面和玻璃板之间一会儿上、一会儿下地跳动起来。我失神地看着看着,不禁鼓掌叫起好来!

"这个游戏叫做'跳舞人形'"爸爸认真地对我说,"绸布和玻璃摩擦就产生了电。电的吸引力使纸人儿向上往玻璃板蹦!"

那天,爸爸趁兴给我讲了许多关于电的故事。他使我懂得,物体不仅和玻璃摩擦可以产生电,和干燥的毛皮、纸张等的摩擦也都可以产生电,人们还用专门制造的发电机发出电来点灯、取暖以及开动有轨电车和无轨电车等等。电的用处可真大呀!

爸爸的"游戏",在我幼小的心田里种下了一颗爱科学、学科学的种子。后来在老师们的辛勤教导和指引下,这一颗科学之种,终于发芽滋长起来,使我逐步走上了科学之路。回顾过去,我深深感到少年儿童时期的教育,对人的一生所具有的极为深刻的意义。为了"科教兴国",为了祖国美好的未来,我国的科学家、作家和教育家真应为少儿科学文化的教育与创作多贡献力量才是!

爸爸陈伯吹作为一个儿童文学作家,当然非常希望我能继承他的事业。我上中学时,他曾不断地鼓励我翻译儿童故事、创作儿童小品,期待我能成长为一个作家。但是,他同时酷爱科学,认为现代的文明一刻也离不开科学的发明创造。因此他十分重视少儿科学文化的教育与创作。凡是有好的科技文艺作品,他都建议我去阅读或观看。记得有一年,上海电影院放映《发明大王爱迪生》的片子,他当即放下手头的工作,带我去看。还有一次,他冒着大雨把我从寄宿制的位育中学中接出来,带我去观看电影《居里夫人传》。他勉励我要学居里夫人,不畏艰险、奉献科学,做一个对社会有贡献的人。他曾多次让我阅读《月球旅行记》,他自己也写了《十一个奇怪的人》等科学故事鼓励广大的少年儿童爱科学、学科学。如果说我这一生中在科技方面还有一丁点儿成绩的话,我首先要感谢我父母和父辈的作家和科学家们,他们曾给少年时期的我馈赠无比宝贵的精神食粮!